MEASURING PLANT DIVERSITY

MEASURING PLANT DIVERSITY

Lessons from the Field

Thomas J. Stohlgren

OXFORD
UNIVERSITY PRESS

2007

OXFORD

UNIVERSITY PRESS

Oxford University Press, Inc., publishes works that further
Oxford University's objective of excellence
in research, scholarship, and education.

Oxford   New York
Auckland   Cape Town   Dar es Salaam   Hong Kong   Karachi
Kuala Lumpur   Madrid   Melbourne   Mexico City   Nairobi
New Delhi   Shanghai   Taipei   Toronto

With offices in
Argentina   Austria   Brazil   Chile   Czech Republic   France   Greece
Guatemala   Hungary   Italy   Japan   Poland   Portugal   Singapore
South Korea   Switzerland   Thailand   Turkey   Ukraine   Vietnam

Copyright © 2007 by Oxford University Press, Inc.

Published by Oxford University Press, Inc.
198 Madison Avenue, New York, New York 10016

www.oup.com

Oxford is a registered trademark of Oxford University Press

The format and design of this publication may not be reproduced by any means
without the prior permission of Oxford University Press. The text was produced
by the National Institute of Invasive Species Science, Fort Collins Science Center,
U.S. Geological Survey, and is in the public domain.

Library of Congress Cataloging-in-Publication Data
Stohlgren, Thomas J.
Measuring plant diversity : lessons from the field / by Thomas J. Stohlgren.
    p. cm.
1. Plant diversity.   2. Botany—Methodology.   I. Title.
ISBN 13: 978-0-19-517233-1
QK46.5.D58S86 2005
581.7—dc22   2005008060

9 8 7 6 5 4 3 2

Printed in the United States of America
on acid-free paper

To my lovely wife, Cindy,
and
to our charming children, Connor, Shannon, and Michael

Foreword

There is no substitute for high-quality data in the scientific process. Hypotheses and models, as well as experiments, also play a central role in the advancement and refinement of scientific understanding. But neither have scientific value independent of quantified observations (data) on the behavior of natural systems. This is true in all fields of science, but particularly so for those disciplines in which the major phenomena cannot be experimentally manipulated, such as astronomy and cosmology (e.g., the formation and dynamics of stars, galaxies, and the universe as a whole), geology (the history and dynamics of the earth's solid structure), and ecology (the evolution and dynamics of the living organisms distributed over the earth). In spite of its central role in scientific progress, the collection of data on the properties and dynamics of ecological systems is often looked down upon as "soft science," or as the early, primitive, "descriptive" phase of the science of ecology.

The prestige of experimental science is epitomized by the emphasis on an experimental approach in research funding decisions. This is exemplified by the fanfare that accompanied the recent publication of ecological experiments heralded with the announcement from Peter Karieva that "ecological science has entered the big leagues, the world of particle accelerators and orbiting telescopes."

The status and credibility accorded such experimental results have led some ecologists to question or dismiss field observations that were inconsistent with the new experiments. Two types of field observations have recently been called into question. First, the widely observed pattern of plant diversity being lower in areas with high plant productivity, often associated with fertile soils or the addition of fertilizer, than in areas with intermediate productivity was called into question by experimental results that seemed to show that plant productivity increased monotonically as the number of plant species increased. Various efforts have attempted to explain or dismiss the "anomalous" patterns found in the field data.

The second type of field data criticized for being inconsistent with experimental results was the observations (including some on which this book is based) that areas with high numbers of native species also have high numbers of exotic species. These observations conflicted with new experimental results that seemed to show that areas with high numbers of native species were more resistant to invasion by exotic species, and consequently had lower numbers of exotic species than areas with fewer native species. The field data not only failed to match the experimental results, but were also inconsistent with an old and widely held belief (hypothesis) that high diversity conferred resistance to invasion, as we learned from Charles Elton in 1958. The fact that this familiar hypothesis was supported by the results of several new experiments called the validity of the field observations into question and led to efforts to explain how the field studies could produce such misleading results.

While on first inspection the new experimental ecology on the model of "particle accelerators and orbiting telescopes" seemed to invalidate the ecology of quantitative data based on field measurements, this initial conclusion has been reversed. My extensive reevaluation and analysis of these experiments has led to the discovery of numerous problems with the experimental design, analysis, and interpretation of both the diversity-productivity experiments and the diversity-invasibility experiments. In two such examples, the inconsistency between the field data and the experiments reveals the flaws in the experiments, not vice versa. In the final analysis, it is quantitative data based on observations of "real-world" systems that represent the truth to which both hypotheses and experimental results must be compared. The limitations imposed by experimental controls produce results that are typically relevant to only a small portion of the full range of conditions (e.g., the mean, range, and spatiotemporal variability of multiple environmental factors) experienced by natural ecological communities.

This book addresses the collection and analysis of field measurements of plant communities and includes data that are relevant to and that have played a role in the resolution of the two controversies discussed above. The ecological patterns analyzed and discussed here illustrate the power of high-quality ecological data to the advancement of scientific understanding.

While some recent ecological experiments have failed to live up to the high standards of particle accelerators, the methods and data discussed in this book illustrate the commonality between quantitative field ecology and the high-tech science based on orbiting space telescopes. Two parallels merit brief elaboration. Just as deep-field telescopes allow the detection and analysis of individual stars and galaxies of different ages, different stages of cosmogenesis, and differing initial or current interstellar conditions, so the selection of field sampling locations allows analysis and comparison of plant communities at different stages of development, with different initial and current environmental conditions. A second similarity relates to the collection of cosmological data at multiple frequencies along the electromagnetic spectrum, from X-rays to visible light to infrared radiation. Just as these multiple frequencies allow sampling of different types of cosmogenic processes across a range of spatial resolutions, so the multiscale sampling methods described in this book allow the detection of biological patterns caused by ecological and evolutionary processes that occur at different spatial (as well as temporal) scales.

It is critical that quantitative field ecology be based on the recognition that different types of processes (e.g., competition versus dispersal) operate at different spatial and temporal scales. The inevitable consequence of failure to make this distinction is that the effects of a given process can be undetectable in samples collected at an inappropriate scale. The danger of ignoring the process-scale relationship is that processes that remain undetected by an inappropriately scaled sampling design are often assumed to be unimportant for regulating ecological phenomena. Such errors misdirect the scientific process and may lead to serious errors in resource management and conservation.

Tom Stohlgren has done an excellent job in raising these issues in the context of quantitative field ecology, building on the strong tradition of Frederick Clements, Robert Whittaker, John Curtis, and other great plant ecologists of the 20th century. The sampling methods that are developed, evaluated, and applied in the pages of this book provide a sound foundation for a quantitative field ecology that can advance our understanding of fundamental ecological and evolutionary processes. They also provide a standard for testing hypotheses and for evaluating the relevance and validity of hypotheses and experimental results. This book provides a good introduction to field sampling for beginning plant ecologists, as well as a fascinating set of results and analyses that will stimulate and challenge experienced ecologists.

—Michael Huston
San Marcos, Texas

Preface

Most textbooks on measuring terrestrial vegetation have focused on the characteristics of biomass and cover, and on the density or frequency of dominant life forms (trees, shrubs, grasses, and forbs), or on classifying, differentiating, or evaluating and monitoring dominant plant communities based on a few common species. Sampling designs for measuring species richness and diversity, patterns of plant diversity, species-environment relationships, and species distributions have received less attention. There are compelling, urgent reasons for plant ecologists to do a far better job measuring plant diversity in this new century. Rapidly invading plant species from other countries are affecting rangeland condition and wildlife habitat, placing more plant species on threatened and endangered species lists and increasing wildfire fuel loads. Attention has shifted from the classification of plant communities to accurately mapping rare plant assemblages and species of management concern to afford them better protection. More ecologists, wildlife biologists, and local and regional planners recognize the value in understanding patterns, dynamics, and interactions of rare and common plant species and habitats to better manage grazing, fire, invasive plant species, forest practices, and restoration activities. Thus revised and new sampling approaches, designs, and field techniques for measuring plant diversity are needed to assess critical emerging issues facing land managers.

This book offers alternatives to the approaches, designs, and techniques of the past that were chiefly designed for dominant species and other purposes. I focus on field techniques that move beyond classifying, mapping, and measuring plant diversity for relatively homogeneous communities. This book complements methods for measuring the biomass and cover of dominant plant species. Most species are sparse, rare, and patchily distributed. It empowers the reader to take an experimental approach in the science of plant diversity to better understand the distributions of common and rare species, native and nonnative species, and long-lived and short-lived species.

There are six parts. Part I introduces the problem: plant diversity studies are difficult to design and conduct, in part because of the history and baggage associated with the evolution of plant ecology into a quantitative science. Issues of scale, resolution, and extent must be effectively commandeered. Part II implores the practitioner to take an experimental approach to sampling plant diversity with a clear understanding of the advantages and disadvantages of single-scale and multiscale techniques. Two case studies (actual field investigations) demonstrate how to test and assess various field techniques. Part III focuses on scaling plant diversity measurements from plots (or local field observations) to landscapes. Here, five detailed case studies are used to arm the practitioner with model applied studies of plant diversity from issue formulation, through methods and statistical analysis, to cautiously interpreting results. The case study methods employ the select multiscale plot design with advantages and disadvantages, but alternative multiscale methods could easily have been employed. Regardless of the specific study methods selected, the general approaches might be used and improved by others to measure plant diversity to meet various management needs. Part IV provides a brief introduction to modeling plant diversity in relation to environmental factors. Examples of common nonspatial (correlative) and spatial analyses are explained. Part V introduces the concept of measuring temporal changes in plant diversity at landscape scales followed by a case study designed to collect the necessary baseline data to monitor plant diversity. Part VI discusses research needs to better understand changes in plant diversity in space and time.

The intended audience for the book includes upper division college students, graduate students, field ecologists, resource managers, landowners, range conservationists, and others. I specifically target observational studies of plant diversity and patterns of plant diversity at landscape scales, the survey and monitoring of forest understory and grassland plant species, species-environment relationships and species-area relationships, the role of rare habitats in maintaining plant diversity patterns, typical statistical analysis techniques, and modeling spatial patterns of plant diversity.

This book is not a complete text on comparative field methods or biometry. It doesn't cover controlled vegetation experiments or theoretical ecology. It is particularly weak on aquatic vegetation techniques and nonvascular plant sampling techniques, though many of the proposed techniques might be easily adapted for those uses. The book is limited on the many theories and causes of plant diversity, since these topics go well beyond observational studies.

Measuring Plant Diversity is a logical extension and synthesis of a body of work that spans about 10 years, including more than two dozen peer-reviewed journal articles, book chapters, and lecture notes that I have written, with the help of many colleagues. I cite and acknowledge my more than 60 coauthors who have contributed to jointly published papers over the years. My coauthors, in alphabetical order, include Craig Allen, Richard Bachand, Bill Baker, Dave Barnett, Jill Baron, Michael Bashkin, Jayne Belnap, Joe Berry, Dan Binkley, David Buckley, R. Busing, Tom Chase, Geneva Chong, Scott Collins, Michael Coughenour, Nolan Doesken, Charles Drost, Paul Evangelista, Maurya Falkner, Leandro Ferriera, Curt Flather, Paula Fornwalt, Jim Grace, Sue Grace, Laurie Huckaby, Mohammed Kalkhan, John Kartesz, Merrill Kaufmann, Margot Kaye, Kate Kendall, Tim Kittel, Michelle Lee, Jesse Logan, John Moeny, Dennis McCrumb, Greg Newman, Yasuhiro Onami, Paul Opler, April Owen, Bill Parton, Bruce Peterjohn, Betsey Pfister, Roger A. Pielke, Jr., Roger A. Pielke, Sr., Kelly Redmond, Robin Reich, Kelly (Bull) Rimar, J. Rodgers, Mike Ryan, Yuka Otsuki, James F. Quinn, Michael Ruggiero, Lisa Schell, Sara Simonson, Yohon Son, Jason Stoker, Ken Stolte, Kuni Suzuki, Hidiru Suzuki, Brian Vanden Houvel, Tom Veblen, Cynthia Villa, and Gary Wagonner. I have received their full support and encouragement throughout this process. I am the storyteller, but it is *our* story, and I will never be able to thank them enough for their guidance and training.

I tell the story of *Measuring Plant Diversity* by relying heavily on examples. The "lessons from the field" aspect comes from hiking and observing, testing and evaluating, and trial and error. I provide general techniques and methods that can be used directly or easily modified by most students to meet their needs, and case studies that have both general and specific applications. I provide copious tables and figures adapted from previous publications because real field data are important in the learning process. I supply many stand-alone case studies, and "lessons" in the form of photographs and figures to pass on a few hints, raise questions and issues, and stimulate discussion and thought. I believe that "example is the best teacher" —and that the "teacher isn't always right." Wise students will use the lessons in this book as a starting place to address a few questions rather than as a place to look for all the answers to some final exam.

My specific objectives are to (1) provide a basic understanding of the history of design considerations in past and modern vegetation field studies; (2) demonstrate with real-life case studies the use of single-scale and multiscale sampling methods, and statistical and spatial analysis techniques that may be particularly helpful in measuring plant diversity at landscape scales; and (3) address several sampling questions typically asked by graduate students and budding field ecologists. My ultimate goal is to encourage additional studies of plant diversity—arming the next generation of ecologists with a better understanding of the history of the development of vegetation science as it relates to plant diversity, a rationale for taking more of an experimental approach in designing large-scale plant diversity studies (we're all very new at this), and the early beginnings of a better toolkit for future advances in the field of measuring plant diversity. I welcome all feedback, comments, and suggestions.

Acknowledgments

In addition to my valued coauthors, equally valuable field and laboratory assistance over the past 10 years has been provide by Krista Alper, Marcell Astle, Steve Bousquin, Debbie Casdorph, Jean Marie Ederer, Rick Edwards, Helen Fields, Emily Galbraith, Randy Griffis, Michele Hart, Kate Healy, Catherine Jarnevich, Amy Johnson, Susan Klimas, Tom and Jeane Leatherman Alicia Lizarraga, Max Medley, John Moeny, Stephanie Neeley, Lisa Nelson, Anne Overlin, Nate Pierce, Amy Randell, Dan Reuss, James Self, Rick Shory, Elizabeth Smith, Sean Stewart, Connor Stohlgren, Laura Stretch, and Alycia Waters. In addition, the book's final references, tables, and figures received help from Nate Alley, Dave Barnett, Geneva Chong, Paul Evangelista, Mohammed Kalkhan, Deb Guenther, Greg Newman, Sara Simonson, and Alycia Crall—and I can't thank them enough.

Advice from Indy Burke, Norita Chaney, Phil Chapman, James Detling, Bill Gregg, Alan Hastings, Bruce Van Haveren, Michael Huston, Linda Joyce, John Kartesz, Bill Laurenroth, Mark Miller, Dennis Ojima, Michael Palmer, and Marcel Rejmánek has been greatly appreciated.

Logistical support was provided by the U.S. Department of the Interior, National Park Service, National Biological Survey, National Biological Service, Bureau of Land Management, U.S. Geological Survey (the Midcontinent Ecological Science Center, now the Fort Collins Science Center), and Natural Resource Ecology Laboratory at Colorado State University. To all, I am eternally grateful.

Contents

PART I

THE PAST AND PRESENT

1

Introduction

Despite several solid attempts to compile standard methods for vegetation sampling (Bonham 1989; Daubenmire 1968; Elzinga et al. 1998; Mueller-Dombois and Ellenberg 1974), many sampling techniques dealing specifically with plant diversity have not been widely accepted or generally applied. Any review of methods sections in plant ecology journals will reveal a myriad of approaches for selecting sampling sites, plot shapes and sizes, and sample sizes, and many different patterns of sampling. Few studies have collected directly comparable data on plant species richness or diversity because of the many subjective decisions made during the course of each study, from the sampling design phase, through field-testing techniques, through statistical analyses, modeling, and design modifications to application.

There are two primary reasons why standard methods for plant diversity sampling have not been as widely accepted in the same way that standard methods for water sampling and analysis or for soils analysis have been adopted. First, specific study objectives may vary greatly among investigators of plant diversity. Some studies are designed to address site-specific management or research needs, such as vegetation mapping, wildlife habitat analysis, or specific grazing or postfire effects, etc. Specific sampling designs and field techniques might be required for each objective. Second, as noted by Hinds (1984, p. 11), "it is a world of unique places"—it has been difficult for plant ecologists to design off-the-shelf vegetation sampling protocols or monitoring techniques for general use. Vegetation types vary from desert scrub to alpine tundra, from tropical rainforests to boreal forests, and from aquatic lily pads to giant sequoias. Understory plant diversity can vary from sparse and patchy under heavily canopied forests to rich and productive in wet meadows and riparian zones. Vegetation structure at each site may be affected by the topography, geology, soils, hydrology, land use and disturbance history, and the history of dispersal, establishment, growth, and survivorship of the cornucopia of species at each site. It is no wonder that generalized techniques might fail for specific purposes and unique sites, or that specific techniques might fail for various vegetation types and general applications.

This seemingly chaotic status of vegetation science should not prevent plant ecologists from improving the level of standardization among field sampling techniques for evaluating plant diversity. Understanding the nature of the problem is an important first step in designing solutions. I begin with a brief evaluation of the nature of the problem.

Why Conducting Plant Diversity Studies Is Difficult

There are several well-known difficulties in conducting plant diversity studies, including taxonomic difficulties, phenology difficulties, and problems of species rarity. These difficulties are often deterrents to graduate students and young ecologists, and they shouldn't be. They should be viewed as challenges with many potential solutions.

Taxonomy Difficulties

There are recognized problems in identifying plant species in the field. There are large numbers of species in the species pools of most landscapes. For example, Rocky Mountain National Park in Colorado (108,000 ha) contains more than 1000 plant species, and more than 100 plant species from other countries (i.e., nonnative or exotic plant species). The State of Hawaii contains about 1900 species of angiosperms (flowering plants), with half of the species coming from other countries (http://hbs.bishopmuseum.org/hispp .html). The Jepson manual of higher plants in California recognizes 5862 native and 1023 nonnative naturalized species from other countries (Hickman 1993). Some species (many examples), genera (e.g., *Carex*), and families (e.g., Poaceae) are particularly difficult to identify in the field because of small or variable plant parts.

Many species contain subspecies and varieties. This explains the wide range of plant characteristics that investigators are expected to differentiate. A species named "polymorphous" is particularly discomforting to field

---

Hint: It is important to record the finest taxonomic resolution possible in the field—you can always lump species, but you can't split after the fact.

ecologists. Phenotypic variation is common for broadly distributed species. For example, the silver sword in Hawaii varies so much in form from island to island and habitat to habitat that only the best field naturalists can differentiate them. The Hawaiian silver sword alliance contains about 30 species in three genera (*Argyroxiphium, Dubautia,* and *Wilkesia*) with a phenomenal range of anatomical, morphological, and ecological adaptations, but are very closely related based on biosystematics and molecular studies (http://www.botany.hawaii.edu/faculty/carr/silversword.htm).

Another general problem is that taxonomic keys are far from perfect. Many taxonomies are outdated. Local floras often differ from regional and national floras (e.g., the Flora of North America Project). There is often confusion and disagreement over two floras in the same state [e.g., Munz and Keck (1959) and Hickman (1993)]. Additional taxonomic problems occur when species, subspecies, or varieties are lumped or split, or when they are renamed as previously named taxa or new taxa. New names can be a daily occurrence. A newly recognized problem is that many floras are incomplete due to the rapid invasion of nonnative species into nearly every landscape. Discriminating between native and exotic congeners can be very difficult.

The major problem with taxonomy, however, is a lack of investigator expertise in the field (Figure 1.1). Many plant ecologists receive little formalized training in plant systematics, relying on one or two undergraduate classes in taxonomy. Typically these classes involve a survey of plant family characteristics and the identification of 50 to 100 species in full bloom and with all the distinguishing characteristics. Thus many young ecologists may be unprepared for studies in species-rich environments or for landscapes and regions with large and complex species pools.

There are many possible solutions to these taxonomic issues. The most obvious is to better train plant ecologists in basic field taxonomy. There are ongoing efforts to computerize taxonomic keys, including "polyclaves," which can identify plant species from a few field characteristics. Polyclaves are already available for the plant families of the world (http://www.colby.edu/info.tech/BI211/PlantFamilyID.html or http://www.discoverlife.org), and many are being created for state floras. Note that computerized taxonomic keys often have many of the imperfections of written manuals. Still, there is hope that future computerized keys will be more quickly updated and will provide rapid searches for "synonyms" (all historically used names for a given species) and remote computer assistance via satellites, including photographs, line drawings, and online taxonomic assistance.

Phenology Difficulties

Once in the field, there are several phenological difficulties that must be overcome. Surveys and monitoring are expensive, with the most expensive aspect being the cost of getting field crews to and from the sampling sites. In large landscapes, investigators can rarely afford to visit the same

Figure 1.1.   Lesson 1. Plant diversity studies require well-trained, well rewarded taxonomists. The greatest sampling designs, field methods, and statistics are worthless without botanists, taxonomists, and natural history experts. Always indicate the "authority" you are using when collecting so that the meaning and synonymies of the name can be reconstructed. Befriend taxonomists!

sampling site twice. A common concession made in designing plant diversity field studies is to survey sites while most plants are, or have recently finished, flowering (often termed "peak phenology"). Of course, plant species do not perfectly synchronize flowering at any given site, so the plan is generally to capture as many plant species with flower parts as possible. Extensive botanical experience and observations of seasonal climate are helpful, but a naturalist's eye and extensive hiking are adequate substitute skills in many cases.

In some ecosystems, multiple blooms are common. The shortgrass steppe in Colorado and many arid ecosystems contain some species that flower in early spring and some that flower in the late summer or fall, associated with periods of intense precipitation. In such cases, sampling sites have to be visited at least twice a year. In areas protected from intensive grazing, fire, and other disturbances, sampling in the fall often allows for identification of most early (dried flower parts) and late flowering species. The last caveat deserves more attention. Plant parts in nearly all areas are seriously compromised by herbivory from domestic livestock and wildlife. The role of insects and pathogens that attack plant structures also cannot be

> Hint: The solution to finding many partially eaten plants is to search
> outside plots for more complete plant specimens. Plants play a great
> game of "hide and seek" from herbivores, and it is usually possible to
> find more complete plant specimens nearby. Where all the specimens
> of a species are immature, returning to the site at a later date is
> important. It is wise to collect and press partial specimens in a
> working herbarium to compare later to complete specimens.

underestimated. It is not uncommon to have the majority of plants grazed
or browsed in a study area.

Difficulties with Species Rarity

Another well-documented problem with plant diversity studies is the uneven distribution of the number of individuals in each species in an area,
which complicates field measurements and sampling designs. Species abundance curves are almost always inverse J (or negative exponential) curves,
with a few species having many individuals and the vast majority of species
represented by very few individuals. In several field studies in the Central
Grasslands, Rocky Mountains, and Colorado Plateau, we have documented
that about 50% of the species encountered in 0.1 ha plots have less than 1%
foliar cover (Stohlgren et al. 1997b, 1998d, 2000b).

Most species are locally sparse. Like species abundance curves, species
frequency curves tend to have an inverse J shape (Figure 1.2). Of the 550
understory plant species recorded in the 309 plots (0.1 ha each) distributed
in the 850,000 ha Grand Staircase-Escalante National Monument in Utah,
189 appeared in only 1 or 2 of the plots, while only 36 species were found
in 50 or more plots.

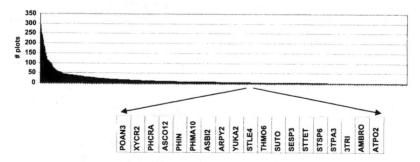

Figure 1.2. Data from 309 0.1-ha plots in Grand Staircase-Escalante
National Monument in Utah (unpublished data). Species codes (PLANTS
database; http://plants.usda.gov/) are shown for selected species.

A consequence of, or correlate to, rarity in species abundance curves and species frequency curves is that most plant species have very patchy distributions. In another example, in a random selection of four 0.1 ha plots in four prairie types in the Central Grasslands, only one of nine species, on average, was found in all four plots in each vegetation type (Stohlgren et al. 1998c). Many species in a vegetation type occurred in only one of the plots. In 1 m² subplots in the same prairie types, a few species were commonly found together, but few of the many locally rare species could be consistently found in the same subplots.

Patchy distributions and natural spatial variation are compounded by temporal variation (Huston 1999; Huston and McBride 2002). Temporal variation can be caused by immigration, emigration, seed germination, extirpation, speciation, and extinction. Immigration is very common, especially considering invasive exotic species. Plant species can germinate soon after field crews leave a site, with seed banks providing sudden species enrichment. Conversely, established species can desiccate to unrecognizable forms just prior to a field crews visit (i.e., local extirpation). Emigration, speciation, and extinction are generally much slower processes that are beyond the scope of most plant diversity studies, but they may be important in some areas and should not be discounted outright.

The solutions to problems of local rarity are to search larger areas, search well, and search often. I will discuss later ways to evaluate how large of an area needs to be searched and how often one should revisit sampling sites. Evaluating how well a site is searched is also important.

Why Designing Plant Diversity Studies Is Difficult

The difficulties just described in conducting plant diversity studies make it difficult to design such studies. Designing plant diversity studies for large landscapes is difficult for many reasons. The difficulties result from high spatial and temporal heterogeneity, which occurs at multiple spatial scales—from the rarest and tiniest of seeds in the seed pool to global patterns of diversity—with many species of plants, animals, and diseases rapidly dis-

---

Hint: The most important question a plant ecologist can ask is, "What did we miss by sampling?" Ecologists often report "what they've captured" (e.g., the number of species in a plot or group of plots) without evaluating the completeness of the work. There are many ways to evaluate completeness by extending the size of plots or using multiscale plots, or by adding time to searches for species. These topics are described later in the text (Figure 1.3). Lesson 2—Always ask, "What did we miss?"

Figure 1.3.   Photograph by Paul H. Evangelista. Used with permission.

persing throughout the biosphere. Imagine designing plant diversity studies in light of the complexities described below.

Plant-to-Plant Neighborhood-Scale Variability

At the neighborhood scale, two plant species can be (1) direct competitors, competing for the same resources; (2) dependent species, such as mosses that grow on particular tree species; or (3) complementary species, which use resources at different times or in different spaces (Huston 1999; Walter 1964, 1971). Furthermore, these dependencies may overlap at various growth stages, or in various habitats (Mueller-Dombois and Ellenberg 1974). When even the smallest plant dies, some resources may be released for a replacement by the same species, another species, or neither. Plants die for many reasons: age, disease, competition, and herbivory are typical causes. The cause of death may be drastic or slow. Life spans vary for plant species living in the same vicinity. Annual grass species can share a site with perennial herbs, moderately long-lived shrubs, and very long-lived trees. Local plant replacements can be sporadic over time, resulting in complex patterns of diversity at many spatial scales. Large numbers of individuals at a site may reduce the chance of local extirpation of a plant species, but it is often difficult to get accurate estimates of the number of spindly annual grasses or diffuse bunch grasses with dead or partially dead centers. It is equally difficult to identify individuals of sod-forming grasses or resprouting shrub and tree species. Defining "individuals" sometimes bothers plant ecologists. The abundance, diversity, variation, and viability of seed banks, and the timing of germination can also complicate the study of local plant species diversity.

Plot- to Landscape-Scale Variability

We can envision a landscape filled with thousands to millions of asynchronous plant neighborhoods superimposed on complex topographic variation (slopes, aspects, elevation), physical factors (soil texture, geology), and environmental gradients (moisture, soil nutrients, solar radiation). Small-scale, frequent disturbances (e.g., tree falls, small mammal excavations) can be superimposed on larger-scale, less-frequent perturbations (e.g., volcanism, fire, insect outbreaks, rapid climate change; see Figure 1.4) that vary spatially and temporally (Huston 1999; Huston and McBride 2002; Pickett and White 1985; Stohlgren 1994).

Spatial variation in plant diversity can be enormous: plant species lists in 0.1 ha plots just 100 m apart in grasslands in the Rocky Mountains overlap 30–80% (60% on average; Stohlgren et al. 1999b). Most species occur in multiple vegetation types (Figure 1.5).

Landscapes represent a mix of species that have migrated to the site over thousands of years under various climate and disturbance regimes. For example, one 0.1 ha area in the middle elevations of Rocky Mountain National Park, Colorado, might include relict limber pine (*Pinus flexilis*) from the last

Figure 1.4.   Lesson 3. Previous vegetation patterns influence seed banks and fire, which in turn influences the complex patterns of plant diversity at multiple scales in the post-burn stand, landscape legacies help to create complex spatio-temporal patterns of plant diversity in most natural ecosystems. Photo reproduced with permission of Geneva Chong.

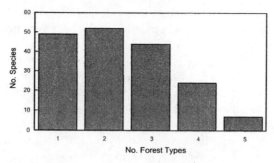

Figure 1.5.   Data from 24 0.1-ha plots in five vegetation types in Rocky
Mountain National Park, Colorado. Most plant species were found in
more than one vegetation type. A more detailed investigation would find
very few species strictly limited to one forest type.

ice age (10,000–12,000 years ago), ponderosa pine (*Pinus ponderosa*) that
arrived approximately 6000 years ago, species of cacti that may have migrated to the site during a warming period (3000–5,000 years ago), various
native plant species that arrived intermittently or in groups over the past
12,000 years, and up to 100 nonnative plant species that arrived from other
countries in the past 150 years. In the same area of the park, there is overlap
in species composition (typically 20–40%) among vegetation types and
complex ecotones between vegetation types, making it impossible to delineate exact boundaries of vegetation types. Spatial and temporal variation in
plant turnover, seed banks, seed dispersal from very distant sites (by wind,
birds, currents, etc.), and the timing of germination add to the complexity
in patterns of plant diversity in most landscapes.

A vegetation survey is sometimes thought of as a snapshot in geologic
time. It might be more realistic to think of a vegetation survey as a collage
of species from multiple geologic times experiencing potentially rapid
changes. The changes can occur in large areas from flooding, fire, or insect outbreaks. Alternatively, the changes can occur in small areas from
small mammal excavations, tree fall, or any small-scale disturbance. It is
obvious that conventional sampling strategies that randomly select a few
small study plots in relatively rare homogeneous study units may not be
adequate to describe plant diversity at landscape scales (Hinds 1984;
Stohlgren 1994).

Landscape- to Regional-Scale Variability

We also can envision a region filled with hundreds to thousands of asynchronous landscapes superimposed on complex topographic variations (longitude, latitude, elevation), climate factors (precipitation, temperature),
geology, and longer environmental gradients (moisture, soil nutrients, solar radiation) than seen at landscape scales. The physical environment and

living organisms are rarely uniformly or randomly distributed. Instead, at all scales, resources are aggregated in patches, or they form gradients or other kinds of spatial structures (Legendre and Fortin 1989).

Measuring plant diversity at landscape to regional scales is further complicated by extremely long lag times in evolution, migration, and speciation. Ancient riverbanks may still contain *Equisetum* (horsetails), a primitive vascular plant that evolved hundreds of millions of years ago, sharing the habitat with newly arrived modern angiosperms from other countries [e.g., the invasive purple loosestrife (*Lythrum salicaria*)]. Migration patterns have been vastly accelerated by modern transportation such as freightliners, airplanes, railroads, and trucks. Many geographic barriers to migration and speciation are easily breeched by the horticultural industry. Reichard and White (2001) report that between 57% and 65% of the flora of Australia was intentionally introduced for horticultural reasons.

Meanwhile, modern land use, primarily agriculture and urbanization, often leaves small islands of natural vegetation in a sea of transformed land. The small refuges may support smaller, more isolated populations of plants and animals, setting the stage for additional phenotypic variation, speciation, or extirpation. Plant ecologists have long realized that plant populations and species are clumped (aggregated) in distribution rather than randomly located on the landscape (Ashby 1948; Greig-Smith 1964). Recognizing the patchy distribution of plant species from the neighborhood scale to the regional scale is the first step in addressing sampling design considerations.

Long-Term Changes in Plant Diversity

Climate and weather history and patterns provide an analogy for understanding long-term changes in plant diversity. Dominant plant species in a region may reflect long-term seasonal patterns in temperature and precipitation. Other species in the region may have been established during various cycles (e.g., El Niño years, severe droughts) and lingered through suboptimal climate years. Still other species may have been established after extreme events such as floods or in high fire years as a result of extreme drought and high winds. Thus understanding long-term mean climatic conditions alone may not necessarily aid in the understanding of long-term changes in plant diversity. Many other factors, such as dispersal, migration, herbivory, disturbance, edaphic characteristics, competition, and seed bank viability, among others, affect long-term changes in plant diversity. Subtle changes in plant diversity at landscape scales cannot be detected without a long-term record (Strayer et al. 1986). Because long-term studies are rare, ecologists have only a cursory understanding of coupled spatial and temporal variability (Haury et al. 1978; Levin 1992; Powell 1989; Steele 1989). Thus, far more long-term studies, innovative sampling and monitoring strategies, and new analytical and modeling tools may be needed to evaluate long-term changes in plant

Figure 1.6. Lesson 4. Plant diversity (northwestern Wyoming) is patchy at multiple spatial scales in all natural landscapes–get used to it!

Figure 1.7. Lesson 5. Tamarisk (*Tamarix spp.*, salt cedar) invading an arid landscape in southern Utah. Plant invaders like salt cedar can radically alter vegetation structure, water and nutrient cycles, soil chemistry, and local plant diversity. More invaders are on the way! (Photograph by Paul H. Evangelista, used with permission.)

diversity (Stohlgren 1994). Plant species are rapidly invading from other countries (Dukes 2002; Floerl et al. 2004; Gilbert and Lechowicz 2005; Mack et al. 2000; Stohlgren et al. 2005a,b,c). The obvious solution is to become extremely clever about evaluating plant diversity at multiple spatial scales and over time.

2

History and Background,
Baggage and Direction

In this chapter I provide a very brief history of plant ecology to focus on how previous ecologists have influenced the ways we typically measure plant diversity today. I draw particular attention to the authors of two textbooks, Rexford Daubenmire, and Dieter Mueller-Dombois and Heinz Ellenberg, because they seem to best reflect the development of many current plant diversity field methods (Barbour et al. 1999). Lastly, I discuss the general direction provided by past plant ecologists, and on the "baggage" of older ideas—how inertia developed and persists in modern plant ecology regarding measuring plant diversity.

History and Background

In the Beginning

To understand modern approaches to the measurement of vegetation, it is important to provide a brief history of plant ecology. Many writers have provided more detailed histories of the people, places, culture, and schools of thought emanating from our scientific ancestors (e.g., Barbour et al. 1987; McIntosh 1985; Tobey 1981). I focus instead on the influence early naturalists and plant ecologists had on paradigms and sampling designs.

One concept binds ecologists: the quest to understand species-environment relationships. Earlier peoples recognized species-environment relation-

ships in everyday life. McIntosh (1985) noted that Theophrastus, from ancient Greece, described mangrove forests related to saltwater habitat. Charles
Darwin's *The Origin of Species* (1859) is the defining volume of observations, theory, conjecture, and wonder about species-environment and evolutionary relationships. Evaluating the primary effects of Darwin and his
predecessors on sampling designs is difficult. In carefully documenting both
broad-scale patterns of vegetation and wildlife in several countries, and the
inseparable links between biotic and abiotic causes and effects, Darwin
saddled future ecologists with the prospects of complex, multivariate explanations for the patterns and processes of plant ecology.

The major force in early plant ecology was Friedrich Heinrich Alexander
von Humboldt (1769–1859) from Prussia (Figure 2.1). Humboldt and others collected 60,000 plant specimens from throughout Central and South
America from 1799 to 1804, later producing 14 volumes on the botany,
physiognomy, and "associations" (vegetation types) of the region (Barbour
et al. 1987). He described many associations with respect to interrelationships among latitude, elevation, and temperature. He believed that all things
were connected by linkages of causes and effects. He published a vegetation classification system in 1806 based entirely on growth form, ignoring
taxonomy for the most part. Humboldt established two important paradigms
in plant ecology: (1) that vegetation can be described in "associations" based
primarily on dominant species, and (2) that species and associations can be
understood in terms of measurements of relatively few topographic and
environmental factors. How did Humboldt influence vegetation sampling
procedures? Most naturalists and plant geographers in the early 1800s con-

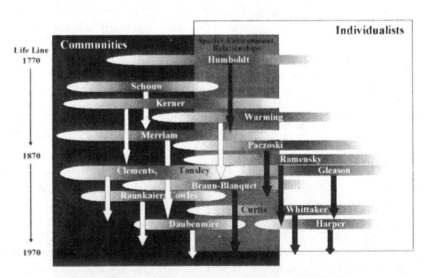

Figure 2.1.   Schematic diagram of selected historical figures who influenced modern plant diversity research.

tinued to use "searching" techniques—collecting and cataloging plant specimens, and taking careful notes of key environmental factors. Humboldt drew
attention to the need for careful, systematic observations ("biological arith-
metic") to show, for example, the vertical zonation of vegetation. Ecological plant geography was born (McIntosh 1985) as Humboldt recognized the
need to accurately measure and sketch natural phenomena.

Influenced by a subset of Humboldt's ideas on plant associations, J. F.
Schouw (1789–1852) emphasized temperature as the single most important
factor in determining plant distributions. He described vegetation associations by attaching the suffix "etum" (meaning "community of") to genus
names of the dominant species. The suffix is sometimes used today. Although
many taxa may have been recorded at each site, attention may have been
inadvertently drawn away from the understory and locally rare species with
an overemphasis on the few dominant species (and their relationship to
temperature) at each site.

Anton Kerner von Marilaun (1831–1898) described plant life of the Danube
Basin with detailed descriptions of vegetation associations and orderly succession. Kerner von Marilaun emphasized order: "The horizontal and vertical structuring of large plant communities is by no means accidental . . .
Every plant has its place, its time, its function, and its meaning. In every
zone, plants are distinct groups which appear as either developing or as finished communities, but never transgress the orderly and correct composition of their kind" [Conard (1951) as quoted in McIntosh (1985)]. Kerner was
typical of most 19th-century plant geographers and the prevailing beliefs of
order in nature and a world filled with distinct plant associations. Plant
geography influenced many future plant ecologists in the design of studies,
the development of quantitative field methods, and the interpretation of
results.

Danish botanist Johannes Eugenius Warming (1841–1924) recognized the
linkage of physiology and morphology. He drew attention to the important
role of soil moisture in vegetation patterns. He summarized vegetation in
Brazil with data on climate, soils, and geology, emphasizing the important
role of moisture and temperature. His textbook(the first) on ecological plant
geography, *Plantesamfund,* published in 1895, created new terms for wetland plants (hydrophytes), moist habitat plants (mesophytes), and dry habitat
plants (xerophytes). He emphasized dominant and subdominant plants
of various vegetation types as a matter of convenience for discussion, but
he criticized the notion that "causes" could be applied to such entities
(McIntosh 1985). He described the effects of fire and succession and clearly
emphasized individual plant species and many gradual gradients of vegetation, soil, and moisture. He did not view succession as necessarily unidirectional. In fact, Warming wrote that plant communities were never stable, not
in equilibrium, and frequently changing due to interactions with animals,
fungi, and competition among plant species (McIntosh 1985). Warming also
noted that competition among species could force a species to grow in a less-

preferred habitat. This could lead one to misinterpret that a species "prefers" a given soil, when it was really displaced by competition in optimal locations. This finding warned future plant ecologists that species-environment relationships might include complex biotic and abiotic components and interactions.

Jozef Paczoski (1864–1941), perhaps the first phytosociologist, realized that plants both respond to and create their own microenvironments in competition with other species. He discussed the interrelationships of species in plant communities. Paczoski's major contribution, although it was overlooked by most plant ecologists, was the concept of gradients between plant communities. Vegetation boundaries are often continuums, affected by microenvironments, disturbance, and various patterns of succession (Barbour et al. 1987).

Leonid Ramensky (1884–1953) fully developed "the individualist concept" before Henry Gleason (Barbour et al. 1987; see below). He demonstrated before Josias Braun-Blanquet (see below) that species independently followed environmental gradients and created tables of species by foliar cover. Ramensky also developed terms for various plant strategies similar to competitors, stress-tolerators, and ruderals [before Grime (1977)], and similar to r and K strategies [before Pianka (1980); see Barbour et al. 1987]. Ramensky (1924) made several other contributions to vegetation sampling. First, he championed the need for a "quantitative and methodologically substantiated registration of facts" to support plant ecology. Second, he proposed and used a standardized square quadrat (with a diagonal of 2 m) for recording the "horizontal projection of ground shoots" (now termed foliar cover). Third, he used the standard deviation to compare variation among plots. Lastly, Ramensky ingeniously noted where plant species were absent in a plant association—a concept that would later be used to help predict probable distributions of rare plant species at landscape scales.

Clinton Hart Merriam (1855–1942) formalized the concept of elevation zones of montane vegetation. His detailed descriptions of the Cascade Mountains of Washington, which focused on temperature (growing season) as the major determinant of vegetation life zones, led to his life zone maps of North America. The focus then returned to the geography of dominant species. Merriam's influence on vegetation can be seen in Bob Bailey's ecoregional maps today, and the concept of biomes (http://www.fs.fed.us/institute/ecolink.html).

Frederick Edward Clements (1874–1945) is the most well-known historical figure in American plant ecology (Barbour et al. 1987; McIntosh 1985; Tobey 1981). Clements drew heavily on the plant geography views of Oscar Drude (1902), who wrote about the communal life of species that caused an ever-changing dynamic factor in changing the face of the world (McIntosh 1985). Clements expounded on these views to insist on the organism-like "plant associations," "regional formations" of vegetation, and orderly processions of succession from early "seral stages" to "climax." He developed a broad background in plant geography by describing the distribution of every

plant species in Nebraska. His study sites ranged from southwestern des-
serts, Death Valley, and coastal California, to Pike's Peak and alpine regions
in Colorado. He knew that some species were environmental indicators for
documenting succession and stressful sites.

Clements (1916, 1936) was dogmatic about his ideas on plant associations
and succession. He would often classify discrete plant associations based
on one or two dominant species. However, he recognized "open" and "closed"
communities, where changes in species composition were likely or less
likely, depending on the availability of limiting resources such as light, water,
minerals, and air, and the use of those resources by a variety of plant species. His views on plant associations as holistic superorganisms were widely
accepted in his time. The concept of stable, self-sustaining communities
controlled by climate are the basis for "potential vegetation maps" which
are widely used today in climate modeling, ecosystem delineation, and
biome mapping.

With his colleague Roscoe Pound, Clements began standardizing the area
of vegetation sampling by quantifying species composition in 1 square mile
areas in Nebraska (Pound and Clements 1898a,b). "Frequency" was documented: the percentage of square mile areas occupied by given species. Then,
within 5 m × 5 m quadrats, they enumerated the abundance of individuals
of each species. Quantitative, multiscale vegetation sampling was born.
Clements later reduced the size of the small sampling unit to 1 m^2 and the
size of the larger sampling unit to 16 m^2, greatly reducing the costs of sampling (Weaver and Clements 1938).

Pound and Clements also created an index of abundance (A) that combined information on frequency, estimated cover, and abundance in quadrats (or smaller units). The index of abundance, $A = (t \times e \times a)/T$, where t is
the number of sample units that contain a given species, e is the estimated
mean cover of the species throughout the sample area, a is the number of
individuals in quadrats, and T is the total number of sample units. Since e,
the mean cover of the species throughout the sample area, was often a pure
guess, and frequency was affected by quadrat size, the abundance index was
criticized by many and not used often, even by Clements. However, the
concept of combining information on species abundance at local scales with
occurrence at landscape scales was a very clever addition to quantitative
plant ecology.

Clements had the most profound effects on the way vegetation is sampled
today. He contributed five indelible staples to the measurement of plant
diversity. First, Clements, like Ramensky (and Gleason, see below), insisted
on the use of a "quadrat" in vegetation sampling. Having a standard, repeatable, statistical measure in various plant communities was revolutionary in
terrestrial plant ecology. Clements stated that experience and simple observations without enumeration was invalid (McIntosh 1985). Second, viewing the world as largely homogeneous vegetation units separated by other
distinct, homogeneous units, Clements guided the use of quadrat sampling

into recognizably homogeneous "representative" communities and formations. In this way, order could be detected from the chaotic approaches and
descriptions of others in the past. Third, he emphasized an inductive procedure to test multiple working hypotheses—primarily with quantitative
sampling. He used quadrats to describe current conditions (species richness
and composition, frequency, cover), experiments (removal and regrowth
experiments), and monitoring (assessing change over time). Fourth, the concept of an index of abundance, or combining information across spatial scales
and vegetation characteristics, set the stage for "importance values"—a way
to measure the dominance of particular species at particular sites—and
"diversity indices"—a way to combine species richness and abundance values. These concepts and contributions are discussed later. Fifth, Clements'
new direction included studies of vegetation change rather than descriptions of the status quo. This moved the discussion from simple surveys
and plant species inventories to monitoring. The hope was that this would
provide a better understanding of the underlying causes of orderly community change—succession.

Danish plant ecologist Christen Raunkaier is best known for life form classifications of plants such as phanerophytes, chamaephytes, cryptophytes, etc.,
describing plant architectures and leaf traits. More important, he strongly
supported quantifying plant "formations" and published several important
works between 1908 and 1934 on the use of statistics in plant ecology. Raunkaier sought to "improve upon the uncertain picture we obtain by subjective estimates of plant communities" (McIntosh 1985). He noted that all
species should not be weighted equally when comparing vegetation types.
Raunkaier influenced vegetation sampling in several ways. First, he advanced the use of quadrats, going beyond the use of cover classes (e.g., by
Clements) by recording individuals by height class to gain insights on plant
species dominance. Second, he took an experimental approach to sampling
by evaluating the effects of quadrat size on species richness, always looking
for new ways to quantify plant frequency and dominance.

Henry Chandler Cowles (1869–1939) focused on vegetation change over
time—succession—on dunes near Chicago. He discussed the classification of
"vegetation societies" and "vegetation cycles." Cowles was an "observer" and
often doubted the utility and meaning of "natural experiments" (McIntosh
1985). Cowles, a contemporary of Clements, noted that succession was not a
straight-line process with a fixed ending. He recognized that past conditions
and processes could greatly influence current vegetation conditions.

Henry Gleason (1882–1975), unappreciated in his day, is now influencing plant diversity sampling in profound ways. Gleason's paper on "The
Individualistic Concept of the Plant Association" (1926) strongly challenged
Clements' well-accepted body of work. Gleason was driven by a strong desire to understand the distributions of individual species. He noted that most
"communities" were in the eyes and minds of investigators. Gleason thought
that plant species were clumped in distribution on the landscape, but not

as tightly associated with other species as they were with environmental factors. He used quadrats to test whether individuals and species were randomly distributed, finding that most were distributed in clusters or patches (Gleason 1920). This work launched decades of studies on theoretical ecology and the ideas about random versus nonrandom distributions in natural populations (Greig-Smith 1957; Hutchinson 1953; Pielou 1969).

Gleason had little effect on the sampling design and methods development of early plant ecologists. So strong was the Clementsian paradigm of a landscape filled with largely homogeneous communities that most early plant ecologists were preoccupied with describing, classifying, and mapping communities (Whittaker 1962). Lost was Gleason's emphasis on the distribution of individual species. Two advances by Gleason were particularly ignored by most plant ecologists in the 20th century. First, like Raunkaier, he was well aware of the effects of quadrat size on species richness. More important, Gleason recognized the importance of quadrat sampling throughout a large area rather than confining quadrat sampling to small, "representative" areas of supposed communities. He tried, in vain, to shift the emphasis of vegetation science from dominant species and artificial communities to individual species—the primary unit of evolution as noted 75 years earlier by Darwin.

Josias Braun-Blanquet (1884–1980), influenced by Kerner von Marilaun, developed methods for association measurement, classification, and nomenclature (Barbour et al. 1987). The approach is sometimes referred to as the Zurich-Montpellier, or Sigmatist, school of taxonomic natural history, in which communities are typed by characteristic species, not necessarily the dominant species (McIntosh 1985).

Braun-Blanquet simplified the collection of field data. An abbreviated description of the Braun-Blanquet or relevé method is as follows [see Barbour et al. (1987) for a more complete description]. An extensive hike of the study area is completed and specific vegetation communities are selected for more detailed survey. In each community, several stands are subjectively selected and thoroughly surveyed to develop as complete a species list as possible. Although this widely varies with investigators, the most typical or representative area of a stand in each community type generally is subjectively selected for further study. Detailed information is recorded in a fairly large plot (determined by a nested quadrat technique and species area curve, or the investigator's experience; see chapter 5). In each plot, the foliar cover of each species is recorded in vegetation cover classes ($<<1\%$, $<1\%$, 1–5%, 5–25%, 25–50%, 50–75%, 75–100%). The "sociability class" of each species is also recorded [from 1 (growing singly) to 5 (growing in large, almost pure stands)] to differentiate between clumped versus scattered species with the same cover. The seasonal importance of species, or "periodicity," and environmental factors might also be recorded. The Braun-Blanquet relevé method allows for classification of plant communities with more species in the database than those that record only dominant species at each site (e.g., the way Clements typically classified associations). The Braun-Blanquet relevé method was and remains one of the most widely used approaches in the world, outside North America (Barbour et al. 1987).

A. G. Tansley and T. F. Chipp wrote *Aims and Methods in the Study of Vegetation* (1926). The preface of the book begins with an analogy that store owners and managers of large estates each must be aware of the nature of their stock and extent of supplies. The emphasis of the methods was to quantify the types of vegetation that could be "directly exploited by man," and to enhance a knowledge and interest in vegetation. The authors cautioned readers "one can acquire a considerable floristic knowledge and yet know next to nothing about vegetation." Likewise, investigators could learn about vegetation without having complete information on floras.

Tansley and Chipp (1926) also recognized "stable" and "unstable" vegetation as communities that were and were not in permanent equilibrium, respectively. They not only recognized the preponderance of "climax" communities [in strong agreement with Clements (1916)], but they also recognized many other subjective classifications of plant assemblages. "Society" was an association of subordinate (usually not dominant) species. "Consociations" were typically found groups of dominant species that might grow separately in pure stands in some areas. "Clans" were defined as small associations of single subdominant species in limited areas. Seral stages were subdivided into primary seres (or piseres), subseres, and habitat-specific seres. Few investigators used these terms frequently.

Tansley and Chipp (1926) recommended belt transects (66 ft × 6600 ft; 20.1 m × 2012 m) to sketch the main boundaries of vegetation types, with aerial photography as an aid to mapping. Sadly the authors noted "there are few, if any, countries, at least within the confines of the British Empire, which still are mainly covered by a virgin vegetation that can be mapped and studied" (p. 48). Still, they noted the value of general observations and note taking, and the mapping of vegetation zones and ecotones (or "transition belts") that are "almost universal" between zoned communities. Charting small areas of vegetation could use quadrats (square areas) up to 100 m² in forested types. Permanent quadrats could be used to assess vegetation change. Transects were said to be "useful in homogeneous vegetation as a method for recording the composition of the community," and indispensable for "representing zonation and transitions." Line transects were used to describe "really homogeneous" vegetation. The authors suggested using quadrats in conjunction with line transects to get a two-dimensional view of vegetation composition.

Robert H. Whittaker (1920–1980) noted the uneven treatment in the history of plant ecology given to the "community-unit theory" of Kerner, Clements, and early plant geographers. He questioned whether the plant community was an appropriate basic natural unit. Still, he developed ordination and gradient analysis techniques to quantify the classification of plant communities and associated environmental variables. However, Whittaker vehemently challenged Clements' ideas and understood the complex pro-

cesses affecting succession, productivity, and species diversity. His approach
to gradient analysis continues to influence many vegetation studies today.
But it was his insights and multiscale field methods in species diversity
which may prove to be his most useful and long-lived contributions to the
study of vegetation. As we shall see, multiscale field sampling, pioneered
by Whittaker, is the foundation of modern quantitative approaches to measuring plant diversity (see chapter 5).

Oosting (1948) recognized the difference between the qualitative synthe-
ses of many early plant ecologists and the descriptive analytical statistics of
specific stands or sites in plant ecology. The analytical statistics included the
frequencies of species, cover, abundance, density, and biomass. He also recognized that these common measurements were made on plots that varied in
shape, size, and number—so comparisons among studies were difficult.

To avoid sample area issues, some methods were developed to estimate
cover and abundance from lines and points along transects (Cottam and
Curtis 1949), or from distances of plants to selected sampling points (Greig-
Smith 1957). Many early plant ecologists relied on assumptions on the importance of measuring predominantly dominant species (or life forms) in
relatively homogeneous units (McIntosh 1985). In Europe, this generally
translated to single, large plots (approximately 100–4000 m²), subjectively
located in the "most representative" site of a dominant vegetation type.

Many others contributed to the advancement of quantitative plant ecology. McIntosh (1985) points out that in 1903, an observant animal ecologist, C. B. Davenport, noted five profound generalizations of animals on the
beach:

1. The world contains numberless kinds of habitats.
2. Each organism has its own habitat consonant with its own structure.
3. Dispersal distributes organisms to better and worse habitats.
4. Those reaching a habitat for which they are better fitted will thrive
   and multiply, and the converse.
5. This process goes on until the organism is found primarily in suitable environments.

Davenport saw the process as complementing natural selection (McIntosh
1985). These ideas reopened the lines of reasoning by Paczoski and Ramensky,
and may have influenced Gleason (1917)—drawing attention to the individualistic approach to plant ecology.

W. S. Cooper (1913), a student of Cowles, describes vegetation as a con-
tinually changing mosaic caused by a myriad of small-scale disturbances.
In 1924, British ecologist A. S. Watts described gap-phase dynamics in beech
forests, forcing future ecologists to recognize predictable small-scale disturbances embedded in large-scale, mixed-species, heterogeneous forests. The
works of Cooper and Watts remain largely ignored by broad-based plant
geographers, vegetation classifiers, and mapmakers. Their works are well

known to gap dynamics modelers, individual-based plant modelers, and those studying biodiversity.

A. S. Watt formally introduced "gap" dynamics in woodlands that focused research at substand scales in 1947. While Watt used a population demographic approach, his emphasis on microenvironments drew attention to habitat heterogeneity at multiple scales.

Bray and Curtis (1957), Curtis and McIntosh (1951), Goodall (1957), and Greig-Smith (1957) developed mathematical techniques to assess data on multiple factors affecting species niches or quantifying similarities in species composition among plots (i.e., the birth of clustering and ordination algorithms).

It is this well-blended, historical mix of people, ideas, and methods that formed the early textbooks on vegetation measurement by Rexford Daubenmire (1968) and Dieter Mueller-Dombois and Heinz Ellenberg (1974).

Rexford Daubenmire

Rexford Daubenmire so greatly reflects modern vegetation measurement that his textbook on plant communities (plant synecology) (1968) is still on the bookshelves of many ecologists. The text has been cited more than 300 times in primary journals since 1968 according to the Web of Science (http://isi1.isiknowledge.com/portal.cgi). I focus on Daubenmire because of his enormous impact on rangeland sampling in the western United States. His 20 cm × 50 cm quadrat and transect sampling technique (chapter 7) has been used in hundreds (if not thousands) of studies, is the most widely used technique among federal land management agencies, and is still recommended in U.S. Department of Agriculture (USDA) Forest Service rangeland sampling manuals. Modern textbooks still describe his approaches (e.g., Barbour et al. 1999) and many government and nongovernment organizations rely on this or similar approaches to assess riparian vegetation, grazing effects, and fire effects. While not solely attributed to Daubenmire or any one plant ecologist of his time, the emphasis of sampling vegetation in "homogeneous areas" may have inadvertently led to the common use of small quadrats and small sample sizes still evident in many vegetation studies today (Kareiva and Anderson 1988; Stohlgren et al. 1998c). Daubenmire serves to illustrate how far we have come, and how far we need to go, in the science of measuring plant diversity.

Daubenmire was an exceptional naturalist and ecologist who recognized interactions from the level of genetics and individual plant "behavior" to communities as the building blocks of ecosystems. Influenced perhaps heavily by Braun-Blanquet, he strongly believed that homogeneous communities formed because of the different environmental tolerances of various taxa and the large-scale patchiness of the environment. Therefore each kind of community could be distinguished by the relative abundance of a certain few plant species from the regional species pool that can establish and persist in the specific environment, and coarse-scale patchy environments create a mosaic of homogeneous communities. However, he had a very elastic definition of "community." It could be any assemblage of organisms, large or small, containing two or more species. Community could be used in an abstract or concrete sense. Still, Daubenmire steadfastly claimed that all organisms in a community were ecologically interrelated, though they might differ in the degree and kind of interrelationships. Species could have "obligative relationships" (e.g., mycorrhizal fungus and conifer seedlings) or very indirect relationships that might be difficult to measure because locally rare species may have low functional significance. Thus the large, common species dominate the biotic character of the ecosystem, and many species may be relatively inconsequential. This exemplifies a strong tendency of early plant geographers to disregard low-biomass species (i.e., those assumed to have low functional significance) because they did not contribute much to forage production or commercial timber volume.

Daubenmire cleverly observed habitat heterogeneity at several scales. At the microsite scale, he noted that plants differing in height or seasonal development could coexist by varying their timing of resource use. At the landscape scale, he noted that ecotones (i.e., gradients where two communities meet) often exist. He noted that most species of plants are shared across ecotones. Oddly, these insightful observations are obscured by far more emphasis on mostly homogeneous communities with defined boundaries in space and time. For example, he was adamant about avoiding ecotones for experiments on the response of vegetation to burning, grazing, or fertilization.

Daubenmire influenced the measuring of plant communities by reinvigorating the focus on the level of "plant associations," following Braun-Blanquet and traditional European approaches to plant ecology. Associations were defined as a kind of plant community of definite composition, representing a uniform physiognomy, though he acknowledged that "definite" and "uniform" were relative terms that meant different things to different people (Daubenmire 1968, p. 27). Most of the field techniques in Daubenmire's book are presented in the context of describing plant communities, associations, and stands. A "stand" was defined by Daubenmire as "each piece of vegetation that is essentially homogeneous in all layers and differs from contiguous vegetation types by either quantitative of qualitative characters." As an extension of the Schouw, Kerner, and Clementsian views of the world, he reasoned that an "association" is a portion of climax stands ("in which the dominants of corresponding layers are essentially the same;" p. 27) that differ slightly in chance dissemination or a transitory historical factor. So groups of similar stands make up an association. Daubenmire then defined a "habitat type" as all the area that now supports, has supported, or will support one plant association. I mention these various definitions because a landscape is often stratified based on stands, associations, community types, or vegetation types prior to sampling. Difficulties in stratification (or classification) can arise when (1) terms are defined differently by different workers, (2) the levels of homogeneity within or among various classes differ among workers, or (3) the locations of boundaries differ among workers. It was underappreciated that Daubenmire introduced the topic with a plea for improved skills in taxonomy, realizing that knowledge of more than a few dominant species was necessary. He also recognized that contemporary vegetation was an indicator of site condition—a point lost on many mappers and modelers of "potential vegetation."

Below are various attributes of vegetation, Daubenmire's recommended approaches and methods in 1968 to assess them, and various potential problems with each of the approaches:

Density   For surveys of large areas with little cost, rough estimates of density (i.e., the number of species per unit area) can be made with five abundance classes: very rare, rare, occasional, abundant, and very abundant. The potential problem here is that these classes are subject to various interpretations by investigators.

Frequency   For more detailed work, frequency (i.e., the percent occurrence in a number of plots) is a more valuable statistic. A species that appeared in 40 of 50 plots would have a frequency of 80%. The potential problem here is that frequency is influenced by plot size. Back in 1909, Raunkiaer noted that with plot sizes as small as 0.01 m^2 (i.e., small enough so that two individuals of the same species would rarely be contained in a given plot), frequency estimates approximated density estimates (Raunkiaer 1934). Daubenmire, himself, noted that very small plots, 0.75 inch (2 cm) diameter loops, and even points (0–5 mm diameter, sharpened metal rods) were used by various investigators to measure frequency.

The specific problems of point methods have been discussed by several investigators. Daubenmire noted that descending rods often used in point methods contact large leaves such as broadleaf herbs better than small, thin-leafed grasses. Several thousand points might be needed to accurately capture locally rare species. Small plots may accurately describe the frequency of common species, but may miss locally rare species or species with patchy distributions, resulting in lower frequencies for those species. Large plots capture more species, generally increasing the frequencies of locally rare species. Frequency values cannot be compared among studies that use different plot sizes. Raunkiaer assessed the frequency in 20 cm × 20 cm quadrats, and with "frequency classes" of 0–20%, 21–40%, 41–60%, 61–80%, and 81–100%. He noticed that frequency data collected this way often resulted in an inverse J-shaped curve, but in some cases, a dominant species overwhelmed the plot, resulting in a flattened, even distribution of species by frequency class.

Daubenmire also clearly recognized the need to record plant species missed by smaller quadrats. In the exercise of classifying ecosystem types (or community types) he cautioned that "one should make sure the sample is adequate, and always those species which are present but do not fall in the sampling system should be listed" (Daubenmire 1968, p. 77). Perhaps overlooked, or at least unreported by Raunkiaer and Daubenmire, are the more complex challenges of sampling plant diversity when it is greatly effected by the size of the quadrat and the spatial distribution of common and rare plant species in the study area (as we shall see later in chapter 5).

Dominance   Plant "coverage" (now commonly called foliar cover; the sum of the shadows cast by stems and leaves for each species) provides a common estimate of dominance. Small plots or quadrats, like the still commonly used 20 cm × 50 cm "Daubenmire quadrat," are better suited for precise measurements of cover in grasslands and herbaceous communities. These methods appear in many modern textbooks and manuals for measuring vegetation (e.g., Barbour et al. 1987, 1999). Sometimes cover estimates are made to the nearest one-tenth percent, nearest percent, or in cover classes (Vestal 1943). Potential problems are well known to most plant ecologists. Overhead canopy cover of tall trees is difficult to estimate, leaves of different species overlap, cover measurements can change drastically over the season, and various growth forms such as skinny grasses, compound leaves, and grazed plant parts create difficulties in accurately estimating cover. Many plots are needed to estimate the cover of most species. First, the small quadrats that aid in the precision of cover estimates for dominant species miss many locally rare species around each quadrat, so they underrepresent local diversity and cover estimates. Second, because most species have patchy distributions across a landscape, diversity and cover are underestimated at landscape scales.

Line interception (Hormay 1949, p. 228) was reviewed as an alternative means to measure plant cover by species. The length of cover is recorded for each plant along a given length of outstretched tape, then summed by species and converted to a percentage. Despite the recognized problems caused by the nonuniform treatment of taxa with different growth forms (overhead trees versus skinny grasses), Daubenmire touted the technique as a cost-efficient means to monitor changes in plant communities. The underlying assumption of the use of line intercept methods for monitoring community-scale changes is that the methods actually capture a significant number of species and accurate estimates of cover of species to be representative of vegetation changes in the larger portions of the unsampled community. Means to assess the accuracy of the method, the adequacy of sampling intensity, and the representativeness of the samples are not addressed by Daubenmire (1968).

Sociability   Sociability was described by Kershaw (1963) and Greig-Smith (1964) as the frequency of occurrences of two species in a given size quadrat. For example, if plant species were randomly distributed, and species A was reported in 50% of the plots and species B was reported in 50% of the plots, then the two species would, by chance alone, occur together 25% of the time (0.5 × 0.5 = 0.25). If the value greatly exceeded 25% or was significantly less than 25%, then the species distributions might be positively correlated or segregated, respectively. Such measurements are heavily dependent on plot size.

Daubenmire (1968, p. 58) clearly recognized the contagious distribution of plant species on a landscape. He used examples of offspring near parents, concentrations of plants in locally favorable microenvironments, nurse tree or nitrogen-fixing species creating more favorable environments for other species, allelopathy (toxic) effects of some species inhibiting other species, and repeat uses of favorable sites after adults have died. Left unsaid was that patterns of contagion can operate at multiple spatial scales, and plant diversity sampling schemes must accommodate the peculiar spatial patterns of various species assemblages under a variety of environmental conditions.

Importance Values   Importance values try to relate the relative contribution (or dominance) of a species in a plant community. The concept was developed by Curtis and McIntosh (1951) to synthesize plot data for species in a prairie-forest ecotone in Wisconsin. Importance values were the summation of relative productivity, relative density, and relative frequency (Table 2.1).

Penfound (1963) focused on dominant species, as evidenced by the concept of importance. His examples were based on the point-centered quarter method, a means of surveying for forest and rangeland plant species. At random points along a transect, the nearest species in each of four quadrats (NW, NE, SE, SW) was recorded. Thus the common species close to each point was recorded and the "importance value" calculation ranked the most dominant of the common species. Daubenmire (1968) noted that the size and shape of a quadrat will influence frequency measures, and summing density and productivity (or cover) values may be compromised in a single importance value. In addition, the point-centered quarter method tends to greatly underrepresent rare entities that are clustered (aggregated) on the landscape (Ludwig and Reynolds 1988). And as Barbour et al. (1987) point out, any difference in importance values among species are "submerged in the addition process." Thus the high relative frequency of species B may be overlooked—and the species may be a sod-forming invasive grass [e.g., Ken-

Table 2.1.   Hypothetical example of importance value calculations for three species in a plant community.

Species	Relative density (1)	Relative cover (2)	Relative frequency (3)	Importance value (1 + 2 + 3)
A	25	75	30	130
B	5	10	70	85
C	40	20	40	100

Species A contributes most to the community, while Species B contributes least, using these criteria

tucky bluegrass (*Poa pratensis*)] that could soon alter relative density and cover of species A and C in the community. Of course, importance values were not intended to assess future importance, but synthetic measures should be viewed with caution because relative biological importance may be quite different from relative statistical importance for any set of attributes.

Constancy   Constancy is the frequency of species in given-sized plots in a community. If a species occurs in 6 of 10 plots, each placed in a different stand in the same vegetation type, the constancy is 60% (Barbour et al. 1987). In discussing constancy, Daubenmire (1968) noted that the same area must be considered in all stands. However, species may not be distributed evenly, randomly, or similarly in all the stands, so the same sample area in each stand could yield dissimilar results. Daubenmire (1968, p.77) clearly recognized the individualistic distributions of species:

> Since all species are individualistic with regard to habitat tolerances and requirements, relatively few can be expected to fall into the pattern of the group showing high constancy that distinguishes an association. Furthermore, all members of an indicator group are not invariably present, and an occasional stand may warrant inclusion even when all the usual indicator plants are absent, providing other aspects of the ecosystem suggest such a placement.

He believed chance was one factor governing plant species distributions.

Homogeneity   Daubenmire realized that most techniques and mathematical tests of homogeneity imply that "(1) every plant occurs wherever the environment is suitable for it; (2) floristic lists alone are sufficient for judging degrees of similarity; (3) all species, dominant as well as accidentals, have equal indicator value; and (4) biological classifications can be rigorously objective" (1968, p. 77). He added that "the results of lengthy computations often do little more than verify vegetation discontinuities that are evident to a trained synecologist by careful inspection." I interpret this to show Daubenmire's disdain for complex multivariate statistical approaches and perhaps a slight bias toward qualitative observations by trained naturalists and vegetation ecologists (e.g., Clements and Küchler). However, Daubenmire worried that "two communities sharing the same species list may not be so closely related ecologically as their taxonomic lists suggest. Clearly abiotic components need to be given some weight, and plant lists should not be relied upon entirely" (1968, p. 77). On the one hand, he recognized problems with species and community taxonomy, largely due to chance and the individualistic distributions of species, yet the field methods described in his book often were far better at capturing common species for classification than for capturing rare species. He clearly recognized that "every species has a distinctive ecological amplitude, and nearly all are found in two or more associations" (p. 79)—so communities are really more reflected by environmental conditions than by species.

Plant Species Fidelity   Daubenmire defined fidelity as the extent to which a species occurs regularly in every stand of an association (1968, p. 78). Generally this was a qualitative assessment from 1 (low fidelity) to 5 (high fidelity) as follows:

- F1: rare in the association, common in at least one other.
- F2: not showing species affinities for the association ("companion species").
- F3: better represented in this association than in others, but common elsewhere.
- F4: rather uncommon in other associations in comparison with this ("selectives").
- F5: completely, or almost completely confined to this association ("exclusives").

He noted that many plants fall into classes 1–3, few into class 4, and extremely few into class 5 because most plant species occur in two or more associations. Daubenmire claimed that for classification of associations, "one must depend on a group of species that: (1) have high constancy; and (2) occur as a group only within a distinctive range of environment" (1968, p. 79). However, this seemed in direct conflict with his findings that few species had high constancy.

Sampling Tree Volume or Forage Biomass in a Landscape   On random sampling of landscapes for the volume of timber or forage biomass, Daubenmire stated: "One cannot synthesize associations from samples taken at random over a landscape. Widespread associations are needlessly over-sampled, uncommon types are under-sampled, and many samples fall astride ecotones so that 'associations' lists include incongruous combinations" (1968, p. 80). However, Daubenmire did not advocate plant diversity studies in rare vegetation types and ecotones (if they serve as unusual sites). For local inventories, Daubenmire liked stratified sampling, but he stopped short of suggesting "stratified *random* sampling" (1968, p. 81). Europeans often favored a single 10 m × 10 m plot (Daubenmire 1968, p. 81), while some used much smaller plots that averaged the data for each stand. Several small plots allow for an analysis of frequency, sociability, variance, and sample size requirements. Cover estimates are more precise in smaller plots (i.e., there is more agreement between observers and time periods).

Heterogeneity   Daubenmire admitted that landscapes contained some level of heterogeneity. He made innumerable astute observations about potential problems in species composition studies: problems that related directly and indirectly to sampling vegetation in heterogeneous areas. He also observed considerable overlap in species composition at ecotones and the contribution of species from adjacent communities. He acknowledged many factors influencing spatial variation in vegetation, such as soil characteristics, heterogeneity of grazing impacts due to differences in palatability, soil disturbances from small mammals, changes in the animal community, seed bank and viability characteristics, soil chemistry and biology variation, introduced plant species, evolution of species, competition, and species-specific habitat requirements. Measuring that heterogeneity would not be easy!

Although Daubenmire (1968) recognized both spatial and temporal variation, he emphasized patterns of temporal variation (i.e., succession). He noted several recurring vegetation stages and patterns that occur in long periods following disturbance, which still hold true in many places today. Two-thirds of his book was devoted to succession, changes in vegetation communities through time after fire, flood, abandonment of agricultural fields, etc. He identified temporary communities ("seres"), some with floristically and structurally distinctive stages.

Daubenmire (1968) found that between disturbances, small, short-lived plant species are replaced by large, long-lived plant species (i.e., that annuals may be replaced by perennials, and short-lived perennials may be replaced by long-lived perennials, so sample designs for long-term monitoring must be able to accommodate a variety of taxa). Daubenmire found that there is usually a diversification of life forms over time (i.e., from increasing numbers of interspecific dependencies, energy flows, biomass, dead organic matter, soil development, and nutrient concentrations in foliage). This suggests that plot sizes appropriate for mostly grasses in early succession areas may not be appropriate for shrubs and trees later in succession on the same site. Daubenmire also recognized that some communities persisted as repeating early successional seres ("disclimaxes"), while climax communities could contain small areas of different sere communities. He noted various microhabitats created by small mammals or physical forces at the scale of individual plants or larger areas. This suggests that multiscale investigations and many replicate plots are needed in large areas to adequately capture spatial and temporal variation. Daubenmire noted the frequent destabilization and changes in sand dunes, erosion (removing soil from some sites, depositing it on others), lake and pond siltation, and organic matter buildup. For example, for temporal variability, he noted disappearing relics of earlier stages of succession, while some invasive species become established but are unsuccessful and soon perish (Daubenmire 1968, p. 105). Thus Daubenmire clearly established a precedent for plant communities that were not perfectly homogeneous, although there was little emphasis in his book on precisely measuring the exact spatial extent of homogenous and heterogeneous areas, and he did not emphasize quantifying the level of heterogeneity in any stand, community, or landscape.

Hint: Typical patterns of vegetation change, noted throughout Daubenmire's book, should be considered when stratifying vegetation types in the design phase of plant diversity studies. He provides hundreds of very general and accurate observations of landscape ecology.

Macroplot Nested Designs   Some nested plot techniques were designed to measure large plants (e.g., mature trees) in large plots, while measuring small plants (e.g., seedlings and understory herbaceous layers) in small plots within the macroplot. If properly designed and tested in various habitats, this has the distinct advantages of colocating measurements of overstory and understory species without oversampling the smaller species. One example comes from Daubenmire (1968, p. 87, Figure 26), where a 15 m × 25 m macroplot is established (Figure 2.2).

It includes three contiguous 5 m × 25 m plots for recording the diameter of trees more than 1 m tall. Trees less than 1 m tall are counted in two 1 m wide belt transects along the long axis of the center 5 m × 25 m plot, and the frequency and cover of shrubs and herbs (by species) are recorded in 20 cm × 50 cm quadrats spaced at 1 m intervals overlaying the belt transects. It was suggested that understory species not encountered in the quadrats also be recorded (but there is little evidence this was done often by Daubenmire or those that followed him). Assuming the site is completely homogeneous, one soil profile is examined and assumed to be representative of the entire macroplot. In placing small quadrats within a larger study plot, Daubenmire strongly suggested placing quadrats along a line and spaced one quadrat width apart (systematic sampling), such that the entire series of small quadrats fall in a homogeneous unit. Oddly, while Daubenmire recognized that systematic sampling would oversample common habitats and undersample rare habitats at the landscape scale, these same concerns were not expressed at the plot scale. The motivating theory was that systematic sampling was unbiased and gives greater accuracy, but the costs of oversampling were not considered, nor was there an expressed concern for missing locally rare or heterogeneous habitats. The only disadvantage of systematic sampling noted by Daubenmire was when there were obvious linear patterns in vegetation (e.g., grazing terracettes).

Vegetation Type Mapping and Classification   Daubenmire (1968, p. 245) thought vegetation type mapping, even if successional stages were included,

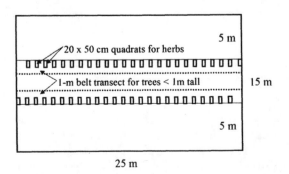

Figure 2.2.   Re-draft of Daubenmire's macroplot-nested design.

was of limited utility, too temporary, and not economically justified. Of course, this was in an age before fairly inexpensive, readily available satellite imagery. Daubenmire saw mapping "potential" climax communities (including symbols for various cover types and subareas of various seres and disclimaxes) as enduring, as it would represent topography, climate, soil, and flora. This assumes that current plant communities can be clearly distinguished and delineated, and single, deterministic, successional trajectories are known for each community. How might the paucity of climax stands affect such a mapping effort? "The rarity of climax stands is not such a drawback as the uninitiated might imagine, for much of the disturbed vegetation can show in its population structure the nature of the climax state from which it was derived and to which it can revert" (Daubenmire 1968, p. 257). However, few details are presented on how to accurately estimate the potential climax communities in large, complex landscapes and regions, or how the methods might work outside the Pacific Northwest, with which Daubenmire was most familiar. Daubenmire realized that most classification systems are based on small plots (1–10 m²) in homogeneous stands, thus typically only common species generally are recorded. The small plots reduced the potential for high species composition overlap among plots, except where the sampled stands were extremely similar in every respect (soils, microclimate, slope, aspect elevation). In fact, Daubenmire (1968, p. 267) dismissed the importance of rare habitats and landscape features:

> A classification can be satisfactory without including all the vegetation in the landscape. There is little reward in greatly increasing the complexity of a classification in an attempt to make it include the motley array of one-of-a-kind ecosystem scraps associated with unusual combinations of environmental conditions which commonly occur but occupy little area even in the aggregate.

Still, he suggested a complete species list was needed for each of the major vegetation types identified in the ecosystem.

Daubenmire stressed the need for accuracy assessments of vegetation maps. He suggested that at least 100 objectively located points in a landscape should be visited to see what percentage of these is accounted for by the classification. It was unclear what levels of accuracy might be acceptable.

Sample Site Selection   Daubenmire's comments on sample site selection probably were intended for classifying vegetation rather than for quantifying plant diversity. However, the selected passages below may have unduly influenced vegetation ecologists in the decades that followed. Daubenmire knew that many of the vegetation attributes above were influenced by plot size and shape, sample size, and particularly by sample site selection. Daubenmire (1968, p. 82) recognized that selecting plot locations subjectively (e.g., picking "typical" or "representative" plot locations) was subject to bias and criticism. Still, he claimed: "In a zealous effort to avoid subjectivity, it is easy to commit the worse sin of violating common sense" (p. 86). He stressed that "one of the most fundamental requirements for a valid statistic is that the stand which is sampled must be homogeneous, for a fraction of an area cannot be relied upon to represent the entire area unless the latter is homogeneous" (p. 79). "Our problem is one of elimination of as much heterogeneity as possible, especially variability attributed to differences in intrinsic habitat factors and history of disturbance" (p. 80). He suggested rules to reduce heterogeneity: (1) no sample may overlap even part of an ecotone, and (2) never overlap two soil types—and embrace a minimum variation in topography. He noted that "any significant heterogeneity in areas accepted as (homogeneous) stands rapidly obscures the differences between associations" (p. 80). Therein lies the unstated goal of synecology—to maximize differences between associations by sampling homogeneous and "representative" stands. Daubenmire (1968, p. 80) stated that: "Since homogeneity of environment decreases rapidly with increasing area, every attempt should be made to correlate *specific* biotic assemblages to *specific* soil and microclimate conditions, the smallest sample that is adequate is desirable." Daubenmire (1968, p. 268) did not stress the need for additional sampling in heterogeneous habitats: "Although hybrid stands may be very abundant, the best use of time is to replicate analyses of the pure types." Clearly the emphasis on describing vegetation was placed on measuring dominant species in homogeneous units.

In short, Rexford Daubenmire epitomized the science of vegetation ecology in the 1960s (and for many plant geographers and ecologists today) by espousing that several types of information were needed to assess the organization of communities, including floristic composition, structural and functional characteristics, environmental relationships, successional status, geography, and classification. Daubenmire championed observational studies. He seemed to distrust small-scale ecological experiments, suggesting that there are many factors weighing against the success of ecological experiments, including (1) complex relationships among different factors; (2) different life spans of species; (3) climate change and slowness to equilibrium (i.e., nonequilibrium at the time of study); (4) inconsistent or ephemeral species-environment relationships; and (5) natural spatial variation (Daubenmire 1968; Stohlgren 1994).

Daubenmire also championed correlative studies to summarize relatively accurate environmental records in relation to both community and individual species with a narrow range of working hypotheses as to the factors determining community and species limits. However, it is somewhat difficult today to imagine how information on community and species limits could be gained by restricting studies to relatively pure, homogeneous stands in complex, heterogeneous landscapes. As a testament to the staying power of Daubenmire's methods for measuring foliar cover in 20 cm × 50 cm quadrats along a line consider this—his paper published in 1959 describing these methods has been cited about 575 times according to the Web of Science (http://isi1.isiknowledge.com/portal.cgi).

Insights on Daubenmire's Philosophy and Practice of Vegetation Sampling and His Influence on Sampling Plant Diversity

We all have preconceived notions that influence the way in which we see the natural world. Insights into how Daubenmire viewed vegetation science are easily gleaned by his own observations and criticisms of Clements' "supra-organism" concept and Gleason's "individualistic concept.." It demonstrates all the struggles and volleys of a close tennis match. In a brief review of Clements' philosophy, Daubenmire obviously agreed with the basic notion that species form tightly interacting, easily recognized plant communities: "F. E. Clements referred to the plant community as an 'organism.' This use of this term seems entirely justified, for it is applied even to whole galaxies" (1968, p. 242). *Advantage Clements.* However, Daubenmire also noted that "species aggregations can show every degree of organization from a highly integrated community to groupings with negligible integration" (p. 243). *Advantage Gleason.* Still, Daubenmire largely dismissed Gleason's views because "to accept the hypothesis of complete independence of species with respect to their distribution over the landscape is to repudiate the thoroughly documented principle of competitive exclusion, for many species can invade a habitat following disturbance only to be completely eliminated by competitive pressures as equilibrium is restored" (p. 243). *Advantage Clements.* However, Daubenmire argued that because no two species have identical distributions across environmental gradients, "vegetation lacks objective discontinuities" such that communities (i.e., along steep environmental gradients) are unclassifiable except by wholly arbitrary means (p. 243). *Advantage Gleason.* He stated that "Succession leading to stability is a major unifying principle in synecology" (p. 99). *Advantage Clements.* But despite this general belief, he noted that "owing to the long life span of trees, most forest stands do not have time to reach complete equilibrium" (p. 169). *Advantage Gleason.* While Daubenmire weakly asserted "this phenomena embraces the individualistic concept," there is far more condescension in his review: "No organism lives in a biological vacuum, as implied by the 'individual' concept" (p. 244). He saw a "remarkable degree of similarity in species composition and structure" within communities (p. 244), such that the "biotic community is much more than a group of independent species." *Advantage Clements.* In the end, Daubenmire saw *floras* as a continuum, supporting Gleason, but he saw *vegetation* as largely discrete homogeneous units with significant discontinuities. *Game-Set-Match Clements.*

Dieter Mueller-Dombois and Heinz Ellenberg

In 1974, Dieter Mueller-Dombois and Heinz Ellenberg published *Aims and Methods of Vegetation Ecology,* a text that has probably been used by more English-speaking plant ecologists than any other. The book was strongly influenced by Ellenberg's earlier book *Aufgaben und Methoden der Vegetationskunde* (1956), published in German. Ellenberg was a classically trained phytosociologist who focused on detailed descriptions of Braun-Blanquet's relevé sampling methods. He was a plant geographer, studying the distributions of taxa and their evolutionary relationships (Ellenberg 1958). Mueller-Dombois added theoretical and practical experience from central Canada and the Pacific Northwest. Mueller-Dombois and Ellenberg recognized that the classification of plant communities "is no longer considered an end in itself. Instead, environmental investigation of plant communities and the sociological interaction of plant species have become the main concerns." The primary task was to study vegetation structure, systematics, geographic patterns of communities, and community development, change, and stability. Clearly the vegetation groups of concern ranged from biomes (tundra, forest, grassland, desert) to associations (defined by subdominant species) with "plant communities" in between, anchoring the science.

Sample Site Selection   Some aspects of the text are puzzling in light of the current understanding of plant diversity patterns. For example, Mueller-Dombois and Ellenberg (1974) saw the advantages of stratified random sampling, but only for the more abundant plant communities and in more or less equal-size subsegments (i.e., strata; p. 39). Important, locally rare plant communities harboring unique species assemblages or high in productivity could be underrepresented by such an approach, but this was not addressed as a significant concern. Subjective placement of plots (without preconceived bias) was seen as an opportunity to take full advantage of an investigator's knowledge and common sense, and rapidly progress the science and state of knowledge.

Pattern of Sampling   Mueller-Dombois and Ellenberg (1974) related that a 100% survey of even one vegetation type would be too costly and time consuming, so sampling was necessary. They suggested four essential steps in vegetation sampling: (1) stratification of vegetation cover types (they termed it "segmentation" and "entitation"—an attempt to include most or all significant cover types); (2) sample site selection in each strata; (3) evaluations of the shape and size of plots; and (4) deciding what to record in each sample.

According to the authors, in the first phase, the sampling locations within strata might be selected in three ways. The sites might be subjectively selected with preconceived bias by picking "representative" or "typical" sample site locations, as is commonly done in Europe (Stebler and Schröter 1892), Australia (Goodall 1953a,b), and elsewhere. Stebler and Schröter (1892) used very carefully placed 1 ft² plots in representative grassland types in Swit-

Insights on Mueller-Dombois and Ellenberg's
Philosophy and Practice of Vegetation Sampling
and Their Influence on Sampling Plant Diversity

Mueller-Dombois and Ellenberg (1974) represented a "modern summary of vegetation science," not only to classify communities, but to
quantify how communities relate to environmental factors. Their focus
was on "sociological geobotany," the study of composition, development, geographic distribution, and environmental relationships of
plant communities (p. 8). There was an introductory section (p. 4) on
"how to recognize plant communities." The authors suggested that
"spatial changes in species composition are often obvious even to the
casual observer" and "communities that are nonevident to the unac-
quainted may be self-evident to the experienced investigator" (p. 5).
"Wherever the cover shows more or less obvious spatial changes, one
may distinguish a different plant community" (p.4).

The authors' understanding was that communities were comprised
of stands with only some frequently recurring species and that
communities were not entities like organisms in the Clementsian
sense. The authors noted that they would not ignore contrary views of
vegetation science, most notably from Gleason (1926) and Whittaker
(1970), who mistrusted the easy delineation and description of plant
communities. Still, the authors emphasized as current trends in
vegetation science such topics as the geographic description orienta-
tion of life forms and communities, dominant vegetation dynamics,
statistics, experimentation, ecosystems, and ecosystem classification.
They presented little information on individual species distributions
or patterns of plant diversity. Reminiscent of Daubenmire's book six
years earlier, the emphasis was on dominant species in homogeneous
areas.

zerland. Krebs (1989) calls this "subjective sampling," which thwarts any hope
of using probability statistics to extrapolate results to the remainder of the
unsampled landscape. Sample sites also might be selected in a subjective
manner without preconceived bias, with a strong inference used to evaluate
one aspect of the system (say, vegetation cover), leaving many other aspects
(e.g., understory diversity, soil characteristics, etc.) to be sampled with greater
opportunity for surprises. The last and most serious suggestion was selecting
sample locations in a random, systematic, or other unbiased manner (i.e.,
according to chance) (Mueller-Dombois and Ellenberg 1974, p. 32).

Mueller-Dombois and Ellenberg (1974) recommend random, systematic,
or stratified random sampling to evaluate broadly defined dominant communities. Random sampling (p. 39) can be achieved by creating a grid over

a map of the study area and selecting random numbers as grid locations for sample sites and locating them in the field. Random sampling is required for probability statistics. The authors also state that if strata are too large, results will be influenced by within-type variation.

Baggage and Direction

Clearly the baggage and directions of most of the early plant ecologists were strongly (perhaps too strongly) influenced by Braun-Blanquet, Clements, Daubenmire, Kerner, Merriam, Mueller-Dombois and Ellenberg, and Schouw. Clements was clearly dominant in North America. While the citation index is hardly indisputable evidence, Clements' book on plant succession has been cited about 875 times since 1945, while Gleason's paper on the individualistic concept of the plant association has been cited about 500 times in the Web of Science (http://isi1.isiknowledge.com/portal.cgi). Perhaps influenced most by Braun-Blanquet, vegetation science in Europe and elsewhere consisted largely of community classification, vegetation type mapping, and habitat mapping, relying on single-scale measurements of species richness and cover in subjectively selected, relatively homogeneous areas. The maturing science of remote sensing (especially photographic interpretation and coarse-scale satellite images), the personal computer, and the development of geographic information systems have provided new, readily available tools for vegetation classification and type mapping. The creation of sharply delineated colorful paper maps and computer displays of vegetation were an obvious passion of many vegetation scientists and land managers, and produced countless theses and dissertation topics for graduate students (Figure 2.3). Researchers who model vegetation distributions at landscape, regional, and global scales have relied heavily on maps of vegetation type, community classification, and functional (or physiological) groups or biomes.

Baggage

The prevailing paradigm remains that plant associations and communities are unambiguously real. That is, the landscape is dominated by combinations of strongly interacting species in easily delineated patterns and distributions, with few environmental factors needed to maintain these fairly stable vegetation patterns. The concept of vegetation associations and communities was so readily and broadly accepted that the underlying assumptions of associations and communities have not been routinely challenged. Many current plant ecologists acknowledge that plant associations and communities are largely intended as a convenience for describing the general locations of groups of dominant forest, shrub, and grassland species rather than the cumbersome and unwieldy task of describing the locations and thousands of individual species and variations. Still, the simplified views

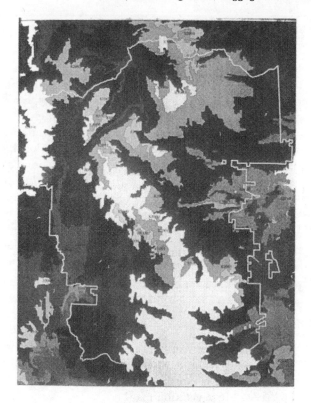

Rocky Mountain National Park Vegetation

Figure 2.3.   Lesson 6. Vegetation type maps and community classification projects have led to an over-emphasis on the distributions of a few dominant species, while underplaying the importance of habitat heterogeneity, and rare habitats and species.

of vegetation description, classification, and mapping have strongly influenced sampling designs in vegetation science, and in particular, the study of plant diversity (Figure 2.3). Sampling designs for a simplified landscape (Figure 2.4, top) might be quite different than sampling designs for complex landscapes (Figure 2.4, bottom).

However, there may have been several unintended consequences from the overwhelming emphasis placed on vegetation classification and dominant plant species. For vegetation type mapping, plant ecologists need not be taxonomic experts. Identifying a few dominant trees, shrubs, or grasses requires far less expertise and training than someone who can identify most of the plant species in a species pool. I'm curious as to whether enrollment in plant taxonomy classes has decreased, and if enrollment has increased in classes on geographic information systems by plant ecologists. Another unintended consequence of vegetation classification is the oversimplifica-

---

On the Value of Vegetation Classification and Type Mapping

The important aspects of vegetation classification and type mapping cannot be denied. Such endeavors are critically important in understanding patterns of biomass and productivity. Relatively few species account for the vast amount of biomass in a plant community. As such, a few species typically monopolize many of the available resources on a site, control nutrient cycling, and modify local environmental conditions. Studies of vegetation classification and analysis of the geography and ecology of vegetation types provide valuable information on the species-environment relationships of dominant species, wildlife habitats, fuel accumulation patterns, and many aspects of landscape diversity. Vegetation type maps often are used to stratify landscapes for further studies of plant diversity and detailed ecological studies on grazing, disturbance, climate, and land use change, as we shall see in later chapters.

---

tion of plant geography, where vegetation gradients, ecotones, and heterogeneous areas are largely ignored. Sharp delineations between vegetation rarely exist, but this myth is consistently maintained for the majority of boundaries represented in vegetation maps.

The most dangerous consequence of our fetish for dominant species mapping is that the assumption of homogeneous communities has had tremendous effects on vegetation sampling designs and field methods. Field sampling was subjectively directed to typical stands, supposedly representative of the homogeneous community, so few vegetation plots would be needed and vegetation quadrats could be small. Variation wasn't particularly expected or sought. There would be no attempt to sample the ends of important environmental gradients. Data recorded commonly included the presence or cover of dominant species, the frequency of a species in a number of stands, density measures of dominant species, estimates of biomass, or basal area of dominant trees. Plots sizes and shapes could vary among investigators because common species are difficult to miss with even modest sampling designs.

How did early plant ecologists further the study of plant diversity? Many of their detailed, naturalist observations provided valuable insights into the complex nature of plant species distributions. Heterogeneity was frequently observed and reported, even though many sampling strategies may have attempted to minimize within-type variation and maximize among-type variation. This could be accomplished by (1) subjectively selecting plot locations in "representative" stands, associations, and communities; (2) mini-

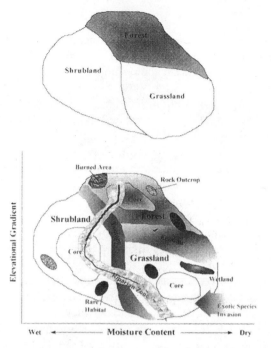

Figure 2.4.   Top: A simplified, classical view of vegetation, consisting of
nice homogeneous communities. Bottom: A more realistic characterization of vegetation including moisture and elevation gradients, core areas
of vegetation types with adjacent more heterogeneous renditions of the
types, ecotones (sharp boundaries) and ecoclines (gradual boundaries)
between types, riparian zones, rock outcrops and rare habitat types,
disturbed areas, and areas of exotic plant invasions. Lesson 7. Few, if any,
natural landscapes are composed mostly of homogeneous communities,
so sampling designs based on the notion of homogeneous units may be
poorly suited for more complex, heterogeneous landscapes.

mizing sampling in heterogeneous areas and rare habitats; and (3) using a
few small plots in each representative site.

How much baggage accompanied the emphasis on the identification,
characterization, delineation, and mapping of plant communities? The sampling design above would tend to reinforce the notion of stable communities controlled by few species and environmental factors: the Clementsian
view. The theory drives the field methods, the field results support the
theory. Issues about plant species diversity, the rarity of certain species, and
individual species-environment relationships were not being addressed by
many plant ecologists. Future directions in measuring plant diversity would
develop from previously underappreciated concepts.

Direction

Gleason (1926) seemed to have very little influence on vegetation science in the 75 years after his landmark paper on the individualistic concept. If he had, questions would have been raised about sampling designs and field methods, and many opposing theories might have been more completely explored. Robert Whittaker (1970) drew attention to multiple environmental gradients controlling the distributions of dominant and subdominant species, resurrecting many of Ramensky's and Gleason's individualistic concepts. John Harper's (1969, 1977) focus on plant population biology resurrected Darwin's, Humboldt's, and Warming's appreciation of species and populations as primary units of evolution. These concepts and the many detailed observations of general patterns of earlier ecologists are reinforced by contemporary ecologists James Grace (1999), Michael Huston (Huston 1994, pp. 15–63), and Michael Palmer (1994), who provide excellent "future directions" by summarizing general patterns of species diversity. These patterns, combined with the individualistic theory of species distributions, provide essential background reading for plant ecologists. Each introduction (best stated by Huston) rightly claims that "no single process or theory can explain a phenomenon as complex as biological diversity." A variety of patterns have been shown, each challenged by contradictory data, but some generalities that have emerged and help in planning plant diversity studies are as follows:

- Latitudinal gradients. Diversity is lowest at the North and South Poles, from nearly monospecific in subarctic areas, and it increases toward the topics, culminating in the species-rich tropical rain forests (see Currie 1991; Dobzhansky 1950; Glenn-Lewin 1977; Monk 1967; Reid and Miller 1989; Stohlgren et al. 2005a).
- Stress. High environmental stress (e.g., severe drought conditions) usually leads to low plant species richness (Grime 1979; Stohlgren et al. 2005b).
- Productivity gradients. Low productivity (high stress) sites usually have low diversity. Sites intermediate in productivity usually have high diversity, but very productive sites exhibit competitive exclusion, resulting in low diversity. Hence a "hump-backed" model of species richness is often seen along a long productivity gradient (Bond 1983; Grace 1999; Grime 1979; Huston 1994; Stohlgren et al. 2005a,b).
- Altitudinal gradients. Diversity decreases with increasing elevation, corresponding to decreases in air temperature (Kikkawa and Williams 1971; K. Yoda, noted in Whittaker 1977). In the Santa Catalina Mountains of Arizona, the highest plant diversity in 0.1 ha plots was at low to intermediate elevations (Whittaker and Niering 1975). However, in all of these studies there was no attempt to account for species overlap among plots at similar elevations, so little can be said about the richness of floras in various elevation zones.
- Heterogeneity gradients. High species richness can result from local heterogeneity in the ratios of resources and many species with low

resource supply rates (Tilman 1982). Or, more simply, environmental heterogeneity leads to increased species richness (Grubb 1977; Harper 1977; Huston 1994; Palmer and Dixon 1990; Pollock et al. 1998).

- Long-term fertilization. Fertilization decreases plant diversity over time. The Parkgrass experiments from 1856 to 1949 clearly show that species richness declines in fertilized plots over time (DiTommaso and Aarssen 1989; Kempton 1979; Willems 1980).
- Area. Species richness increases with sample area (Gleason 1922, 1925; Preston 1960, 1962a,b, 1969). The primary reason for this is that environmental heterogeneity increases as area and the number of habitats also increase, and rare habitats continue to be included. Contiguous samples (high spatial autocorrelation). Contiguous quadrats in homogeneous areas will record fewer cumulative species than contiguous quadrats in heterogeneous areas (Shmida and Wilson 1985).
- Age of the substrate. At least in the early stages of primary succession, such as following the eruption of Krakatoa (Doctors van Leeuwen 1936) or the formation of a new island (Fridriksson 1975), plant species generally increase with substrate age.
- Evolution, endemism, and lag effects. The number and diversity of plant species has generally increased with geologic time (Niklas et al. 1983) and geographic isolation, then lag effects (persisting on site despite climate and other changes).
- Infrequent massive disturbances. Glacier advances, huge floods, and intense wildfires can greatly decrease local plant diversity.
- Frequent, less severe disturbances. High plant diversity is said to accompany intermediate levels of disturbance. This claim is moderately supported in the literature (Brooks and Matchett 2003; Hobbs and Huenneke 1992).

Of course, many factors may be correlated and cross-correlated with these patterns. Sun angle, day length, mean temperature and temperature extremes, moisture regimes, and glacial history, among others, were noted by Huston (1994). The main direction provided by early and current plant ecologists is one of guided wonder. This has created an atmosphere of interest and enthusiasm to make many more measurements of plant diversity.

Magurran (1988) wrote a primer on ecological diversity and its measurement, providing a rationale for measuring diversity. The text includes detailed definitions and measures of alpha diversity (local), beta diversity (differentiation diversity among sites), and gamma diversity (overall diversity in a group of areas across environmental gradients) and several examples reproduced many times in plant ecology texts and in university courses. Krebs' (1989) text on ecological methodology and the Barbour et al. (1999) text on terrestrial plant ecology complement Magurran's book and provide essential and broad introductions to statistical ecology and plant ecology, respectively. Krebs targets measures of abundance, simple spatial patterns, basic sampling and experimental design, similarity and species diversity indexes, and exceptional examples from plant and animal sciences. Barbour et al. provide a complete text on the history of plant ecology, plant life history and population dynamics, classification and succession, and environmental factors—all targeting the relationships between vegetation and the environment. These three essential references form much of the foundation for the text that follows.

Two related texts add additional direction and describe the challenges in future studies of plant diversity. Michael Rosenzweig (1995) provides a general examination of species diversity in space and time. Rosenzweig synthesizes much of the published literature on broad-scale patterns of plants and animals, including effects of latitude, elevation, species-area relationships and the different forms of species-area relationships on islands versus continents, habitat diversity, diversity in geologic time, and population dynamics that effect diversity. Issues of scale permeate the book. The strength of the text lies in the more than 100 examples where the number of species (or log number of species) is regressed against area (or log area), latitude, island age, time, and many other factors in many areas. The taxonomic completeness of datasets is sometimes, but not always questioned when richness values are presented for islands, counties, and countries of various sizes. The use of 0.1 ha plots to assess the relationship of plant species richness to latitude [original data by Gentry (1988)] was criticized as being a "standard" size for plant diversity studies, but probably too small to census tree diversity. Thus the accuracy and completeness of the published relationships are less emphasized in the text relative to the general patterns presented. Still, the relationships provide the basic hypotheses for understanding patterns of plant (and animal) diversity in space and time. Rosenzweig's book is an essential primer for population and landscape ecologists.

The second text providing direction for measuring plant diversity is by Stephen P. Hubbell (2001). Hubbell proposes a unified neutral theory of biodiversity and biogeography based on many long-accepted and often repeated mathematical patterns found in nature, including (1) negative exponential distributions of species abundances (with very long tails of rare species with low abundances); (2) consistent patterns of species rank abundance curves for communities with a given number of species and neighboring communities; and (3) consistent species-area relationships in three phases from local to regional to global scales. The theory presumes that "no species can increase in abundance in a community without a decrease in the collective abundance of all other species." The theory appears to hold particularly well for tropical regions, where a zero-sum change in tree species abundance has been carefully measured at multiple spatial scales. It is uncertain how the theory would hold under a wider variety of natural circumstances. For example, areas under primary succession could be strictly additive in diversity and species abundance for decades to millennia (e.g., after the eruption of Krakatoa). Plant species invasions tend to increase local and regional plant species richness without immediate corresponding extinctions, and fire and other disturbances can radically alter species richness and abundance patterns on the same sites that supported fewer species and lower abundances in the predisturbance condition. Likewise, invading plant pathogens such as white pine blister rust or sudden oak death can greatly reduce overstory trees without replacement by other trees species (recognizing as the author did that extinction processes may simply result in replacement by other species from the species pool). These critiques aside, Hubbell (2001) draws attention to the majority of species, the tails of the distribution, and issues of scale in population attributes and community diversity.

Both Rosenzweig and Hubbell clearly recognized the contribution of rare species and habitats to patterns of plant diversity, but neither author provides techniques or suggestions for actually measuring diversity at multiple spatial scales or over time. This begs for the need for better measurement of locally rare species at multiple spatial scales in many biomes, and for many years to test the assumptions of the neutral theory and species-area calculations.

3

A Framework for the Design
of Plant Diversity Studies

Based on the frequency of commonly asked questions from students, professional ecologists, and colleagues, it is evident that a basic framework is needed in the design of plant diversity studies. There are obvious gaps in the information provided by plant ecology textbooks. Newcomers and even seasoned veterans to vegetation sampling often ask the same series of questions.

Throughout the course of this book, I attempt to offer simple answers whenever possible, and not-so-simple answers where applicable, and logical, systematic approaches to those questions without obvious or immediate answers. My general philosophy is to take an experimental approach to vegetation sampling in general, and to plant diversity sampling in particular. I assume that most available sampling techniques are like hypotheses that must be proven—they must be accurate, precise, complete, and cost efficient. "Problems" in measuring plant diversity often lead to a better understanding or appreciation of diversity. Only after careful observation, repeated trials, and comparisons with other techniques can the hypotheses (methods) be accepted or rejected.

A framework for sampling plant diversity includes initial decisions on goals, objectives, scale, and sampling design. Sampling design is further complicated by decisions on plot size and shape, sample size, sampling intensity, and sampling pattern, which interact and affect the results of plant diversity studies. Using the generalized framework that follows may help in planning landscape-scale plant diversity studies and in evaluating the strengths and weaknesses of alternate study designs and field techniques.

Partial List of Commonly Asked Questions

- How should I measure plant diversity?
- What specific questions can I adequately address with the funds I have available?
- What plot shape and size are best for the task at hand?
- How many samples (plots) do I need?
- How should I select plot locations (or how should I array the plots)?
- How do I know if I've sampled enough?
- How do you sample for extremely rare plant species?
- When should I sample to record the greatest number of plant species?
- What ancillary environmental data should I collect?
- How do I relate the species data to the environmental data?
- Should I stratify my samples? How should I stratify? What map resolution is needed?
- How does spatial scale affect my sampling results?
- How should I measure and interpret species-area relationships?
- What do diversity indexes tell me, and which diversity index works best?
- What if some plant specimens cannot be identified to the species level?
- How does my sample relate to the population?
- How do my sample sites reflect the broader, unsampled landscape?
- How do I estimate the number of plant species I missed?

Goals

First and foremost, the goals of a study must be articulated clearly. An age-old question in ecology is *what to measure* (Krebs 1989; Risser 1993). What is the grand question being asked? Several examples follow:

- Where are the hotspots of native plant diversity in the landscape?
- What is the affect of large ungulate grazing on plant diversity in the study area?
- How much overlap is there in species composition between recognized vegetation types?
- Are there more species of invasive nonnative plant species in riparian zones or upland sites in this landscape?
- How will species composition change in this set of old fields over time? Or how will it change in a forest, shrubland, or grassland after fire?

Often, the primary goal of a study is to reasonably answer an important, general question. Identifying important questions is often a difficult first step—and it should not be underestimated. Resolving the basic questions in a survey, monitoring, or research program will help define study objectives and incorporate the practical and theoretical constraints.

Objectives

The objectives of plant diversity studies detail what is doable in the field to approach the stated goal. Jones (1986) describes the process as "scoping" and "problem definition," where general problems are reduced to specific ones, where specific issues are identified, and where priorities are set for specific data needs. Particularly important when setting objectives for landscape-scale studies of plant diversity is that existing data (and auxiliary data collected by others) must be evaluated fully for spatial and temporal completeness, accuracy, and precision. Note that this is usually no small task, but it helps to identify the types and levels of data that are needed (Stohlgren 1994, 2001).

Narrower objectives will usually reduce the scope of a study by eliminating extraneous data (MacDonald et al. 1991). Priorities for data collection are based on the objectives and the practical and theoretical constraints and limitations of conducting a study. For example, the questions directing the "goals" of a study include very broad terms such as "plant diversity" and "species composition." Plant diversity in the broadest sense would include genetic diversity, plant diversity (species richness, abundance, frequency, cover, composition, and functional diversity; Petchey et al. 2004), vegetation type/community diversity, and seed storage and viability. Typically funding may only allow an evaluation of species richness and cover, and the specific objectives must state this limitation. Species composition is a broad term that may encompass species richness, abundance, cover, biomass (above and below ground), frequency, seed storage, etc. Again, specific study objectives might be far more limited. Both terms—plant diversity and species composition—imply a specific area and time under investigation, and a level of taxonomic specification (subspecies, variety, species, genus, or functional group), and each must be clearly specified in the objectives.

At this point in designing a plant diversity study, it is helpful to visualize data products (e.g., tables, figures, maps, model outputs). It is important to acknowledge the potential limits to which study results can be extrapolated spatially and temporally (see Berkowitz et al. 1989). Specific objectives also help in the selection of appropriate spatial scale, sampling design, and field methodologies, which are briefly described below. Investigators must periodically evaluate whether specific objectives are addressed directly by the field data. This feedback loop ensures that the overall study goals are met or approached. By articulating clearly the goals and objectives of the study, the ecologist can begin to address issues of scale.

Scale, Resolution, and Extent

Three main geographic features of plant diversity studies are scale, resolution, and extent. Scale is a confusing, multiuse term that generally refers to the dimensions of observations, both spatially and temporally (Levin 1992; Peterson and Parker 1998). Levin (1992) also states that large-scale ecological studies require the interfacing of phenomena that occur at very different scales of space, time, and organization. The scale selected for plant diversity studies sets the limits on which ecological phenomena can be studied and results interpreted. Scale is comprised of two components, extent and resolution (or grain). In simplest terms, extent refers to the size of the area in question, while grain refers to the level of detail to which natural features and habitats are identified (Figure 3.1).

For example, a 754 ha area may seem too large for an investigator to evaluate plant diversity, so a 75 ha area might be selected for detailed study (i.e., the fine-scale study area in the top of Figure 3.1). This changes the extent of the study area. While this may provide more detailed information about the 75 ha area, two key vegetation types (wet meadow and burned pine) would be missed in the finer-scale area. The two vegetation maps in Figure 3.1 are derived from the same aerial photograph, but they have two different minimum mapping units (or different resolutions or grains). The minimum mapping unit is the smallest area recognized by the photointerpreter for a

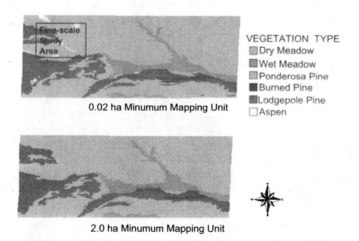

0.02 ha Minumum Mapping Unit

VEGETATION TYPE
▨ Dry Meadow
▧ Wet Meadow
▨ Ponderosa Pine
■ Burned Pine
▧ Lodgepole Pine
☐ Aspen

2.0 ha Minumum Mapping Unit

Figure 3.1. Vegetation maps of a 754-ha area of Rocky Mountain National Park, Colorado. The top map uses a 0.02-ha minimum mapping unit and a hypothetical fine-scale study area. The lower map uses a 2.0-ha minimum mapping unit (adapted from Stohlgren et al. 1997a). Lesson 8. Many plant diversity studies rely on existing vegetation maps to guide surveys and select sample sites. If the maps are too coarse in resolution, rare and important habitats could be easily missed.

particular map. Larger minimum mapping units have larger grain sizes and they are less expensive to produce. The potential drawback to larger grain sizes in plant diversity studies is that key vegetation types may be missed. In the example, most of the burned pine vegetation type and all of the aspen stands would be missed with a 2.0 ha minimum mapping unit. In this case, the burned pine and aspen stands contributed several unique plant species to the local species pool (Stohlgren et al. 1997a).

As I show later, many plant diversity studies take many measurements at relatively few sites, thus recording information on relatively few species in the local species pool. Unique events and peculiar site characteristics and land use histories may be captured or missed if the extent is too small or if the resolution (grain size) is too coarse, so observed plant diversity patterns may not be extrapolated confidently to other areas.

In contrast, regional- or national-scale field surveys of plant diversity may collect relatively few measurements at many scattered sites. These measurements may adequately assess spatial variation across large areas, but may not detect high-resolution variability or temporal trends because of the few measurements recorded at each site. This illustrates the major trade-off between local/landscape-scale and regional/national-scale plant diversity studies. There is an inevitable loss of information (or accuracy) as study extent increases and less ability to extrapolate results as study extent decreases.

Issues of scale, extent, and resolution are easily overcome with infinite funding. However, in the real world, funding levels and cost efficiency are important constraints and considerations in sampling designs. Decisions must be made on whether to use several small plots that cover wide environmental gradients, or fewer large plots that capture local environmental variation. Alternatively, a nested-intensity sample plot design could be used as a compromise in some situations (see chapter 17). Here, many measurements taken at a few small sites are nested in larger areas where fewer measurements are taken. In this way, results of intensive studies on temporal trends may be extrapolated to larger areas with known estimates of accuracy and precision (Haslett and Raftery 1989). A series of extensive plots and fewer intensively studied plots is one way ecologists can gain an understanding of the "representativeness" of the intensive study sites.

Ecologists often use a simplified definition of the term "scale" to describe either plot size (e.g., 1 m², 10 m², 1000 m² plot) or the area of inference (e.g., plot scale, landscape scale) once the area of a particular study has been defined. Sampling at multiple spatial scales is also an important component of plant diversity studies (Shmida 1984; Whittaker et al. 1979). Because it is virtually impossible to inventory an entire landscape for vascular plant diversity (much less for biodiversity), sampling species richness at multiple spatial scales may allow for mathematical estimates of total diversity. Mueller-Dombois and Ellenberg (1974) recognized that "truly, every square meter of habitat and plant cover shows certain differences when compared on an absolute basis" (p. 24).

Sampling Design

Sampling designs for plant diversity studies revolve around issues of plot shape and size, sample size, and the pattern of sampling. These factors greatly influence results.

Plot Shape and Size

Decisions must be made about plot shape and size, the parameters selected for study, and the frequency, precision, and accuracy of measurements, which all affect the value of the long-term dataset (Zedaker and Nicholas 1990). Consider three plot (or quadrat) shapes commonly used in vegetation surveys, a circle, a square, and a rectangle (0.5 m × 2 m), each with an internal area of 1 m² (Figure 3.2). Theoretically, in a homogeneous area with randomly distributed plant species, the circular plot should record the fewest number of species, because it has the smallest perimeter:area ratio. Few if any areas are strictly homogeneous, so larger diameter (or wider) plots would tend to cross more environmental gradients than smaller plots. A square plot should recover more species than a circular plot and a rectangular plot should recover more species than either a circle or a square. In a field test in a montane meadow, the rectangular plot did recover more vascular plant species than the circular or square plots (Figure 3.2). The square and circular plots behaved fairly closely to theoretical expectations. The rectangular plot behaved as expected. Even in this simple study, there are many potential sources of error. The sample size was small relative to the heterogeneous study site with complex environmental gradients and patchily distributed species. I, the observer, could have contributed "observer error"

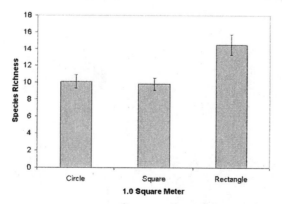

Figure 3.2. Effects of quadrat shape in measuring vascular plant species richness in Beaver Meadow, Rocky Mountain National Park, Colorado. For Figures 3.2 to 3.4, n = 30 randomly selected plots in the same 1 ha area. Vertical bar is the standard error.

with decisions on which plants were located in or out of the plot, or in the identification of some species in a grazed landscape. The study site may be atypical. In general, the field study conformed to theory, showing that rectangular plots may be superior to circular or square plots in this area (but see Keeley and Fotheringham 2005).

Obviously, in most natural habitats, more plant species would be recorded in larger plots than in smaller plots (Figure 3.3). This is the underlying concept behind species-area curves. In this case study, roughly 30% more species were found when plot size was increased by three times. However, there is a cost associated with increasing plot sizes: more time is required to identify more species. If cost efficiency is a concern, then time can be factored into the equation (Figure 3.4). Very small plots recover few species and take little time searching for unrecorded species. Moderate-size plots contain more species and require more searching time. Larger plots take considerably more time searching for species, additional taxonomic expertise to identify species, and more decisions on which species are inside or outside the plots longer borders. Thus, in the study case, it was more efficient (in terms of the number of species recorded per unit time) to establish many 1 m² plots than to establish the smaller or larger plots tested.

Many decisions on plot shape and size are not this easy. The case example assumed the study objectives were to maximize cost efficiency and species richness information at a single (1 m²) plot scale. In some cases, larger plots might be preferred for monitoring changes in species richness in fragile habitats where trampling between smaller plots could cause damage. Or smaller plots might be monitored to determine the effects of invasive non-native species on individual native plant species.

Plot size can affect species accumulation (or species effort) curves. In a preliminary test of quadrat size in a shortgrass steppe in Colorado, first grade students (supervised by graduate students) established five 1 m² quadrats and five

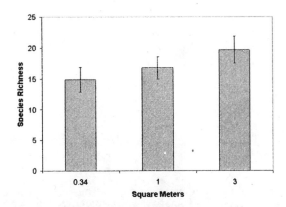

Figure 3.3.   Effects of quadrat size in measuring vascular plant species richness in Beaver Meadow, Rocky Mountain National Park, Colorado.

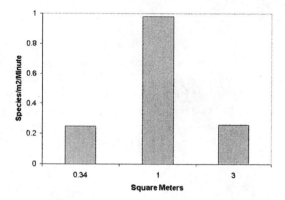

Figure 3.4.   Cost efficiency viewed as the number of species recorded per m² per minute of sampling when measuring vascular plant species richness in Beaver Meadow, Rocky Mountain National Park, Colorado.

0.1 m² quadrats and simply counted the number of "new" plant species encountered (Figure 3.5). The smaller quadrats kept capturing the most common species, missing the patchy, locally rare species. The larger plots captured some locally rare species. This student survey is far from perfect. Still, the obvious conclusion was big plots catch more than little ones. Less obvious is that extrapolations of the species accumulation curve from small plots may grossly underestimate the actual number of species found in larger areas of the habitat (Figure 3.5; also see Barnett and Stohlgren 2003; Stohlgren et al. 1995b).

Sampling with one plot size may not tell the whole story. The investigator may wish to evaluate the effects of fire or grazing on plant diversity—and evaluate the effects of scale on the results. In a case study detailed later in chapter 11, we asked whether long-term grazed areas differed in plant species richness

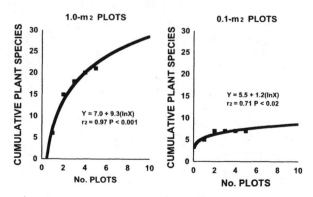

Figure 3.5.   Cumulative species recorded by Ms. J. Martin's first grade class in short grass steppe in Colorado (May 1999).

compared to long-term ungrazed sites. In a four-state area, 260 pairs of 1 m² plots and 26 pairs of 1000 m² plots were established in long-term grazed and ungrazed sites (Table 3.1). In this study, there was a significant increase in plant species richness in the 1 m² plots in grazed areas, but no significant difference on the 1000 m² plots. This suggests an increase in species density (or packing of species) on small plots, but without significant differences on larger plots. Investigators relying only on small plots or single-scale designs may have assumed that the results were independent of scale (i.e., similar across scales).

In some studies, effects may not be so scale dependent. However, many investigators fail to even ask the question, "How does sample scale affect my results?" As a consequence, various results may be entirely accurate and statistically sound, but misleading or ecologically benign. Throughout this text I will explore the many reasons to collect plant diversity information at multiple spatial scales.

Sample Size

One of the most commonly asked questions by graduate students conducting plant diversity studies is "How many sample plots will I need?" The unsatisfying short answer is "Sample size determination is tricky business, and it is strongly dependent on specific study objectives, habitat heterogeneity, statistical considerations, and cost constraints. Next question?" A study to evaluate the species richness (numbers of species only) in two grazed and ungrazed pastures will likely require fewer plots than a study whose objectives are to evaluate species richness, species diversity, composition, and biomass differences in the same two pastures. The number of samples needed to assess species-area relationships in homogeneous areas will be less than in heterogeneous areas with patchily distributed species (Figure 3.6). Theoretically, in a homogeneous area, most species would be encountered in the first few samples. In most heterogeneous areas, where small species-rich microhabitats are patchily distributed in large expanses of species-poor habitat, few species would be recorded by chance in the initial samples, with more species recordings as the rare species-rich habitats are encountered, followed by more common species-poor areas.

Table 3.1.   Mean species richness (standard error in parentheses) from 260 pairs on 1 m² plots, and 26 pairs of 1000 m² plots that were established in long-grazed and long-ungrazed sites in Colorado, South Dakota, Wyoming, and Montana.

Plot size	Grazed plots	Ungrazed plots
1 m² plots	8.3 (0.3)	7.1 (0.3)
1000 m² plots	32.6 (2.8)	31.5 (2.5)

Adapted from Stohlgren et al. (1999b).

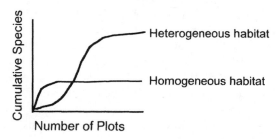

Figure 3.6.   Hypothetical relationship of cumulative species found in heterogeneous habitats (with rare patches of very high plant diversity), and homogeneous habitats (with evenly distributed species).

It might also be difficult to assess the completeness of species richness sampling because of the linear or uniform placement of quadrats within a larger plot. Quadrats placed along a line likely would miss species that are patchily distributed (Figure 3.7; also see chapters 6 and 7). Randomly placed quadrats would generally cover more heterogeneous areas and more complex environmental gradients.

Assessing differences to an 80% level of confidence will require fewer samples than assessing differences to a 95% level of confidence. A study budget of $4000 may produce fewer plots than if the budget is $40,000. Obviously there are trade-offs between many of these factors. Small budgets or high heterogeneity may preclude a high level of confidence. It's a tricky business.

Typically, appropriate sample sizes in vegetation biomass studies are determined after evaluating between-plot variance from initial field tests. Krebs (1989, pp. 229–260) provides a complete description of sample-size determinations, though the various approaches and formulas are primarily for species abundance, population size, or a mean value (e.g., biomass). For

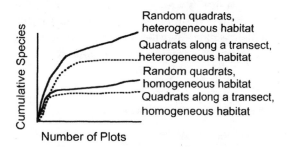

Figure 3.7.   Hypothetical relationship of cumulative species found in heterogeneous habitats (with rare patches of very high plant diversity) and homogeneous habitats (with evenly distributed species) with either random quadrats or quadrats of the same size placed along a transect in the same study area.

example, the effects of plot shape and size are not as apparent in many biomass studies as they are in plant diversity studies. There are many potential sources of error with both biomass and diversity studies. Both can have
systematic errors (errors in measurement devices), such as when very small
quadrats collect only a few common species. Both can have sampling errors
(errors caused by the pattern, timing, or completeness of sampling), such as
when sampling along a line misses rare, but important components of diversity or biomass. Both may have considerable spatial and temporal variability due to complex environmental gradients. Interacting natural processes
on an already heterogeneous landscape further complicate both types of
studies.

Sample size requirements to measure plant diversity are unknown for
many reasons. Methodologies for plant diversity studies are still being developed and tested in many habitats. It is unrealistic to attempt to survey
and monitor all species: most species are rare, and to monitor most or all of
them would be cost prohibitive (Magurran 1988). Statistical techniques to
determine taxonomic completeness are in the developmental stage (Heltshe
and Forrester 1983; Miller and Wiegert 1989; Palmer 1990; Palmer et al.
1991). Last, but not least, sample size, which is determined by sample variance (Krebs 1989, 1999), is unequivocally dependent on the pattern of plot
placement (see below).

Another important consideration is the intensity of sampling (number and
size of plots on the landscape), which influences the amount of variation
accounted for by the study and the ability to accurately detect trends over
time. Hinds (1984) stated: "the major statistical difficulty is specifying appropriate replication standards in a world that is full of unique places." While
the debate continues about replication and pseudoreplication (see the box
below), in most cases there are rarely enough true replicates in ecological studies. In a review of hundreds of studies in top ecological journals, Kareiva and
Anderson (1988) showed that (1) as plot size increased in ecological studies,
the number of replicates decreased; and (2) once plot size reached about
3 m², the number of replicates was consistently less than five. Few of these
studies were likely plant diversity studies, but it is instructive that much of
our knowledge of plant ecology may be based on small sample sizes and small
plots—methods to be remedied in the measurement of plant diversity.

Formula-Based Determinations of Sample Size

The reader is advised that there are no easy ways or off-the-shelf protocols
for determining sample sizes, given that different study objectives, issues
of scale, and various patterns of sampling greatly influence results of plant
diversity studies. Many textbooks offer sections on sample size determination that provide important background reading (Bonham 1989; Cochran
1977; Krebs 1989). These are often followed by critical articles in scientific

Reminder: Precision and Accuracy

Mean Biomass Example: Many textbooks and field manuals have described accuracy and precision in terms of comparing different techniques to measure the mean cover or biomass of a stand of vegetation (e.g., Barbour et al. 1987). It is important to gain some knowledge of the accuracy and precision of field samples. Often the true population mean is unknown, such as the average foliar cover for a large meadow. For this example we will assume that a much greater random sampling of the meadow provided a reasonable estimate of the true population mean (Figure 3.8). In case 1, the three sample plots from relatively homogeneous areas in the meadow provided a very close approximation of the true mean cover value (high accuracy), and the variance among plots was low (high precision). In case 2, the three samples had the same high precision (low variation) as in case 1, but none of the sample values were close to the true mean cover value, perhaps sampled from atypically productive or less productive areas of the meadow. In case 3, the mean value of three samples from heterogeneous areas of the meadow was again very close to the true mean value, but the variance was higher than in case 1 (i.e., high accuracy and low precision). Precision and accuracy were relatively low in the case 4 samples, suggesting atypical and heterogeneous areas of the meadow (Figure 3.8). Larger sample sizes, collected with a random or stratified random sampling design, generally increase accuracy and precision. It is often advisable to collect additional samples in an area, keeping track of improvements in precision as additional samples are collected.

Mean Species Richness Example: In the same vein, the terms precision and accuracy can be applied to studies of mean species richness. A series of samples may be high or low in precision and accuracy (Figure 3.8). However, there is an important distinction to the biomass example above. Species richness values are heavily dependent on quadrat size and shape. The precision of a set of samples for plant species richness may be enhanced by using rectangular plots (see Figure 3.2), but this precision is a trade-off with decreased accuracy scaled to the perimeter:area ratio (Krebs 1989). Rectangular plots, which cover a wider range of environmental gradients, may more accurately describe the variance among plots in heterogeneous areas. Larger, rectangular plots may reduce the variation further, but with increased costs associated with more plant identifications. Generally, fewer plots are needed to assess mean biomass in grasslands than mean species richness.

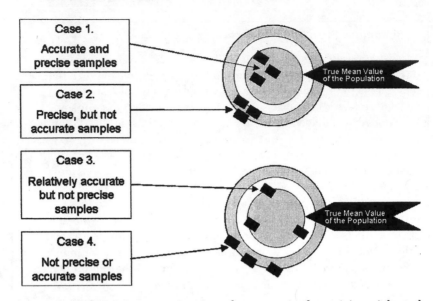

Figure 3.8.   Schematic representation of accuracy and precision. Adapted from Barbour et al. 1987.

journals (Blackwood 1991; Hurlbert 1984; Kupper and Hafner 1989; Thomas and Krebs 1997), which serve to complicate the matter—in a very positive manner—causing the investigator to devote more time and thought to sample size issues. It is easy to fall prey to simple formulas and computer-assisted programs to select appropriate sample sizes. Thomas and Krebs (1997) provide a brief review of some of these computer programs. In any case, it might be best to forget some of the old axioms from Statistics 101, where assumptions of homogeneity and normality are made without question, and where a sample size of 30 is selected because the t-values level off in the t-table.

Replication and Pseudoreplication

Replication and pseudoreplication are bewildering concepts for many landscape ecologists (Hawkins 1986; Hurlbert 1984). Pseudoreplication occurs when samples are not dispersed in space (or time) in a manner appropriate to the hypothesis being tested (Hurlbert 1984). For example, repeated sampling in one (or a few) subjectively placed neighboring plots may not be representative of the forest stand, vegetation community type, or landscape inferred by the hypothesis. Hargrove and Pickering (1992) argue that pseudoreplication can't (and needn't) be avoided in landscape or regional ecology considering the multivariate, spatially, and temporally-complex facet of a landscape.

They suggest greater reliance on combinations of "natural experiments" (i.e., chronosequence studies), "quasi-experiments" (i.e., before/after impact studies), and meta-analysis (i.e., a statistical technique that produces a unified result from diverse and sometimes contradictory studies). However, these study types are not without their problems, as you will see in subsequent sections and chapters.

How might pseudoreplication influence plant diversity studies? Consider the following example. An investigator wishes to evaluate the effects of grazing on plant species richness in a rangeland. The investigator can establish 40 plots in one grazed mesa top and 40 plots in the adjacent ungrazed mesa top (a "natural" experiment; Figure 3.9). If the plots are randomly located on the two mesa tops, they are independent samples and true replicates of the one "paired site." In this case, the effects of grazing are only considered at this one paired site—there are no replicates at the treatment or landscape scale (Guenther et al. 2004).

Alternatively, the investigator can establish 10 randomly located plots in and adjacent to four grazing exclosures at different, similarly treated sites. In the first instance, the investigator cannot extrapolate the study results beyond the one pair of grazed and ungrazed sites. The plots on the two mesas are not independent samples of the population of grazed and ungrazed sites in the rangeland. However, as long as the investigator acknowledges that the true sample size is one ($n = 1$ set of grazed and ungrazed sites), and limits the discussion to that one local site, then the quadrats are not pseudoreplicates. In the second case, the investigator is free to extrapolate to the sample of four pairs of exclosures and adjacent sites. If the landscape contained four exclosures, and the four exclosure sites and paired sites were selected randomly and encompassed most of the rangeland, then the investigator may cautiously extrapolate results to the landscape. This presumes that the combined suite of samples adequately represents the landscape and that the exclosures were sited randomly. If the landscape contained 12 exclosures, and the four sampled sites were selected at random from the suite of 12 sites, then the investigator may cautiously extrapolate results to the "population" of sites. However, if the four pairs of sample sites were subjectively selected, or if the sites were not representative of the population of sites or the landscape, then the results could not be extrapolated beyond the sampled sites: the sampled sites and the quadrats in each are not true, independent replicates of the population. Hurlbert (1984) claims many ecological studies suffer problems of pseudoreplication, generally in the form of extrapolating results beyond the sample dataset and inappropriate to the hypotheses being addressed. Being careful in the design phase and in the interpretation phase can prevent most problems of pseudoreplication.

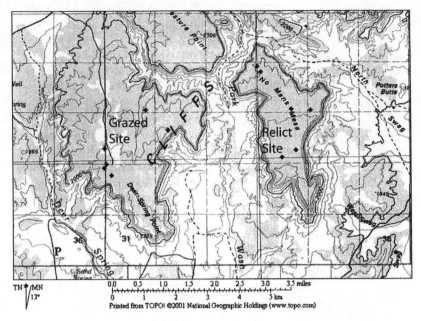

Figure 3.9.   Lesson 9. Most plant diversity studies include poorly replicated treatments or habitats. Don't design the next poorly designed study!

Sample size determination is a process, one that is sometimes complicated
and usually iterative (Krebs 1999).

The first part of this section addresses formula-based techniques for determining sample sizes, briefly summarizing the clear and concise treatments
of population-based sampling by Elzinga et al. (1998, chapter 5 and appen-
dix 7). Subsequent parts address more iterative approaches to sample size
determination for plant diversity studies.

One of the most puzzling aspects in the objective science of plant ecology is the number and impact of several subjective decisions that are made
when determining sample size. Generally the investigator must consider the
acceptable confidence limit, alpha (α) level (type I error rate), beta (β) level
(type II error rate), and statistical power of certain tests. For example, an
investigator may want to determine the number of 1 m^2 plots needed to
compare the average number of plant species per plot in burned and unburned portions of a field. The hypothesis might be that the burned area has
equal plant species richness in the spring following the fire compared to
unburned sites. The investigator subjectively sets a 95% confidence limit
to reduce the type I error rate (i.e., the 5% probability of suggesting a significant difference between the sites when there really is no difference). The
investigator must then subjectively set the minimum detectable change
(MDC), which is the size of the change the investigator wants to be able to

detect (Elzinga et al. 1998). In this example, the investigator wants to detect a 20% difference in mean species richness between burned and unburned areas. The statistical power of a test is a complement of the a level (i.e., $1-a$), so high power is inversely related to making a type I error. The following somewhat complicated example is meant to be instructive rather than dogmatic in the use of formula-based determinations of sample size. In this case, the investigator wants to be 95% certain of detecting a 20% difference in mean plant species richness. In a preliminary random sampling of ten 1 m^2 plots in burned and unburned portions of a field, the investigator calculates the following values (Table 3.2).

Assuming the burned and unburned sites are truly independent of each other, and that the investigator needs to determine the number of samples (n) required to detect a 20% difference in species richness compared to unburned areas (i.e., 20% of 6.5 species/plot = 1.3 species/plot), the following equation may be used:

$$n = 2(s)^2(Z_\alpha + Z_\beta)^2/(MDC)^2$$
$$n = 2(2.1)^2(1.96 + 1.64)^2/(1.3)^2$$
$$n = (8.82)(12.96)/1.69$$
$$n = 67.6 \text{ plots, or 68 plots.}$$

The key point is that there are several caveats to relying on formula-based determinations of sample size. The parameters of interest (e.g., plant species richness, number of nonnative species) are assumed to be normally distributed with homogeneous variances. For very large samples (n = far greater than 30 and with very large plots), the assumptions may be correct in some rare cases, but most frequency distributions of plant species richness are typically inverse J-shaped. For small sample sizes ($n < 30$), Z coefficients generally underestimate sample sizes (Elzinga et al. 1998). A less obvious problem is that the initial sampling of a few to several plots may not represent the true spatial variation in a complex landscape. Increasing sample sizes may extend sampled environmental gradients and include rare or unique habitats. The variance may not be homogeneous with increasing spatial extent. One option is to rerun the sample size determination formulas as new plots are established. Another option is to rerun the tests with different plot sizes. This creates an iterative process that is more computationally expensive, but may yield more reasonable results.

Table 3.2. Hypothetical data on mean plant species richness from ten 1 m^2 plots in burned and unburned areas in the field.

	Burned area	Unburned area	Difference
Mean	8.5	6.5	2.0
Standard deviation	2.4	2.1	1.4

Reminder: The probability that an interval of values contains the true value of a parameter is known as the confidence level. Elzinga et al. (1998) call the α level (also called the type I error), the false-change error rate. This happens when the investigator's data suggest a significant change when no change has taken place. Elzinga et al. (1998) call the β level (type II error) the "missed-change error rate," which is when the investigators dataset suggests no significant change when there has been a change. The MDC greatly affects statistical power. Power analysis (Fairweather 1991) is a function of the standard deviation of a test set of data, the number of sample units, the MDC, and α, the false-change error rate (see Elzinga et al. 1998). Often the sampling design can be changed (increasing the size or number of plots) to improve the power of a test by reducing the standard deviation. Power also can be increased by increasing the minimum acceptable level of significance (α level at 0.10 rather than 0.05) or by increasing the MDC level (e.g., detecting a 10% change rather than a 1% change in species richness). Values for Z_α and Z_β can be gleaned from the appendices of statistics textbooks, with the most frequently used values presented below (Table 3.3):

Confidence level	α	β	Power	Z_α	Z_β
90%	0.10	0.10	0.90	1.64	1.28
95%	0.05	0.05	0.95	1.96	1.64
99%	0.01	0.01	0.95	2.58	2.33

If the investigator can assume that 10 pairs of plots in the burned and unburned areas of the field are "paired" samples, the appropriate equation would be

$$n = (s)^2(Z_\alpha + Z_\beta)^2/(MDC)^2$$
$$n = (1.4)^2(1.96 + 1.64)^2/(1.7)^2$$
$$n = (1.96)(12.96)/2.89$$
$$n = 8.7 \text{ plots, or 9 plots.}$$

In this case, s is the standard deviation of differences between each pair of burned and unburned plots, and the 20% difference is based on the larger of the two means (i.e., $8.5 \times 0.2 = 1.7 = MDC$). Note that far fewer samples are needed to detect differences in the second case by making a change in the sampling design (i.e., the use of a paired-plot design) and the assumption that each pair of plots is nearly identical in every way except for the "main effect"—grazing in the case example. Imagine the real-world difficulty of finding nine pairs of 1 m² plots (or 1000 m² plots) with the same slope, aspect, elevation, microtopography, soil characteristics (texture, nutrients, water-holding capacity, moisture), disturbance history, and plant species composition, structure, and age such that the only difference is the level of grazing. Plot size would greatly affect these results. Picking truly paired sites is difficult in ecological studies.

More Iterative Approaches
to Sample Size Determination

The previously described formulas for sample size determination require preliminary data (and a standard deviation) to estimate how many more samples will be required. This is an iterative approach. There are several other iterative approaches that may be useful in determining sample sizes in plant diversity studies.

The Running Mean

In many habitats, variance often decreases with sample size. It is helpful to plot the running mean of samples as new samples are incorporated (Table 3.4; see Krebs 1999, pp. 256–258).

Plotting the running mean shows that after high initial variance, the running mean eventually approaches the population mean (Figure 3.10). An appropriate preliminary sample size is thought to be when the curve of the running mean levels out. In this hypothetical example, the appropriate preliminary sample size is seven or eight (Figure 3.10).

However, this technique also carries several assumptions. It is assumed that the sample parameters are normally distributed and that the sites are randomly selected. It is further assumed that the samples encompass the full range of environmental gradients from the potential population of samples in the landscape, and that the ordering of the samples into the running mean calculation had no effect on the outcome. Of course, all of these assumptions can be tested to some degree. Several statistical analysis software programs can evaluate the normality of distributions. The random selection of points can be carefully managed while conducting the fieldwork. The range of environmental gradients (e.g., slope, aspect, elevation, soil moisture) can

Table 3.4.   Hypothetical data of the number and mean number of plant species in eight successive 1 m² plots.

Sample	No. of species	Mean
1	8	
2	11	9.5
3	12	10.3
4	7	9.5
5	9	9.4
6	10	9.5
7	11	9.7
8	10	9.8
9	9	9.7
10	11	9.8

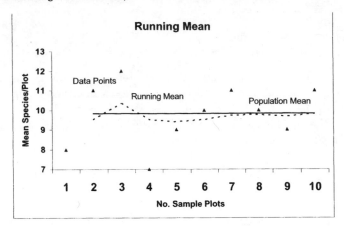

Figure 3.10.   Plot of the running mean (dotted line) from Table 3.4. Data points are triangles with a sample mean (an assumed population mean) represented by the solid line.

be evaluated for subsets of data points as they are entered into the running mean equation. The order of points can be randomized and repeated prior to entering the points into the running mean calculation.

Species Accumulation Curves

Species accumulation curves and species-area curves also can be used to assess the taxonomic completeness of sampling (Figure 3.7). A species accumulation curve plots the cumulative number of species encountered (y axis) against time or effort in a cumulative number of similar-size plots (x axis). The species-area curve plots the cumulative number of species against the cumulative area of the quadrats (or area searched)—the quadrats do not have to be of similar size (see chapter 5). For example, Jaccard (1901) [as discussed in Daubenmire (1968)] constructed species-area curves to assess when "time spent studying more area is wasted." This was generally considered to be the point at which the species-area curve leveled off. This point was seen as an important floristic characteristic of an association.

Upon closer examination, the example in Daubenmire's book (Figure 27; Daubenmire 1968, p. 89) is unfortunate for several reasons. First, for this particular set of data, unlike most I have seen, the curve levels off very quickly. In the two stands of rich steppe vegetation, only 2–3 m² of sampled area were required to record nearly all the species in the stand. It may not be a coincidence that the transect and quadrat method (later known as Daubenmire transects) (Stohlgren et al. 1998c) typically included twenty 20 cm × 50 cm quadrats in a stand (i.e., 2 m² of sampled area).

Second, homogeneous areas may have been preferentially selected for the study so that environmental gradients would be minimized, thus potentially underestimating species-area curves in more heterogeneous areas. Third, high spatial autocorrelation (similarities among adjacent plots and along a transect) may artificially reduce the species-area curve so that it may not be representative of a stand. There usually are small but important patches of diversity in the stand, but not along the transect or captured by small quadrats. The assumptions are that the area being sampled is relatively homogeneous such than new samples do not significantly extend environmental gradients or encompass new habitats or microhabitats. The ordering of the samples is assumed not to influence the results (Figure 3.11).

There are more sophisticated Monte Carlo techniques (Krebs 1999) and computer-assisted techniques for assessing running means and species accumulation curves. For example, Estimate-S (Colwell 1997) will randomize samples and generate smoothed species accumulation curves. However, the underlying assumptions for sample size determination procedures remain. In real life, it is difficult to truly randomly select plot locations and increase the number of plots without extending complex environmental gradients—so species accumulation curves rarely level off for landscapes or regions. Species accumulation curves are also greatly affected by plot size and shape, sample size, and pattern of sampling.

Spatial Pattern of Sampling

A premier challenge in ecological research is to develop field techniques to detect and quantify patterns in space and time and to explicate underlying mechanisms (Lubchenco et al. 1991). Krebs (1989) noted that many studies have used "accessibility sampling," such as areas in flat terrain, close to roads, or near facilities. It is often painfully obvious that these sites may not fully represent the plant diversity in the broader landscape. In this section I briefly describe how the spatial patterning of sampling influences results. I then address the theoretical and analytical limitations of synthesizing information

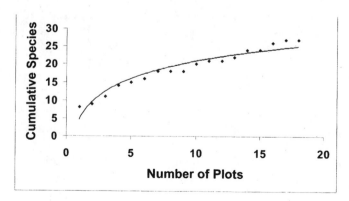

Figure 3.11.   Hypothetical species-accumulation curve and data points.

on spatial trends from various patterns of sampling sites. In a homogenous environment, the pattern of sampling (the spatial arrangement of study plots on the landscape) is largely irrelevant. For example, estimating biomass in a field of corn is accomplished simply with a random or systematic sampling design. In natural systems and landscapes, plot placement is *everything*. Consider the following. Natural landscapes are complex mosaics of vegetation types and structures, soils, geology, and animals. Superimposed on these resources are temporally and spatially variable responses to disturbance, herbivory, competition, and pathogens. The resulting complex landscape is complicated further by spatially and temporally variable land use histories and external threats (e.g., air pollution, invasion by nonnative plants and animals). Determining the optimum pattern of sampling required to describe and understand such a landscape is far from trivial (Figure 2.4).

Within a larger study plot, Daubenmire (1968) strongly suggested placing quadrats along a line and spaced one quadrat-width apart (systematic sampling), such that the entire series of small quadrats fall in a homogeneous unit. Oddly, while Daubenmire recognized that systematic sampling would oversample common habitats and undersample rare habitats at the landscape scale, these same concerns were not expressed at the plot scale (in fact, they were suggested). The motivating reasoning was that systematic sampling was unbiased and gave greater precision, but the costs of oversampling were not considered, nor was there a concern for missing rare or heterogeneous habitats. The only disadvantage of systematic sampling was when there were obvious linear patterns in vegetation (e.g., grazing terracettes).

Recall from chapter 2 that Mueller-Dombois and Ellenberg (1974) suggested that sampling locations within strata might be selected with preconceived bias by picking "representative" or "typical" sample site locations. However, Krebs (1989, 1999) calls this "subjective sampling," which limits the use of probability statistics to extrapolate results to the remainder of the unsampled landscape. Mueller-Dombois and Ellenberg (1974) also said that sample sites might be selected without preconceived bias, but this would not absolve the ecologist from estimating how much bias had been incorporated by her or his strong inference—again the samples collected in such a venture could not be used to extrapolate results to the remainder of the unsampled landscape. Lastly, the authors suggested selecting sample locations in a random, systematic, or other unbiased manner (i.e., according to chance) (Mueller-Dombois and Ellenberg 1974, p. 32; see "Temporal Patterns of Sampling" section below).

Systematic surveys evenly place plots on a grid across the entire study area (Figure 3.12). While this plot layout is designed to produce unbiased estimates of a parameter, it is easy to see how some vegetation types might be oversampled, while small, locally rare habitats might be missed entirely.

Daubenmire (1968) recognized the problems of systematic sampling. It is equally obvious that simple random sampling with small (affordable) sample sizes can also oversample common types, while rare features are

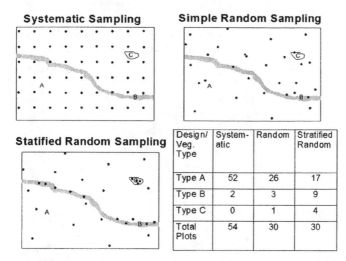

Figure 3.12.   Potential effects of various sampling designs on capturing plant diversity information from three vegetation types in a hypothetical landscape.

undersampled (Figure 3.12). This was also shown in field studies (Fortin et al. 1989; Legendre and Fortin 1989). Extremely rare features on landscapes are very difficult to capture with strictly systematic surveys (Stohlgren 1993). Still, systematic surveys remain the design of choice for very successful monitoring programs such as the U.S. Forest Service Forest Health Monitoring Program (Cline et al. 1995, chapter 9).

Stratified random sampling is said to combine the benefits of random and systematic sampling (Mueller-Dombois and Ellenberg 1974). Strata selected to include rare and common strata or vegetation types, with randomly located sample sites within them, have been efficient in sampling plant species richness in natural landscapes (Figure 3.12) (Stohlgren et al. 1997a,b,c). A disadvantage to stratified sampling, in general, is that the results are restricted to the specific objectives of the study, and they last only as long as the strata remain intact. Vegetation gradients pose particular problems in stratifying areas for sampling plant diversity (Stohlgren et al. 2000b). Statistical approaches by Goodall (1953a,b, 1954) with random and systematic plots failed to detect discrete boundaries between vegetation types, revealing the gradients of populations and vegetation types that would make stratified sampling difficult for that particular study site. Meanwhile, Russian ecologists (Sukachev 1945; Sukachev and Dylis 1964) found discrete vegetation boundaries exactly matching environmental boundaries (e.g., edaphic and topographic edges) that would make stratified sampling easier for their particular study sites. Oceanographers (Colebrook 1982; Magnuson et al. 1991; McGowan 1990), limnologists (Nero and Magnuson 1992), and meteorologists (Haslett and Raftery 1989) may be more advanced than terrestrial vegetation ecologists at synthesizing information from widely scattered sampling sites. However, some research has focused on the problems of spatial autocorrelation and optimizing the spatial arrangements of vegetation plots on forested landscapes. For example, Fortin et al. (1989) and Legendre and Fortin (1989) tested several different patterns of plot placement to evaluate which one was more capable of detecting the spatial structure (i.e., high-density areas) of a sugar maple (*Acer saccharum*) forest. By collecting data from a larger number of plots, they subsampled the data using various spatial patterns and interpolated data from the subsamples (using kriging). They then compared the spatial patterns of each subsample pattern to the population pattern. They showed a systematic cluster design (uniform plots each with a paired offset plot) was superior to a strict systematic or random sampling pattern using the same number of plots. However, stratified random sampling was not compared to the other designs in the study.

I used the systematic cluster design to describe ponderosa pine (*Pinus ponderosa*) stands in Lava Beds National Monument in northeastern California (Stohlgren 1993). My two objectives were to quantify the spatial patterns of high basal area pine stands and to estimate the number of bald eagle winter roost trees (old-growth trees) in the study area. While the sampling design detected the general distribution of stands of large pines, it failed to accurately determine the number of rare, heavily clustered bald eagle winter roost trees in the study site (Stohlgren 1993). It seems that more research in experimental design is needed to optimize the pattern of sampling required to detect rare, but important landscape features.

The pattern of sampling also influences species enumeration for evaluations of biodiversity (Sokal and Thomson 1987). After establishing more than five hundred 0.1 ha plots at uniform 1 km grid intersections in Sequoia and Kings Canyon National Parks in California (0.015% of the land base), Graber et al. (1993) report encountering two-thirds of the vascular plant species known in the parks. Here, the uniform placement of plots likely contributed to missing the rare, but important landscape features in the parks (i.e., species-rich areas such as riparian zones, small meadows, serpentine outcrops, etc.).

Temporal Patterns of Sampling

It is humbling for plant ecologists that generalizations about species richness were stated early and concisely by a limnologist. In the 1920s, Thienemann noted that (1) the greater the diversity of conditions in a habitat, the greater the species richness; (2) deviations from normal conditions result in fewer species, but greater abundance of individuals; and (3) species richness and stability increase the longer the site has been in the same condition (later termed the diversity-stability hypothesis) (McIntosh 1985).

The second major component of pattern is temporal: how frequently should data be collected at each site. The pattern of sampling of long-term study plots determines the ability to accurately evaluate spatial and temporal changes in vegetation at landscape scales. The major analytical challenges are determining (1) the appropriate dimensionless metrics of temporal and spatial variability (Magnuson et al. 1991); (2) the magnitude of variation owing to "location," "year," and "other" (Magnuson et al. 1991); and (3) the number of replicate samples (i.e., measurements, plots, transects, etc.) required to study spatial and temporal trends.

Magnuson et al. (1991) reported that it is difficult to compare diverse ecosystems without "dimensionless parameters." These are parameters that can be compared across a wide array of locations, or for long time periods (e.g., precipitation and temperature, nutrient concentrations, species richness, etc.). The resulting data can be analyzed with a model II analysis of variance (ANOVA) (Sokal and Rohlf 1981) to determine the magnitude of variation owing to "location," "year," and "other" (Magnuson et al. 1991). However, it may be difficult in many plant diversity studies to factor out measurement errors from the interaction between the main effects of "year" and "location." Without adequate replication of study sites, the interaction and error terms cannot be separated.

One constant in plant ecology is the interest in species richness, whether it is the number of species in a quadrat, area of undefined size, or community type. Worster (1977) reported an observation of Gilbert White (an 18th-century naturalist) that the longer one looks in an area, the more species are found. Plants die. Measuring the site by site replacements of plants and species in a carousel of replacements over large areas will not be easy (Palmer and Rusch 2001; van der Maarel and Sykes 1993).

There is still much to learn about the interaction of sampling design, sample size, and the pattern of plot placement in plant diversity studies. Many plant species in the deserts of Death Valley, California, germinate and flower every 3–10 years, depending on the previous fall's precipitation (Hunt 1975). Plant ecologists need creative study designs relying on nested sampling techniques with an unbiased selection of study plot locations and many replicate plots. An experimental approach is needed.

PART II

AN EXPERIMENTAL APPROACH TO SAMPLING

4

Single-Scale Sampling

There are many types of single-scale sampling that have been used by plant ecologists for decades. The simplest and oldest form of single-scale sampling for plant species richness is "searching."

Searching Techniques

Usually the objective of searching techniques is to create or amend a species list or to search for rare plant populations in a meadow, rangeland, park, landscape, or region in a botanical survey. The methods rely on the taxonomic expertise of the field crew, the specific area surveyed, the level of intensity of the survey, and the ability to detect the species in the area. For example, an experienced research team with excellent botanists will probably record and collect more plant species than an inexperienced crew over the same area. Both crews may record more species if they spend more time in the field or visit the site several times over several years. Both crews might be less successful in sites where tiny, prostrate herbs hide under larger, thorny shrubs, or in remote, dangerous microsites.

How well does searching work? Consider herbarium specimen additions over the years from Rocky Mountain National Park, Colorado (Figure 4.1). Collections got off to a slow start until 1925, when the first botanical study in earnest collected more than 450 species (including varieties and subspecies). Other studies in the 1950s, 1970s, and 1980s each added significant

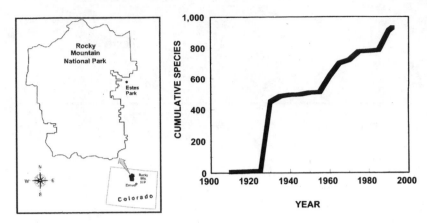

Figure 4.1.   Map and cumulative plant species added to the herbarium at Rocky Mountain National Park, Colorado.

numbers of species to the herbarium. However, it is impossible to estimate the completeness of the combined searches. The field crews may have looked in different or overlapping areas, missed some botanically rich areas, ignored remote locations, or overzealously collected some taxonomic families of greater familiarity while other families were overlooked. Regression models that assume independent samples from random locations would be inappropriate for this type of data.

Searching may be an important technique for developing initial species lists, for targeting taxonomic surveys in suspected species-rich areas, or for augmenting plot-based sampling designs where only a tiny portion of the landscape or region can be affordably sampled (see Magurran 1988, pp. 51–55). A combination of searching and plot-based sampling techniques may be needed to develop near-complete species lists and for monitoring plant diversity in large, complex landscapes. Searching may be the only effective technique to locate extremely rare species, such as threatened and endangered species. Searching methods can be improved, or at least better documented, by carefully reporting the skill levels of the crews, the areas covered, the time spent searching, and the recovery rate of new finds.

Single-Scale Plots

The most common form of vegetation "sampling" (in the true sense of the word) is single-scale sampling. There are many different goals and objectives that can be realized with single-scale sampling of plant diversity. Fortunately current vegetation sampling texts cover this approach almost

exclusively (Daubenmire 1968; Mueller-Dombois and Ellenberg 1974), so I
will discuss these methods only briefly.

Single-scale plots are generally sized to the vegetation (Table 4.1). Small
plots are used for mosses and lichens, while larger plots are used for shrubs
and trees.

The classic example of single-scale sampling is the use of small points or
quadrats in a large sampling area. Cain (1944) suggested that 20 quadrats
0.1 m × 1.0 m in size adequately described alpine vegetation in Wyoming.
Parker (1951) thought that 300 very small (2 cm diameter) plots would capture several common species in rangeland vegetation. Rice (1952) used
twenty 0.32 m × 0.32 m plots to describe tallgrass prairie in Oklahoma. Dix
and Butler (1954) needed 80 quadrats that were 0.35 m × 0.35 m to evaluate
the effects of fire on Wisconsin prairies. Note that a fairly small area is actually measured in small-quadrat studies. These above examples range from
less than 0.1 m² for the Parker (1951) method to 9.8 m² for the Dix and Butler (1954) method.

Historically the relevé method of community classification greatly influenced plot sizes and sampling locations. Remember, the relevé method was
typically used to document the distribution of common, dominant species
in a landscape. Mueller-Dombois and Ellenberg (1974, p. 46) reminded plant
ecologists that a sample stand (or relevé) should be large enough to contain
all species belonging to a plant community, the habitat should be uniform,
and plant cover should be homogeneous. Obviously, small plots would not
contain *all* the species in a plant community, nor would many small plots.
However, in uniform habitats and areas of homogeneous plant cover, small
plots will likely capture dominant species. In addition, if the dominant
species are short in stature and ubiquitous across the landscape, then small

Table 4.1. Suggested quadrat or plot sizes in various textbooks
for routine vegetation sampling.

Vegetation	Daubenmire (1968)	Mueller-Dombois and Ellenberg (1974)
Mosses and lichens	25 m²	
Mosses		1–4 m²
Lichens		0.1–1 m²
Herbs		1–2 m²
Herbs and shrubs	0.1 m²	
Dry-grassland		50–100 m²
Tall herbs-low shrubs		10–25 m²
Tall shrubs		16 m²
Large plants/trees/forest	4–100 m²	200–500 m²
Forest understory only		50–200 m²

plots will suffice. In fact, for the purposes of vegetation classification, Mueller-Dombois and Ellenberg (1974, p. 47) correctly state that "less obvious or rarer species may be dispersed among the more obvious species, and the dispersal pattern of these less obvious species may be considered of secondary importance." So we see that in some textbooks and field manuals, rather small plots are recommended for general sampling, without specific information on sampling efficiency for various possible objectives (Table 4.1). When species richness and diversity were discussed in earlier texts, plot size recommendations were still simplified. Daubenmire (1968), for example, noted that in the "floristically impoverished arctic tundra," 1 m² plots may be adequate, while 2 ha plots may be required in tropical rainforests.

A transect is another form of single-scale sampling. During the past century, many North American plant ecologists used linear transects to assess understory vegetation. Belt transects were typically reserved for analyzing vegetation gradients across ecotones. Daubenmire (1968, p. 88) was enamored with the "superiority" of the elongated plot—or line transect. He saw transect and quadrat (20 cm × 50 cm) techniques as a means to precisely survey many clusters of species, quickly gaining quantitative data on species richness and cover. He recognized that point frequency sampling (e.g., a row of 10 points spaced at 5 cm intervals) may affect the independence of observations: "most

---

Concerning the Use of Cover Classes

Cover classes have often been used to increase precision among investigators (Daubenmire 1968). In foliar cover studies in grasslands, for example, precision can be improved by using smaller quadrats (e.g., 20 cm × 50 cm) and larger cover classes. However, large cover classes may not ensure high accuracy because of the way midpoint averages are often used to sum field data. Two investigators might easily agree that the cover of some herb species was between 0% and 5%, but fewer would agree that cover was 2% versus 3% in a 0.1 m² quadrat or a 1 m² quadrat. However, there is often a trade-off between precision and accuracy where cover classes are used, because broad cover classes may not accurately reflect the actual cover of a species and the midpoints of cover classes (used to calculate total cover) may not accurately reflect the actual cover of a species (Table 4.2). For many vegetation types in the central and western United States, there is an inverse J-shape (negative exponential) distribution of species identity (x) and cover (y), with about 50% of the understory species having less than 1% cover. There will usually be more species on the lower end of each cover class. Thus, using the midpoint average in calculations will usually inflate estimates of total cover and exaggerate the mean cover per species (Table 4.2).

Table 4.2. Hypothetical data on species cover (estimated to the nearest 1%) and cover class data for the same species.

Species	Actual percent cover	Cover class	Cover class midpoint
a	1	1	2.5
b	3	1	2.5
c	1	1	2.5
d	1	1	2.5
e	1	1	2.5
f	2	1	2.5
g	4	1	2.5
h	9	2	7.5
I	1	1	2.5
j	2	1	2.5
k	1	1	2.5
l	1	1	2.5
m	1	1	2.5
n	1	1	2.5
o	1	1	2.5
p	6	2	7.5
q	12	3	17.5
r	28	4	37.5
s	3	1	2.5
t	2	1	2.5
u	1	1	2.5
v	1	1	2.5
w	3	1	2.5
x	2	1	2.5
Total percent cover	88		120
Mean cover/species	3.7		5.0

Cover classes	Range	Midpoint
1	0–5%	2.5%
2	5–10%	7.5%
3	10–25%	17.5%
4	25–50%	37.5%
5	50–75%	62.5%
6	75–100%	87.5%

of the ten points make the same kind of distinctive contacts" (p. 88). Regardless of the presumed minor problems associated with various small-scale techniques, single-scale plots were generally regarded as cost-efficient, information-rich techniques that could be effectively used to describe many vegetation characteristics. Some examples may illustrate effective uses of single-scale sampling in describing plant diversity patterns.

Example 1: Plant Species Diversity
and Succession in Abandoned Fields in Illinois

Single-scale methods have long been used to assess plant species diversity. One typical example is the work of Bazzaz (1975), who examined plant species diversity in a chronosequence of abandoned agricultural fields in southern Illinois. The herbaceous cover of each plant species was estimated (to the nearest percent) in forty 1 m × 2 m plots. Shrubs and trees were sampled in 25 plots: 4 m × 4 m for shrubs, and 10 m × 10 m for trees. Bazzaz used the Shannon-Wiener index to compare species diversity in fields that were 1, 2, 4, 10, 25, and 40 years since abandonment. The Shannon-Wiener index (H) is defined as

$$H = \Sigma Pi \log_2 Pi,$$

where Pi is the proportion of cover of the species within the sample. A low H value generally suggests a site with few species and a few dominant species, while a high H value suggests considerably more species, usually with many species of low cover [see Krebs (1999, p. 410–454) for a complete discussion of species diversity measures]. In this set of old fields, diversity (H) was lowest in the 4-year-old abandoned field and tended to be higher in fields abandoned more than 15 years, seemingly increasing up to 40 years (Figure 4.2).

Rice and Westoby (1983) investigated plant species richness in 0.1 ha plots (20 m × 50 m) in seven sites from northern to southern Australia. The author's recognized that the large plots were needed to capture locally rare species and make meaningful comparisons to data from similar-size plots collected

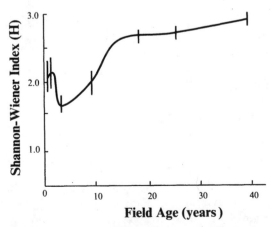

Figure 4.2.   Idealized diversity changes in a chronosequence of abandoned fields in southern Illinois (adapted from Bazzaz 1975 and used with permission from the Ecological Society of America).

Reminder: Chronosequence studies substitute space for time (e.g., Bazzaz 1975). Rather than monitor a set of fields for 40 years or more, an investigator simply finds fields (or forest stands) of different ages and assumes (1) all the sampling sites were similar at time zero (i.e., initial conditions were identical), and (2) that the succession would occur in a similar way if sampling sites were monitored over time. These assumptions are rarely tested (see Picket 1989). Two or more sites are rarely identical in terms of all topographic, biotic, and abiotic characteristics. Species patterns (cover by species, pattern of diversity, productivity, etc.) are rarely identical at multiple spatial scales in any two field or stands. Subtle differences in initial floristics, soil characteristics, and moisture availability, and continuing differences in small-scale disturbances by rodents, insects, and diseases can complicate the interpretation of chronosequence results. Some of these concerns can be minimized by detailed comparisons of the environments of the sampling sites, having replicate sites (e.g., several 1-year-old fields, several 4-year-old fields, etc.), and by following some or all of the fields for a few years to see if the temporal patterns match chronosequence patterns. Cost and adequate replication are always important concerns, but it may be wiser to do fewer chronosequence stages well than to do more stages without adequate replication and evaluations of spatial variation in environmental characteristics.

Bazzaz (1975) effectively used dominance-diversity curves (Whittaker 1967, 1970) to show how species ranked by relative cover displays two important aspects of species diversity simultaneously. In this example, the 1-year-old field contained far fewer species than the 40-year-old field, with high relative dominance in a couple of species (Figure 4.3). The 40-year-old site had many more species and many low-cover species. The 4-year-old, 15-year-old, and 25-year-old fields had intermediate patterns. Bazzaz concluded that (1) species diversity generally increased with succession, (2) microsite heterogeneity and structural diversity may increase diversity, (3) strongly dominant species may reduce diversity, and (4) dominance-diversity curves elongate over time in support of the "niche preemption hypothesis" (Whittaker 1965), where community resources are shared by more species.

by Whittaker and several others. The results clearly demonstrated that temperate, closed-canopy forests had, on average, fewer than 50 species per 0.1 ha, whereas tropical rainforests averaged 140 species per 0.1 ha. Floristic evolutionary history was thought to have a strong influence on patterns of species richness, with present-day climate and soil characteristics contributing to a lesser degree.

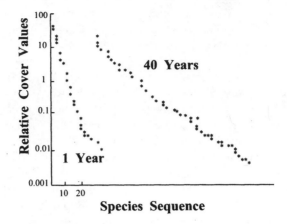

Figure 4.3. Idealized dominance-diversity curves for a 1-year-old and 40-year-old abandoned fields in southern Illinois (adapted from Bazzaz 1975).

Diersing et al. (1992) developed a Land Condition Trend Analysis (LCTA) program for monitoring U.S. Army lands throughout the United States—4.9 million ha of land. For plant diversity, species lists and voucher specimens are collected and herbaceous plant species are recorded each 1 m (point) along a 100 m transect in a 600 m² plot (6 m × 100 m). The cover of woody vegetation is recorded throughout the 600 m² plot, while birds are monitored with modified point counts and small mammals, reptiles, and amphibians are recorded in traps on the periphery of the plot.

Double Sampling

Double sampling is a "nested intensity," single-scale sampling technique where less detailed measurements on many plots are "corrected" with detailed measurements on a few plots. For example, "eyeball" or "ocular" estimates of aboveground biomass might be made on twenty-five 0.1 m² plots, while the vegetation in every fifth plot is clipped, dried, and weighed to quantify and correct for any bias in the eyeball estimates. Ahmed et al. (1983) provide a detailed example of how to correlate many ocular estimates of dry weight to fewer measurements of three rangeland grass species and two herbaceous species. Coefficients of determination (r^2 values) ranged from 0.81 to 0.94 for estimating the dry weights of species from ocular estimates. Lesson: There are ways to cut costs and increase the spatial extent of sampling without sacrificing accuracy. Still, the loss in precision and accuracy must always be measured when less intensive designs are substituted for more intensive designs.

Example 2: Plant Species Diversity
in Floodplain Habitats in Brazil

A colleague and I set out to quantify the effects of river level fluctuation on plant species richness, diversity, and distribution in floodplain forest habitats in Jaú National Park, Brazil (Ferreira and Stohlgren 1999). I provide much more detail in this example to take the reader through a typical single-scale plant diversity study (i.e., from objectives, methods, and data analysis to interpretation of results) because these are commonly used tools of the trade and lessons from the field.

The river levels in the Central Amazonian study area fluctuate up to 14 m annually, with the flooding period ranging from 50 to 270 days between the rising and falling phases. The extensive forests along the rivers contain plant species that are well adapted to annual flooding. This study sought to answer two basic questions: (1) How did the duration of flooding affect tree species richness, diversity, and density at local scales (i.e., plot-level scales)? (2) How did the duration of flooding affect tree species distributions at landscape scales? In another way, we asked how are the various flooded habitats different and how are they similar?

The study area was located in Jaú National Park, located 200 km northeast of Manaus, Brazil (Ferreira and Stohlgren 1999). The floodplain forest is seasonally inundated by the Rio Jaú, a blackwater river. The elevation ranges from 22 to 27 m above sea level. Variation in river water level was markedly seasonal, with the normal rising phase occurring between late December and early July, and the draining phase occurring from the end of July to early December.

All trees larger than 10 cm in diameter [at 1.3 m; diameter at breast height (DBH)] were measured and identified in twenty-five 10 m × 40 m randomly selected plots in lake, river, and stream habitats. Within the plots, six measurements of water level were taken using the previous year's watermarks on the tree trunks.

Two-way ANOVA was used to test for differences between habitats and sites sampled for flooding duration (water depth), species richness, diversity index, tree density, and basal area. Where site effects were not significant, site data were lumped by habitat for one-way ANOVA. Tukey's test was used as a means comparison test where the F-test was significant.

There are many different types of diversity indexes used by plant ecologists [see Krebs (1989), Ludwig and Reynolds (1988), or Magurran (2003) for far more detail on various indexes and their strengths and weaknesses]. In this example, the widely used Shannon-Weiner diversity index was used to assess the diversity of plots in different areas. Jaccard's coefficient (see chapter 4) was used to quantify the species composition overlap within and among habitat types (i.e., between 25 randomly selected plots within each type, and among the complete species lists in each habitat type) (Krebs 1989).

In this example, an ordination technique that clusters groups of similar plots based on species composition and dominance (i.e., nonmetric multi-dimensional scaling ordination method) (McCune 1997) was used to determine the differential distribution of species among the 75 plots in the three habitats, using only abundant species (more than 10 individuals overall) in the analysis. The ordinations were performed on PC-ORD with all default options as recommended (McCune 1997). Full descriptions of various ordination methods can be found in Krebs (1999)

Evaluating Differences Among Habitats

It is important to realize that evaluating differences among habitats is made easier for study designs that rely on stratified random sampling. After all, the strata were defined as having wildly different characteristics—flooding duration in this study. However, it is often unknown "how much" the strata differ. In this case, ANOVA showed that species richness, diversity, water level, and flood duration were significantly different among habitat types (Table 4.3). Plots in the lake habitat had significantly lower species richness and diversity, and greater water level and duration of flooding than the river and stream habitats. Tree density and basal area were also significantly different among the habitats sampled, with the lake habitat having higher tree densities and lower basal areas than the river and stream habitat types (Table 4.3).

Species richness in plots was strongly negatively correlated to the water level [tree species richness ($r = 0.82$, $P < 0.001$) and duration of flooding ($r = 0.84$, $P < 0.001$)]. Diversity indexes in plots were also strongly negatively correlated to the water level ($r = 0.88$, $P < 0.001$) and duration of flooding ($r = 0.88$, $P < 0.001$) (Ferreira and Stohlgren 1999).

Table 4.3.   ANOVA results (df = 2, 72) and mean values (standard error in parentheses) of site characteristics by habitat type ($n = 25$ plots/type).

Characteristic	Lake	River	Stream	F	P	R^2
No. of species	8.9 a	24.8 b	29.4 c	138.9	<0.0001	0.68
	(0.3)	(0.8)	(1.2)			
Diversity	1.62 a	2.67 b	2.79 b	168.0	<0.0001	0.82
	(0.05)	(0.04)	(0.05)			
No. of Trees/plot	45.7 a	33.5 b	37.5 b	4.2	<0.02	0.10
	(2.0)	(4.1)	(2.3)			
Water level (m)	8.1 a	2.3 b	2.4 b	154.5	<0.0001	0.81
	(1.4)	(1.3)	(1.3)			
Flood duration (days)	260.0 a	105.7 b	89.8 b	133.5	<0.00 1	0.78
	(7.1)	(8.1)	(8.6)			

Means with dissimilar letters are significantly different at $P < 0.05$ using Tukey's test.

Adapted from Ferreira and Stohlgren (1999).

At the plot level, or local scale, the variation in water level and duration of flooding strongly negatively influenced tree species richness and diversity (Table 4.3). Plots in the lake habitat, which were flooded about 260 days/year had about one-third the number of species, 60% of the diversity (Shannon-Weiner index), and 35% higher density of trees compared to river and stream habitats. The combination of low species richness and high density in the lake habitat suggests that colonization potential is high, but only for species that can tolerate inundation (Ayres 1993; Ferreira 1997; Keel and Prance 1979). At the plot scale, there was little species composition overlap between plots within each habitat type. There was an average of 15.0%, 13.6%, and 20.4% overlap in species composition in 25 paired plot comparisons in the lake, river, and stream habitats, respectively (Ferreira and Stohlgren 1999).

Nonmetric multidimensional scaling (see Ludwig and Reynolds 1988) partially differentiated the species distribution among the 75 plots in the three habitats sampled. Axis 1 discriminated a cluster of tree species formed by lake plots corresponding to the longest duration of flooding. The second, less-distinct cluster contained plots from the Jaú River margin, which are subjected to intermediate periods of inundation, and stream habitats with the shortest flood duration. Axis 2 partially discriminated some plots in the Jaú River margin and stream habitats (Figure 4.4).

In this example, the ordination scores of axis 1 were strongly positively correlated to water level and flooding duration ($r = 0.77$, $P < 0.001$ for both), while the scores of axis 2 were negatively correlated to water level and flooding duration ($r = -0.40$, $P < 0.001$). The ordination analysis distinguished two plant assemblages in the different flooding habitats, which clearly separated the lake habitat from the river and stream habitats (Figure 4.4). This plot-level analysis suggested that tree "communities" could be determined primarily by differential physiological tolerance of species to flooding duration (Junk 1989; Keel and Prance 1979; Kozlowski 1982). However, the major weaknesses of this ordination analysis were that (1) only abundant species were used, so the separation of clusters likely exaggerated the differences between communities that include common and rare species; and (2) little is learned about species composition overlap among various habitats or about the numbers of species that are unique to different flooding regimes. Keep in mind that the particular methods used in plant diversity studies affect the results and interpretations of the data.

In this example, local-scale responses may merely reflect species-specific tolerances to flooding. Highly flood-tolerant tree species dominating the lake habitat included *Macrolobium acaciaefolium*, *Burdachia prismatocarpa*, and *Eschweilera tenuifolia*, while shrub species included *Coccoloba pichuna*, *Eugenia inundata*, and *Symmeria paniculata*. Corroborating studies by Takeuchi (1962) reported the same species in floodplain species of Rio Negro. Junk (1989) also found that species such as *C. pichuna*, *E. inundata*, and *E. tenuifolia* could withstand up to 10 m of flooding for 280 days.

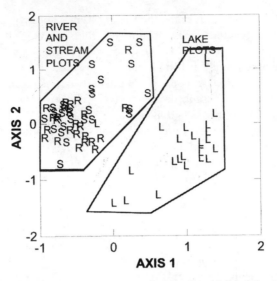

Figure 4.4. Basic ordination diagram of plots from lake (L), river (R), and stream (S) habitats derived from non-metric multidimensional scaling. Adapted from (Ferreira and Stohlgren 1999).

Evaluating Similarities Among Habitats

Scale greatly influenced species composition overlap among strata. At the habitat scale (or landscape scale), there was considerably more overlap in species composition. Combined species lists of each habitat type showed that 60.6% of the tree species are shared between the stream and river habitat types. The lake habitat had lower species composition overlap with river (45.6%) and stream (42.6%) habitats, but species overlap among types was still substantial. Assessing uniqueness is also important. Fifty-four species of trees were found in the stream habitat, with six of those species unique to stream habitats (*Macrolobium suavelens, Naucleopsis caloneura, Gustavia pulchra, Caraipa* sp. 1, *Aspidosperma nitidum,* and *Mauritiella aculeata*). Fifty-two tree species were found in the river habitat, with only three species unique to that habitat (*Eschweilera* sp. 1, *Eugenia gomidesiana,* and *Quiina rhytidopus*). The lake habitat was the most species poor, with only 33 tree species. Three species were unique to the lake habitat (*Buchenavia oxycarpa, Crudia amazonica,* and *Eugenia cachoeirensis*). Other studies have shown that tree species vary in their susceptibility to flooding (Kozlowski 1982, 1984), but we found that classifications of many species into mutually exclusive flood-tolerant, moderately tolerant, and intolerant species groups was impossible. In contrast to the ordination results (Figure 4.4), we were surprised to find that only three species of trees were unique to the lake habitat; 9.1% of the flora in that habitat. Likewise, the river and stream habitats had 5.8% and 11.1% unique species, respectively. Thus the vast majority of tree species can establish, and perhaps persist, in a variety of habitats. We get a similarly complex view of the effects of flood duration when the data are aggregated by habitat type (i.e., landscape scales).

It is increasingly routine to classify or cluster species into "communities" related to some perceived group response to environmental factors (Ludwig and Reynolds 1988), and floodplain forests are no exception. Junk (1989) reported that floristic composition along the flood-level gradient in várzea floodplain forests represent seral stages of succession, from lowland pioneer communities with a few species to upland forests with many species. Campbell et al. (1992) found the same pattern in the várzea floodplain forest of Rio Juruá in western Brazilian Amazonia. Salo et al. (1986) showed that species distributions in the várzea forest of western Amazonia were strongly related to water movement, erosion, and sedimentation rates because these effects alter community structure and habitat stability.

Based on the local-scale results (Table 4.3), it was true that the plots within habitats showed consistent differences related to flood duration. However, this effect is influenced by scale and the plasticity of the ecology of the species involved. Plots in the lake habitat averaged only 30–36% of the species richness of plots in the stream and river habitats, respectively, and complete species lists (i.e., all plots combined by type) showed that the lake habitat had 42.6–45.6% of the species found in the stream and river habitats, respectively, which was not evident from the plot-level ANOVA or ordination results. Thus, as the scale increased, the differences between the habitat types weakened. High spatial variability in species distributions was shown by (1) the low (13–20%) species composition overlap between plots in a given habitat, (2) relatively high species composition overlap (43–61%) among habitats, and (3) low numbers of unique species in each habitat type (three to six species per 25 plots) (Ferreira and Stohlgren 1999).

This example shows that the diversity of these floodplain forests is primarily due to the plasticity of the ecology of the species involved. There was no analysis of soil characteristics, light availability, distance to other habitat types, or other factors that influence species diversity. However, the low species overlap between pairs of plots and significant habitat-site interaction for diversity indicate immense habitat heterogeneity, past patterns of establishment, or other differences between sites. The high species overlap among habitat types suggests that many of the species might be considered habitat "generalists," even if they are more successful in one habitat rather than another. Establishment in a variety of microsites and habitats creates "redundancy" and a hedge against landscape-scale extirpation despite local mortality (Collins 1987; Collins and Glenn 1991; Collins et al. 1995; McNaughton 1979; Stohlgren et al. 1997a,b,c, 2000b).

The lessons learned from this type of single-scale study are many. Plant ecologists may benefit from a shift from a community emphasis to a species emphasis. This entails a greater recognition of the role of habitat heterogeneity in creating and maintaining species and landscape diversity (Stohlgren et al. 1998a,d, 2000b). In search of generalizations, ecologists often lump species into groups, as if each species in the group respond exactly the same to a given disturbance or set of environmental conditions. The way in which a species responds to the range of a disturbance such as flooding (i.e., depth, frequency, duration, and spatial extent) may be more important than our ecological analyses of subjectively defined communities under "average" environmental conditions. This study focused primarily on mature trees, and information on the demography of seedlings and saplings of each species would have complemented this study of species diversity patterns. Still, the wide species distribution of mature trees is impressive. Rather than viewing the landscape as distinct communities, these study results suggested that diverse landscapes are a complex collage of species that persist in a wide range of microsites and habitats (Collins et al. 1995; Stohlgren et al. 1997c, 1998a, 2000b). Rather than viewing long-term flooding as a stressor that reduces local species diversity, it may be that the variability in duration, depth, and spatial extent of flooding serve to increase landscape-scale diversity by increasing landscape-scale heterogeneity and species plasticity to a variety of habitats.

Nested Plot Designs (but Still Single Scale)

There are many examples of attempts to improve the efficiency of sampling vegetation; many are variations on the theme exemplified by Bray and Curtis (1957) and Bourdeau and Oosting (1959). Bray and Curtis (1957) sampled 300 m × 200 m stands of upland, mixed hardwood forests in southern Wisconsin. Recognizing the huge number of trees (larger than 10 cm DBH) that could be measured in each plot, they selected 40 points at random and measured the diameters (and distances) of the two trees nearest the random points in the plots. Mathematical techniques were used to determine tree species density and basal area by species from the distance measures. They also collected information on shrubs and herbs in twenty 1 m × 1 m quadrats within the plots, clearly recognizing that trees could be measured at one scale and shrubs and herbs at another scale. It would be unreasonable and overkill to collect data on all possible 1 m² areas in the stand (i.e., 60,000 quadrats). Remember that this idea was not new. Pound and Clements (1898a,b) assessed frequency in small 5 m × 5 m plots within 1 mile square areas for similar reasons of sampling efficiency. Still, it became more common to see sampling designs where shrubs and herbs were measured in small quadrats within larger plots used to assess forest vegetation.

Bourdeau and Oosting (1959) quantified *Quercus virginiana* forests in North Carolina in a similar fashion. Their 60 m × 100 m plots were divided into six 10 m × 100 m strips. In two of the six strips, alternating 10 m × 10 m areas were sampled for trees (larger than 2.5 cm DBH), and nested in each 10 m × 10 m area was a 4 m × 4 m quadrat to quantify herb cover and fre-

quency. Thus overstory trees were sampled in 1000 m² and understory vegetation in 160 m² of the 6000 m² plot.

Daubenmire (1968) proposed a system of nested plots for sampling conifer forests in western North America because plant sizes can vary by orders
of magnitude from annual grasses to giant sequoias. In that design (see
Daubenmire 1968, Figure 2.2), all trees more than 1 m tall were recorded in
three contiguous 5 m × 25 m "macroplots." Trees less than 1 m tall were
recorded in two 1 m wide belt transects inside the boundary of the middle
macroplot. Shrubs, herbs, and cryptobiotic crusts were evaluated for cover
in two sets of twenty-five 20 cm × 50 cm quadrats spaced 1 m apart on the
same inside boundaries. He also hinted that the macroplots could be searched
for plant species missed by the quadrats.

Nested plot designs form the basis for many sampling designs in use
today (Krebs 1999, pp. 98–102). The USDA's Forest Health Monitoring Program design includes three 1 m² quadrats for herbaceous data and one 17 m²
area for tree samplings (less than 2.5 cm DBH) nested within each of four
168 m² plots (Figure 4.5) (Burkman and Hertel 1992; Palmer et al. 1991).

Forest "reference stands" (1 ha or larger), commonly used for forest demography studies in the U.S. Pacific Northwest and elsewhere, typically
include eight 1 m² herbaceous quadrats in each of nine 5 m radius "seedling and sampling" subplots systematically arranged within 100 m × 100 m
tree plots (Figure 4.6).

The basic concepts of nested designs are the same in terms of sampling
efficiency despite different sizes, numbers, and placement of subplots. Smaller
areas are devoted to plants with smaller stature. Large samples sizes of smaller
specimens can be obtained with small sample units.

The nested plot designs described above are an improvement in plant
diversity sampling relative to single-scale, single-structure-class studies. For
example, forest surveys that may have previously only recorded information
on the diameters of dominant trees on a plot are now recording information
on saplings, shrubs, and understory grasses and herbs on the same plots.

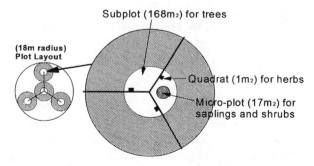

Figure 4.5.   Layout of U.S. Department of Agriculture's Forest Health
Monitoring program plot design.

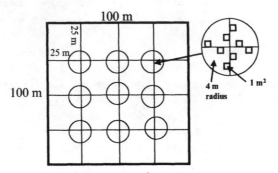

Figure 4.6.   Basic layout of the "Reference Stand" approach to assess long-term demography of forest plant species.

This provides a means to classify communities based on overstory and understory vegetation, assess affinities of overstory to understory plant species, or evaluate species-environment relationships for various size plant species, or in combination.

There are limitations to these nested designs for plant diversity studies. They may undersample many plant species at each site. In many forests, there are many more understory species than overstory species. If small quadrats are used, or if a small total area is sampled, then many plant species could be missed (Stohlgren et al. 1998c). For example, actual sample areas of 5 m² (see Daubenmire 1968, Figure 2.2), 12 m² (Figure 4.5) (Bull et al. 1998), 25 m² (Pound and Clements 1898a,b), 72 m² (Figure 4.6) (Riegel et al. 1988), or 160 m² (Bourdeau and Oosting 1959) may not be large enough to capture locally rare species or species with patchy distributions.

Lessons Learned

Single-scale designs may remain very common in plant diversity studies. In deciding which designs might best meet specific program objectives, it is important to fully understand the benefits and limitations of alternative designs. While aspects may be particular for each case study, some general benefits and limitations might be broadly anticipated and described as follows.

Benefits of Single-Scale Designs

There are three primary benefits to single-scale sampling designs: simplicity, large sample size, and ease of analysis. Single-scale designs are simple to design and conduct. The design phase might include preliminary tests of the appropriate size, shape, and pattern of sampling plots for the vegetation types in question (see chapter 3). Once the tests are complete, training field crews to use a single-size plot is relatively easy. If novel vegetation types or environmental situations arise, additional tests of appropriate plot size and shape can be conducted. If larger plots are needed very rarely in the study to capture local variation, then data can be collected at two spatial scales (i.e., the original plot size and the new plot size) without compromising the comparability of data for all vegetation types.

The second attribute of single-scale plots is large sample sizes. More complex sampling designs, such as nested plot designs, require more setup time and recording time in the field, and usually result in more time spent at each sampling site and thus fewer sampling sites for a fixed budget. There are additional benefits of large sample sizes, including covering broader environmental gradients, establishing more plots in more or rare habitats, and simply observing more of the entire landscape being surveyed. For example, suppose the objective of a particular study is to quickly determine the pattern of native species richness in a given landscape to protect hot spots of native species richness from proposed development. In this example, twenty-four 0.1 ha plots could be randomly distributed throughout the 754 ha study area with simple counts of native species richness in each plot. The fairly large size of the plots ensures that locally rare species are captured at each site. A simple "kriging" spatial analysis program, available in most statistical analysis software programs, can create contour maps of species richness to quickly display the hot spots of native plant species richness (Figure 4.7).

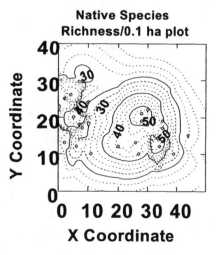

Native Species Richness/0.1 ha plot

Figure 4.7.   Kriging diagram of native species richness in a 754 ha area of Rocky Mountain National Park, Colorado (Adapted from Stohlgren et al. 1999a).

This type of map can be immediately used by project managers to avoid the hot spots of diversity. Other factors may also be important in real-life situations. The locations of rare plants of unique species assemblages or highly erosive soils may also be important, as well as cost and logistical considerations. Still, important information can be quickly gained with single-scale sampling.

Ease of analysis is the third primary benefit of single-scale designs. It is relatively easy to summarize data from single-scale plots. Simple descriptive statistics can be used to quantify vegetation and soil characteristics, and other site characteristics. Spatial analyses are not complicated by different resolutions of primary data (e.g., Figure 4.7).

Limitations of Single-Scale Designs

There are several limitations of single-scale sampling designs. Within-site variation may not be measured. The plot is assumed to be homogeneous, and measurements are assumed to be representative of the entire plot. However, this can mask important information on how species density and overlap change with scale (area). If large plots are used to capture locally rare species, there are increased odds that more complex environmental gradients will be incorporated in each plot. Furthermore, two plots may contain the same number of species, but one of the sites may be far more heterogeneous and have a far greater number of species immediately adjacent to the plot, but there would be no way of determining this from a single-scale design (Figure 4.8).

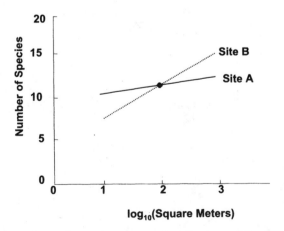

Figure 4.8. Hypothetical species-area curves for two plots with the same number of species at the selected single-scale plot size (12 species in 10 m² sampled).

Figure 4.9.   Lesson 11. Vegetation sampling texts often advise plant ecologists to test various plot sizes in a pilot study before selecting the "correct" or "optimal" sampling frame. They rarely recommend the continued use of two or more plot sizes to evaluate the effects of plot size on study results.

In measuring plant diversity, cost efficiency is often an important consideration. Getting crews to and from a sample sight is often the most expensive aspect of surveys and monitoring, so it may be beneficial to collect a little more information at each site. Single-scale sampling still requires an experimental approach (Figure 4.9). A significant limitation for researchers using single-scale sampling is not knowing which sites (e.g., site A or B in Figure 4.8) might be a higher priority for additional surveys.

5

Multiscale Sampling

Sampling plant diversity at multiple spatial scales may allow for a deeper understanding of relationships between species distributions and composition shifts relative to changing environmental gradients. Multiscale sampling designs are not new to plant ecologists (e.g., Barbour et al. 1987; Mueller-Dombois and Ellenberg 1974), but they are gaining in popularity as a means to measure plant diversity (Brown and Peet 2003; Bruno et al. 2004; Byers and Noonberg 2003; Carrington and Keeley 1999; Davies et al. 2005; Fridley et al. 2004; Keeley et al. 1995), environmental variables (Bellehumeur and Legendre 1998), and genotypic variation (Bell and Lechowicz 1991). Like single-scale techniques, different multiscale techniques have various strengths and weaknesses, and some may be better suited than others for particular study goals and objectives. No single design will be the "be all, end all" for plant diversity sampling. All designs have advantages and disadvantages that must be carefully evaluated. An experimental approach is almost always warranted. The following examples provide but a glimpse of the profound potential of multiscale sampling for plant diversity.

The Nested Quadrat Design

Textbooks on vegetation sampling methods (e.g., Barbour et al. 1987, 1999; Krebs 1999; Mueller-Dombois and Ellenberg 1974) recommend overlaying nested quadrats of increasing size to quantify species-area relationships

(Figure 5.1)—one measure of species richness patterns in an area. The underlying theory is that from any point in a landscape, the cumulative number of species will increase with increasing area around the starting point (Figure 5.1).

Nested circular plots (Figure 5.1) or nested square plots (Krebs 1999, p. 112) would be a quick way to create a species-area curve. However, these designs have not been widely used for many reasons. First, species richness would be expected to be highly variable at small spatial scales (e.g., 1–10 m² scales) in many habitats, and less variable at larger scales. In an analysis of 38 vegetation types, native species richness in more than 7200 subplots of 1 m² had a mean coefficient of variation of 55%, whereas native species richness in the 728 plots of 1000 m² (overlaying 10 subplots each) had a mean coefficient of variation of 33% across the vegetation types (Stohlgren T., unpublished data).

Thus the "anchor" of the species area would not be expected to be very consistent without many replicate nested circular plots in each habitat. Second, these particular designs do not allow for adequate replication at smaller spatial scales, where the variation might be high. Since large areas require more searching time than smaller areas, a disproportionate amount of field time is spent reconciling the most stable portions of the curve. Third, there is considerable nonindependence of points used to construct the species-area curve. The number of species found in rings 1 and 2 is substantially dependent on the number of species found in ring 1. Such a sampling scheme might not provide realistic generalized species-area curves for a given habitat unless replication was extensive.

Still, nested quadrats have long been used to measure species-area relationships (Barbour et al. 1987; Mueller-Dombois and Ellenberg 1974; Stohlgren et al. 1995b). A common design includes a square 1 m² plot that is surveyed

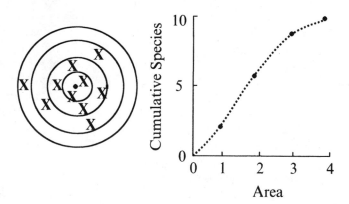

Figure 5.1. Hypothetical concentric plots around a starting point (center dot), with each "X" representing additional species recorded at the site while moving away from the starting point.

for all vascular plant species. Then another 1 m² plot contiguous to the first plot is surveyed, yielding the total number of species in areas of 1 and 2 m² (i.e., the cumulative number of species in a 2 m² area) [see Barbour et al. (1987) for examples]. Then a 4 m² plot is surveyed, and so on, until a 16 m² or 64 m² area is surveyed. The assumptions and problems with this approach are many. First, the data for each "plot size" overlap 50% and are not independent, as in the nested circular plot design above. That is, the cumulative number of species found in plots 1 and 2 (1 m² and 2 m² areas) is strongly dependent on the number of species found in plot 1. The cumulative number of species found in the 4 m² area is dependent on the number of species found in plots 1 and 2. Thus the 50% overlap in area surveyed reduces the independence of data in constructing the species-area curve.

A second problem of the nested quadrat technique, like the nested circle design, is that the resulting species-area curve is heavily dependent on starting location. Assume each "x" in Figure 5.2 represents a different species in a plant community. If, for example, a cluster of species occurs in the location of the contiguous, smaller plots, the species curve will be steep at first, then level off as fewer and fewer additional species are recorded. Conversely, if a cluster of species is located on the largest plot, but not overlapping smaller plot sizes, then the species-area curve would be less steep at the beginning and much steeper at the end of the curve (the 16 m² or 64 m² plot size). If species are patchily distributed at multiple scales, a stair-step function might be seen. Thus the species-area results of one nested quadrat might greatly misrepresent the community (Figure 5.2). Replicate nested quadrats would reduce the starting point limitations of the design, but it would not reduce the problem of nonindependence.

A subtle problem with the nested quadrat design is that the overall shape (and thus the perimeter:area ratio) changes alternately from a square to a rectangle. Changes in the perimeter:area ratio alter the heterogeneity of the underlying survey area. Typically if two plots are the same area, the plot with the greatest perimeter:area ratio will be the most species rich because of the greater environmental gradients underlying elongated plots. Thus by changing the perimeter:area ratio, another factor other than area may be influencing species richness data. Reed et al. (1993) used a nested quadrat design to evaluate the effects of scale on species-environment relationships in a 6.6 ha study site in the Duke Forest (Durham, North Carolina) (Figure 5.3). Presence of plant species in the six smallest quadrat sizes (less than 4 m × 4 m; Figure 5.3) were noted, with trees (larger than 2 cm at 1.37 m; DBH) measured throughout the entire plot. Soil characteristics were collected in the corners of the various nested quadrats. The composition of vegetation was found to be more closely correlated to environmental factors at larger grain sizes. There was no indication that understory composition was influenced by the density or size of nearby trees.

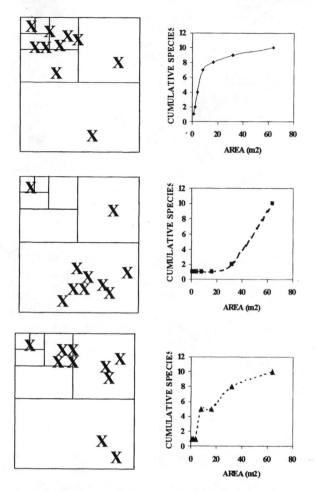

Figure 5.2. Hypothetical use of the nested quadrat design and resultant species-area curves in the same plant community, but with different starting locations.

Some sampling design problems were raised by the authors. The small quadrats were affected by small-scale variations in vegetation and soil. These small quadrats might not have been well replicated in the design relative to the background levels of natural variation in vegetation and soil, especially considering the sizes and locations of many trees, shrubs, and large herbs in the immediate area of the small quadrats. A second issue raised was that the sampling all occurred within a 6.6 ha area that only varied in elevation by 10 m. The authors acknowledged that vegetation samples in a confined area may lead to "lower generality with respect to regional patterns" and that a mix of small and larger plots may be needed to detect locally important processes within larger areas (Reed et al. 1993).

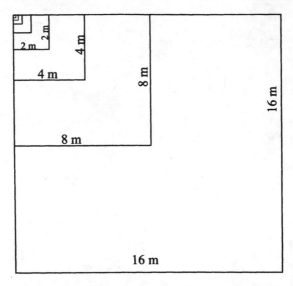

Figure 5.3. Nested-quadrat design described by Reed et al. (1993) to investigate scale-dependence of vegetation-environment correlations in a Piedmont woodland in North Carolina. Square quadrat sizes less than 2 m on a side are 1 m, 0.5 m, 0.25 m, and 0.125 m on a side. The design was replicated in contiguous grid cells over a 256 m x 256 m study area (i.e., sixteen 16 m x 16 m cells).

William Bond (1983) used a nested quadrat design to assess plant species richness in fynbos vegetation on the Southern Cape, South Africa. The plot design was modified from Whittaker et al. (1979) and included five 1 m^2 subplots and a 10 m^2, 100 m^2, and 500 m^2 subplot, nested within a 20 m × 50 m plot (1000 m^2 plot). The patterns of diversity found in the fynbos provide an interesting case study and an excellent example of the many uses of multiscale sampling. The 21 plots showed progressively weaker correlations between species richness at the 1 m^2 and 5 m^2 scales ($r = 0.92$), 100 m^2 scale ($r = 0.78$), and 1000 m^2 scale ($r = 0.65$). Bond attributed this decline to different controls of diversity at large and small scales. Plant species richness (perennial plants only) in 1000 m^2 plots ranged from 21 species on a fynbos island to 104 species in a *Protea*-rich area (Bond 1983). An alternate explanation might be that the small quadrats capture common species better than locally rare species which are more easily captured in larger plots (Stohlgren et al. 1998c). In any case, assessing more complex relationships in plant species richness across scales was just what Bond (1983) and Whittaker (1979) intended of multiscale sampling.

One way to describe these differences in species dominance among sites is with dominance-diversity curves (Figure 5.4). Importance values were cal-

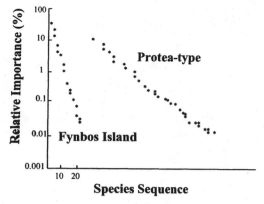

Figure 5.4.   Generalization of examples of dominance-diversity curves for southern Cape woody vegetation (adapted from Bond 1983).

culated as cover × height. In the Bond (1983) example, species are arranged by relative importance values (highest to lowest). The fynbos island site contained 21 species, a few of which dominated the low-diversity site in terms of height and foliar cover. Meanwhile, the *Protea*-type (waboomveld) site contained 104 species, far more evenly distributed in terms of relative importance over far more species (Figure 5.4). In nonfynbos areas, mesic forest sites had a similar pattern to the fynbos island site (Figure 5.4), while xeric succulent thicket sites had a similar pattern to the species-rich fynbos type.

Thus when conditions for plant growth are good, a few species become dominant (high importance values) and fewer plant species coexist in the understory. Species richness in 1 m² plots was positively correlated with foliar cover (less than 0.6 m), total foliar cover, and moisture, but it was inversely correlated to soil fertility. Bond concluded, "few species occur in stressed arid environments of low biomass, but many species coexist in relatively moist environments with high biomass." However, a few plots in areas of very high biomass might lead to the general pattern of a "humpbacked" relationship between foliar cover and species richness (Figure 5.5).

Bond also showed that semilog relationships (i.e., species-log_{10} area curves) described species-area relationships extremely well for all habitats tested. The slopes of the lines clearly differentiated species-rich sites from species-poor sites, and they serve as a description of "pattern diversity." However, Bond concluded that because species richness in 0.1 ha plots is affected by relative patchiness from multiple causes, "few general principles will emerge from large plot studies. Point diversity studies, however, are likely to yield useful information on the mechanisms controlling coexistence of multiple plant growth forms" (Bond 1983, p. 354).

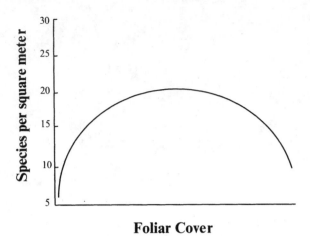

Foliar Cover

Figure 5.5.   Idealized hump-backed curve of the relationship between foliar cover and species richness from Bond (1983).

Whittaker Plot Design

In the late 1970s, Robert H. Whittaker recognized the benefits of measuring vegetation at multiple spatial scales (Whittaker 1977). He designed a sampling scheme that included 10 contiguous 1 m² (square) subplots, two 10 m² subplots (2 m × 5 m), a 100 m² subplot (square), all nested in a 20 m × 50m plot (Figure 5.6).

Shmida (1984) published Whittaker's nested vegetation sampling method as a means to compare different plant communities from different regions of the world. Whittaker long realized that patterns of plant diversity can be elucidated only by systematic surveys and by sampling at multiple spatial scales. Shmida (1984) showed a strong semilog relationship,

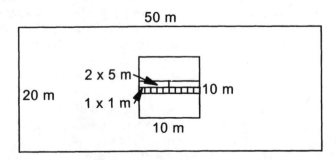

Figure 5.6.   The Whittaker Plot design, a nested plot design (Shmida 1984).

$$S = b + d \log A,$$

between the number of species (S) and quadrat area (A) (b is a constant and d is the slope). This is similar to other reported species-area relationships (Miller and Wiegert 1989; Shafer 1990) and ideally suited for Whittaker's careful design of nested quadrat sizes of 1 m^2, 10 m^2, 100 m^2, and 1000 m^2.

Whittaker's design had three advantages over the nested quadrat design. First, the overall plot size was large, so more locally rare species could be detected in the 1000 m^2 area relative to the typical 16 m^2 or 64 m^2 plot size of most nested quadrats. Second, the spatial scales of measurement increase by a factor of 10 each time (1 m^2, 10 m^2, 100 m^2, and 1000 m^2), increasing the sampling at small scales and reducing the sampling effort for intermediate scales, while efficiently deriving species-area curves (Shmida 1984). Third, there was reduced overlap among subplots and plots at the largest spatial scale (i.e., 10% overlap between the 100 m^2 scale and the 1000 m^2 scale compared to 50% overlap in successive plots in the original nested quadrat design). The most attractive attributes of the Whittaker plot design are that it provides plant ecologists with a standardized, efficient approach to quantify species richness in different plant communities, as well as insights into the effects of quadrat size when determining species-area relationships.

The 20 m x 50 m size of the Whittaker plot (Figure 5.6) was designed initially for more-or-less homogeneous areas so that replicate sites were easy to find (Shmida 1984). Borrowing perhaps from the nested quadrat approach (Figure 5.2), the 1 m × 1 m subplots are overlaid on the 2 m × 5 m subplots, which in turn are overlaid on the 10 m × 10 m subplot. However, the Whittaker plot has three distinct design flaws (Stohlgren et al. 1995b). First, if the habitat is not strictly homogeneous, species richness is influenced by plot shape. Circular or square plots (with a reduced perimeter:surface area ratio) will have fewer species, in general, than a long, thin rectangle covering a more heterogeneous area. Second, plot size and shape interactions may influence species richness. Note that the Whittaker plot design shifts from 1 m × 1 m squares to 2 m × 5 m rectangles to a 10 m × 10 m square, and then back to a 20 m × 50 m rectangle, which confounds the influences of plot shape and size (Figure 5.6). Third, is the problem of spatial autocorrelation. Not only are the ten 1 m × 1 m plots contiguous in one small area of the 20 m × 50 m plot (i.e., high spatial autocorrelation), the successively larger plots are superimposed on the smaller plots (i.e., the plots are not independent in terms of species richness). Thus a species-rich area in one of the 1 m × 1 m plots affects the species richness reported in the larger-size plots. This last problem is ubiquitous among nested (overlaid) quadrats (e.g., Barbour et al. 1987; Mueller-Dombois and Ellenberg 1974), and results may be heavily biased by the starting location in the field (Figures 5.1–5.3). Several replicates for

---

Thoughts About Developing Species-Area Curves

P. Jaccard was the pioneering ecologist of the 20th century who actively sought formulas to explain species-area relationships. Arrhenius (1921) and Gleason (1922) also described species-area curves, but the models greatly overestimated the number of species in very large areas. In the truest sense, developing species-area curves requires three steps. First, unbiased, independent samples of various sizes are collected in the habitat of choice. Then, mathematical formulas are used to "fit" or describe the species-area relationship. Typically, a semilog relationship (where $\log_{10}$ area is correlated to cumulative species richness) has worked well over scales of square meters to thousands of square meters (Shmida 1984), but $\log_{10}$ area to $\log_{10}$ cumulative species relationships have also been common (Stohlgren et al. 1997b,c). Lastly, the curves should be validated with independent data. Tests of the relationship can be conducted in similar habitats or at larger spatial scales to test the robustness of the relationships.

Palmer (1990) and Palmer et al. (1991) compared several methods for estimating species richness, including the number of observed species, extrapolation from species-area curves, integration of the lognormal distribution, and nonparametric estimators. He cautioned that species-area curves may have different forms at different scales; they certainly vary for different habitats (Stohlgren et al. 2005b). Thus investigating species-area relationships will remain a challenging venture.

---

each site or vegetation type are needed in areas where species are patchily distributed (i.e., most areas).

Example Uses of Nested Plot Sampling

This section describes examples of nested plot sampling using a design similar to the Whittaker plot design (Figure 5.6). In both examples, ten 1 m^2 subplots, two 10 m^2, and a 100 m^2 subplot were nested in 20 m × 50 m plots, with four plots in each vegetation type. In the first example, four plots were placed in each of five vegetation types in a 754 ha area of Rocky Mountain National Park, Colorado. The types included ponderosa pine, lodgepole pine, wet meadow, dry meadow, and aspen (Figure 5.7).

At the 1 m^2 scale, the wet meadow, dry meadow, and aspen vegetation types appear fairly equal in plant species richness. However, presenting these data alone might be very misleading, because at the 1000 m^2 scale, the dry meadow type doesn't appear nearly as species rich as the wet meadow and

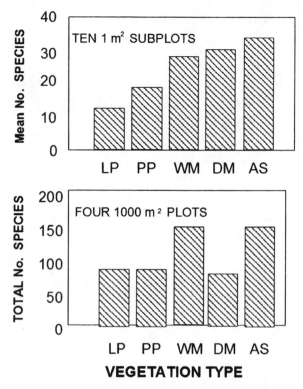

Figure 5.7.   Mean and total number of species in vegetation plots in Rocky Mountain National Park, Colorado. The types include burned Ponderosa pine (BPP), lodgepole pine (LP), ponderosa pine (PP), wet meadow (WM), dry meadow (DM), and aspen (AS). Lesson 12. Imagine how much of the story of plant diversity of various vegetation types would be missed by measuring only 1–m² plots. Plant diversity studies require multi-scale techniques. Adapted from Stohlgren et al. 1997b.

aspen types. At small scales, there were increased densities of common plant species in the dry meadow type, with few additional species added by searching the larger 1000 m² plots. Meanwhile, additional large plots capture many more additional species in the wet meadow and aspen types. The lesson here is that examining species richness at multiple spatial scales is an efficient way to quantify species richness and densities to get a more complete understanding of the patterns of plant diversity in a landscape.

In the second example, four plots each were placed in mixed grass prairie in Wyoming, shortgrass steppe in Colorado, northern mixed prairie in South Dakota, and tallgrass prairie sites in Minnesota. One objective here was to determine which site was more heavily invaded by exotic (nonnative) plant species.

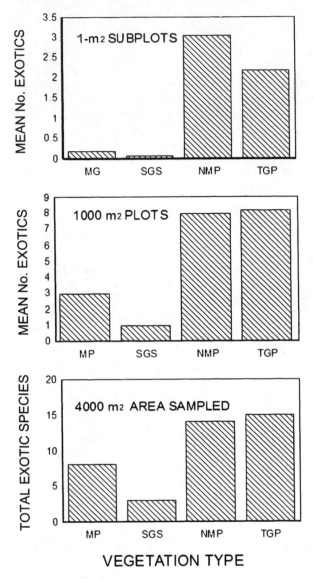

Figure 5.8.   Exotic plant species richness in mixed grass prairie (MG) in Wyoming, shortgrass steppe (SGS) in Colorado, northern mixed prairie (NMP) in South Dakota, and tallgrass prairie (TGP) sites in Minnesota.

At the 1 m² scale, the northern mixed prairie and tallgrass prairie sites appear to be the only invaded sites worthy of much attention (Figure 5.8). However, at the 1000 m² scale, the mixed grass prairie attracts some attention as an invaded site because several of the invading species were not found in the 1 m² subplots, but they were detected in the 1000 m² plots. When the data from the four 1000 m² plots are pooled, even the shortgrass steppe draws some attention for exotic species invasion because detecting some species early may allow for cost-effective containment and restoration. The lesson here is that large plots may be needed to capture locally rare or patchily distributed species of special management concern. Nested plots might have to be fairly large to be effective.

The data collected on four spatial scales (1 m², 10 m², 100 m², and 1000 m²) are well suited for investigating species-area relationships. For example, a comparison of species-area relationships at the tallgrass prairie site and the shortgrass steppe site shows different species-area relationships (Figure 5.9). In this case, it is fairly clear that the tallgrass prairie sites have steeper species-area relationships, suggesting that additional areas sampled in these sites would likely yield many more species than similar areas in the short-grass steppe (Figure 5.9).

Another aspect of measuring plant diversity aided by nested plot de-signs is the analysis of species composition overlap at various spatial scales. There are many possible measures of species composition overlap, the simplest of which is Jaccard's coefficient (see chapter 4). Like species richness, species composition overlap can be greatly affected by the scale of measurements. For example, species composition from Whittaker-type plots (Figure 5.6) could be compared for two vegetation types by comparing species lists from the ten 1 m² subplots in two vegetation types, or by adding in the two 10 m² subplots, or by adding in the 100 m² subplot, or by adding the complete species lists in the 1000 m² plot (Figure 5.10).

Species overlap at various scales can provide insights into the structure and patterning of plant assemblages. In this example, species composition overlap was consistently higher in mixed grass prairie sites at all spatial scales relative to tallgrass prairie sites (Figure 5.10). The mixed grass sys-tem may contain more habitat generalists than the tallgrass prairie systems evaluated here. In 1 m² subplots, species composition is highly variable in the tallgrass sites. However, tallgrass prairie sites became increasingly more similar at larger spatial scales. The nonlinear pattern in tallgrass prairie plots across scales may suggest more complex spatial patterning of species com-pared to the mixed grass prairie site.

Other Multiscale Techniques

Many investigators have used some form of multiscale sampling plots. Wiser et al. (1998) used a series of 100 m² plots with two to seven 1 m² subplots nested within them to assess rare plant occurrences in the south-ern Appalachians.

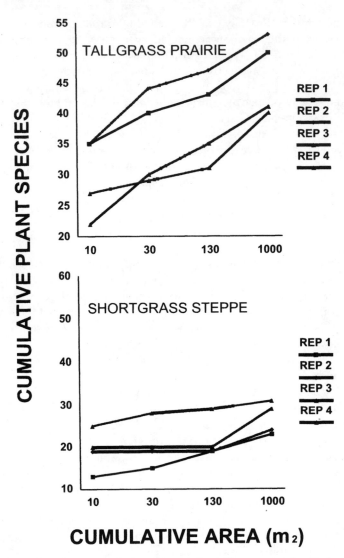

Figure 5.9.   Species-area relationships at the tallgrass prairie site in Minnesota and the shortgrass steppe site in Colorado with four replicates each.

Others have used 20 m × 50 m plots subdivided in various ways. Carrington and Keeley (1999) compared postfire seedling establishment between scrub communities in Mediterranean and non-Mediterranean climate ecosystems (Figure 5.11). Their design was a 20 m × 50 m plot subdivided into 10 m × 10 m subplots with 1 m² subplots placed in the opposite corners of each 100 m² subplot. This yields three scales of measurement with 10 values at 100 m² scales and 20 values at 1 m² scales.

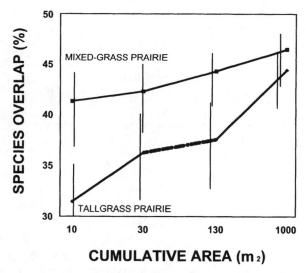

Figure 5.10.   Mean species composition overlap (Jaccard's Coefficient x 100) among four mixed grass prairie plots in Wyoming, and four tallgrass prairie plots in Minnesota, based on cumulative species lists from subplots. Vertical bars represent ±1SE. Lesson 13. Species composition overlap between plant associations is greatly affected by scale. Our understanding of species composition overlap benefits from multi-scale sampling.

The Keeley design provided for many replicates (or pseudoreplicates) at multiple scales, since the seedling data (or plant diversity data) could be regrouped for clusters of contiguous 10 m × 10 m plots. This multiscale nature of the plot has many advantages over single-scale techniques. One value of seedling density in a 1000 m² plot may mask important information on the spatial variation of seedling establishment.

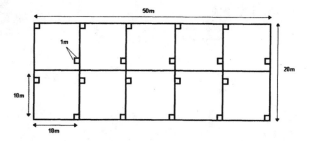

Figure 5.11.   Multiscale sampling design by Keeley et al. 1995, and used in Carrington and Keeley (1999) to assess seedling establishment in shrublands and subtropical forests (redrawn by C.J. Fotheringham from Keeley et al. 1995 with permission).

The Smithsonian Institution published a variation on the nested plot approach for use in Peru (Mistry et al. 1999). In that example, various size plots were used to measure and record seedlings, saplings, and mature trees. A key finding there was that spatial variation was enormous and species rarity and patchiness was striking. About half of the more than 400 taxa recorded were represented as a single individual in one of the ten 1000 m² plots established.

Bell and Lechowicz (1991) used an interesting multiscale design to study the ecology and genetics of fitness in forest plants (Figure 5.12). This design benefited from random selection of subplots (10 m × 10 m, 1 m × 1 m, and 0.1 m × 0.1 m) at each stage within a 50 m × 50 m plot. This provides limited independence of subplots and less dependency on starting location at each scale, but retains overlap and less independence across scales.

For a one-time survey, this design benefited from the randomization of the starting points for the various size subplots. The design may pose some problems for long-term monitoring, as permanent stakes might be required for two corners of each subplot to make remeasurements.

The USDA's Forest Health Monitoring Program is a national program that makes annual evaluations of the condition, changes, and trends in the health of forest ecosystems in the United States (Messer et al. 1991; Riitters et al. 1992) (Figure 5.13). The program, now combined with the Forest Inventory and Analysis Program consists of a nationwide, uniform distribution of sample plots providing a large, unbiased sample of the nation's forests

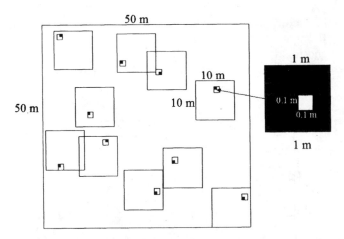

Figure 5.12. A rendition of the sampling design offered by Bell and Lechowicz (1991) to assess genotypic variation in two plant species and environmental heterogeneity in forests in Montreal, Canada.

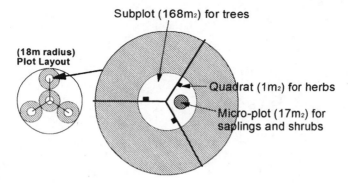

Figure 5.13.   The original USDA Forest Health Monitoring sample plot design (as in Chapter 4).

(1 plot/63,942 ha). The program is designed to provide information for the evaluation of sustainable forests plant diversity as an agreed upon indicator of forest health (Stolte 1997).

As originally designed, the methods called for cover estimates of plant species in three 1 m² quadrats in each of four 168 m² subplots (see Riitters et al. 1992, Figure 4.6). This provided only one scale of measurement for species richness (i.e., the 1 m² scale). The methods were revised to include the searching of each 168 m² subplot, effectively changing the system to a multiscale sampling design with information at the 1 m² scale, the 168 m² scale, and the 672 m² scale (i.e., the four subplots combined). This revised design is also effective in evaluating the variation in species richness at multiple scales with replicate quadrats, subplots, and plots; and evaluating species composition overlap at multiple scales.

Species-Area Curves with Independent Samples

In an ideal world with species randomly distributed in a homogeneous area, sets of small, medium, and large plots would be the preferred way to develop a species-area curve (Figure 5.14). The plots would be randomly located and nonoverlapping, so the plots would be considered "independent" samples. In this example, counting the number of plant species in five 1 m², 10 m², and 100 m² plots provides the data for a simple regression of species and area. Many investigators have found that regressing $\log_{10}$ area to the number of species explains more variance than straight species-area regressions (Shmida 1984; Stohlgren et al. 1997b).

There are many reasons why such an approach would be difficult or undesirable in the real world. First, plant species are rarely randomly distributed. Environmental gradients of soil moisture and nutrients, light, and

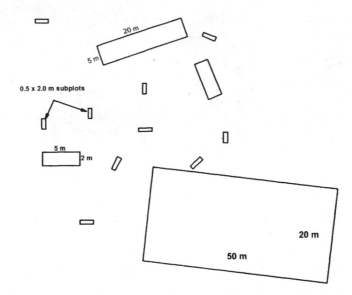

Figure 5.14. Dispersed, independent multi-scale plots.

temperature interact with plant demography, herbivory, disturbance, and other factors to create complex patterns of plant diversity. Second, cost constraints generally reduce the plot sizes used in developing species-area relationships—they are generally small relative to the larger unsampled landscape. Third, it is often difficult for field crews to establish purely random plot locations. Although this has become somewhat easier with Global Positioning System technologies, it is often very costly to have crews locate precise sampling points in a sea of rough topography. Fourth, separate species-area curves might be needed for various vegetation types, and rare, patchy types (e.g., aspen stands in the Rocky Mountains, wetlands or riparian zones in the plains) may be exceedingly costly to sample. Fifth, complex environmental gradients and land uses exist on many landscapes, and they differ for different vegetation types, and the species-area relationships are greatly affected by the magnitude of the environmental gradients captured by the sample plots at the different scales of measurement. Lastly, a series of independent plots would be difficult to permanently mark, relocate, and remeasure.

Lessons Learned

There are several potential benefits to multiscale sampling of plant diversity. First and foremost is that where the underlying patterns of plant diversity are largely unknown for an area (and they usually are unknown or at least poorly studied), then multiscale sampling can be used effectively to evaluate several aspects of diversity, including species-area relationships and patterns of species composition overlap. Several Whittaker plots (Figure 5.6), or Carrington and Keeley plots (Figure 5.11), or Bell and Lechowicz plots (Figure 5.12), or Forest Health Monitoring plots (Figure 5.13), or random plots of various sizes (Figure 5.14) throughout a landscape could provide important knowledge of plant diversity and habitat heterogeneity (in terms of species richness patterns), dominance and rarity, common and unique species assemblages, and insights on patterns of native and exotic species richness (Stohlgren et al. 2005b).

Multiscale sampling also has limitations, the greatest of which is cost. Multiscale sampling is more time consuming than single-scale sampling. Typically more subplots are used and larger macroplots are used in multiscale sampling. Consider the trade-off of putting several 10 m × 10 m relevé plots in an area (where cover by species is recorded for the entire plot) versus fewer Whittaker-type plots (Figure 5.6) for the same cost. The relevé plots might cover a greater geographic range, more rare habitats, and extended environmental gradients. However, the relevé plots may be too small to capture locally rare plant species at each site, and nothing would be learned about within-site variation of species richness, cover, or frequency—and less would be learned about species-area relationships at each site. Cost issues aside, multiscale sampling designs are simply more complex than single-scale designs. Field crews may require more training, data sheets may be more complicated, and analyses are definitely more complicated. In addition, subsampling within a larger plot may provide independent samples, depending on the degree of spatial autocorrelation among subplots and the heterogeneity of the site. However, due to their differing levels of spatial autocorrelation, Lister et al. (2000) required detailed measurements of soil chemistry and vegetation to reveal complex patterns of soil characteristics and species composition.

There are many more examples to illustrate the many positive attributes of multiscale sampling designs. Bellehumeur and Legendre (1998) effectively assessed multiscale sources of variation in ecological variables to examine the spatial structure of environmental variables. Weitz et al. (1993) used nested sampling techniques to determine the scale of variation in soil physical and chemical properties that may be useful in plant diversity studies. Cole (2001) used nested designs (four spatial scales) to identify hierarchical scales of variability in marine environments. Busing and White (1993) subdivided three 1 ha plots into one hundred 0.01 ha subplots and assessed stand descriptors at various grain sizes within the larger plots for an excellent investigation of fine-grain spatial variation in old-growth forests. In these and other ecological studies, it is clear that the scale of measurement affects results and that information from multiple scales is needed in the study of complex environments (Li et al. 2001).

Still, the application of multiscale sampling may be new to many plant ecologists. There is much to be learned by a variety of approaches. How should you test various competing multiscale designs? Again, an experimental approach to sampling is needed. In the next chapter I put two of many possible designs to the test.

6

Comparing Multiscale Sampling Designs

Taking an Experimental Approach

The Issue

Landscape ecologists have not yet agreed on a standardized field methodology for evaluating plant diversity (Krebs 1999; Magurran 1988; Peters and Lovejoy 1992; Soulé and Kohm 1989; Stohlgren 1994; Wilson 1988). In chapter 5, several multiscale sampling designs were presented, but there was little presented in the way of testing various techniques to meet specific objectives.

In this chapter I investigate the process of comparing various multiscale sampling techniques. This was an actual field study, so there were specific sampling objectives (see Stohlgren et al. 1995b), but the objectives were broad and the approach can be generally applied. The goal was to test nested plot techniques that could be used by plant ecologists for unbiased estimates of local species richness and mean species cover, analysis of plant diversity spatial patterns at multiple spatial scales, and trend analysis from monitoring a series of strategically placed, long-term plots.

Background and Sampling Considerations

There are several logical considerations when selecting among potential sampling designs to test. First, it is clear from the goal of assessing plant diversity spatial patterns at multiple spatial scales that single-scale plot techniques might not be appropriate for this exercise. The requirement for unbiased assessments of local species richness might deter the investigator from selecting completely nested designs that are heavily influenced by starting location, such as the nested quadrat design (see Figure 5.2). Data comparability is also an important consideration. Many investigators have used 20 m × 50 m plots (Baker 1990; Peet 1981; Rice and Westoby 1983; Shmida 1984). Thus, conforming to this overall shape and size would have immediate benefits for data comparison. Based on the most commonly used nested designs (see chapter 5), the Whittaker plot had the greatest benefits and least damaging limitations. The design allows for species richness information at four spatial scales (1 m^2, 10 m^2, and 100 m^2 subplots within a 1000 m^2 area), replication of the smallest subplots for cover and frequency data on plant species, and the fairly standard 20 m × 50 m plot (see Figure 5.6). This design was used by Robert H. Whittaker over many years for the measurement of species diversity in different plant communities from different regions of the world. This multiscale sampling of plant diversity allows for (1) evaluations of the influence of spatial scale on local species richness patterns (Podani et al. 1993), (2) better comparisons of community richness than single-scale measurements (Whittaker 1977), and (3) the development of species-area relationships to estimate larger-scale richness patterns (Shmida 1984).

The Whittaker plot has several positive attributes (see chapter 5). Primarily, it provides plant ecologists with (1) a standardized approach to quantify species richness in different plant communities and (2) insights on the effects of quadrat size when determining species-area relationships. However, the Whittaker plot design has three distinct design flaws (Stohlgren 1994), as mentioned in chapter 5. The subplot shapes change from squares to rectangles, confounding the influences of plot shape and size by changing the perimeter:area ratio (see chapter 3). The clustering of subplots in the macroplot may likely hit species-poor areas if species-rich areas are few and patchily distributed. And the successively larger plots are superimposed on the smaller plots (i.e., the plots are not independent in terms of species richness) (see chapter 5). Thus a uniquely species-rich or species-poor area in one of the subplots affects the species richness reported in the larger-size plots. This last problem is common among nested (overlaid) quadrats (e.g., Figures 5.1–5.3, 5.6, 5.11, and 5.12) (Barbour et al. 1999; Mueller-Dombois and Ellenberg 1974; Pielou 1977), the results may be heavily biased by the starting location in the field, and replication is important.

Given the potential problems of the Whittaker design, improvements to the design could be tested in tandem with the original design. We hoped to better describe the average condition of each site with a more scattered distribution of subplots for the long-term monitoring of plant diversity, while maintaining the Whittaker plot's positive attributes.

Methods and Study Areas

We modified the 20 m × 50 m Whittaker plot design (Figure 6.1, top) in several ways (Stohlgren et al. 1995b). We minimized the problems in the original design by using consistent rectangular proportions in the subplots to remove subplot size-shape interactions (Stohlgren 1994). Like the Whittaker plot design, the "modified Whittaker plot" (Figure 6.1, bottom) is 20 m × 50 m. However, the 1 m² and 10 m² subplots were arranged systematically inside the perimeter of the 20 m × 50 m plot. Likewise, the 100 m² subplot is centered in the plot. The three subplot sizes are independent and nonoverlapping, and species richness can be used to construct species-area curves. We realized that the three subplot sizes are nested in and thus overlap the 20 m × 50 m plot, so they are not entirely independent. The subplots combined area represents only 13% of the macroplot. This is a trade-off with the goal to have easily relocatable plots and subplots for monitoring. We hoped that the wider distribution of subplots in the modified design would encounter more species for long-term monitoring.

We compared the Whittaker and the modified Whittaker plot designs by overlaying them in a variety of habitat types in Larimer County, Colorado (n = 13 sites), and in Wind Cave National Park, South Dakota (n = 19 sites). We used the paired t-test (Zar 1974) to compare the number of species recorded in the ten 1 m² subplots, two 10 m² subplots, and one 100 m² subplot of both designs.

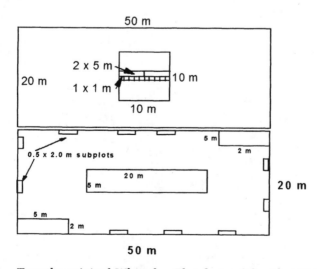

Figure 6.1.   Top: the original Whittaker Plot design (Shmida 1984). Bottom: The Modified-Whittaker Plot design (Stohlgren et al. 1995b).

Results, Discussion, and Lessons Learned

It is obvious that different designs can produce different results. It may be less obvious why this may be the case. The modified Whittaker plot design returned significantly higher ($P < 0.05$) species richness values in the 1 m², 10 m², and 100 m² subplots for the combined dataset (Figure 6.2).

The comparisons of the Whittaker and modified Whittaker designs also showed similar patterns across study sites and in a variety of habitats. Species richness in the vegetation communities tested ranged from 10 species per 0.1 ha to 69 species per 0.1 ha, so the comparison covered a wide variety of sites.

Using data from the 1 m², 10 m², and 100 m² subplots and 1000 m² plots, we found that the modified Whittaker design conformed better to published semilog relationships because the Whittaker plot design, on average, underestimated species richness in the 1 m², 10 m², and 100 m² subplots by concentrating and overlaying the subplots in one small area of the 20 m × 50 m plot (see Stohlgren et al. 1995b).

There are study-specific and general lessons learned by taking an experimental approach to testing alternative sampling techniques. For this specific study, the modified Whittaker nested sampling design looked promising for several reasons. First, the new design minimized the problems of the original Whittaker design. The consistent rectangular proportions remove subplot size-shape interactions, and rectangles may generally perform better than squares at recovering species richness (Stohlgren 1994; Stohlgren et al. 1995b).

The subplots in the modified Whittaker design have less overlap than the subplots in the original design (except for the largest size plot, of course), thus they are influenced less by spatial autocorrelation and nonindependence of observations (Pielou 1977). Because vegetation is often clustered spatially

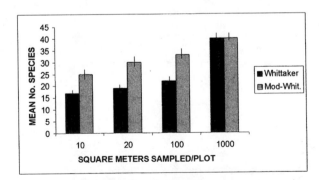

Figure 6.2.   Comparison of the mean number of species recorded by subplot/plot area in the Whittaker and Modified-Whittaker plot designs for different study locations and vegetation types. Vertical bars represent standard errors. Adapted from Stohlgren et al. 1995b.

(Fortin et al. 1989), and most species are rare in cover and abundance (Barbour et al. 1987), the 10 contiguous 1 m² subplots in the original Whittaker design are more likely to miss important patches of vegetation and underrepresent small-scale species richness in the subplots. This is precisely what the test showed (Figure 6.2). In fact, the greatest difference in the two techniques is in the number of species recorded in the 1 m² subplots (Figure 6.2). The ten 1 m² subplots using the modified Whittaker design averaged 26.0 (±1.8; S.E.) species, while the Whittaker design's ten 1 m² subplots averaged only 16.7 (±1.4) species. The original Whittaker design of contiguous 1 m × 1 m subplots (which combined cover only 1 m × 10 m in the center of the 20 m × 50 m plot) probably provides a very biased view of small-scale interactions occurring over limited environmental gradients in a portion of the plot which may exclude many species (Figure 6.2), and because of artificially low variance due to spatial autocorrelation problems (Stohlgren 1994). Particularly for plant species cover estimates conducted in the 1 m² subplots in both designs, the systematic subplots are more likely to give better estimates of mean species cover for the 1000 m² area.

The attractive attributes of the original Whittaker design were maintained in the new design. The nested design allows for analysis of species richness at four spatial scales. The same "sample data sheets" presented in Shmida (1984) can be used. However, species cover abundance data from the 1 m² subplots are improved by recording 9–10 more species, on average, over a broader area of the 0.1 ha plot. Species-area relationships may be more representative of a vegetation type with better estimates of local species richness from a series of distributed subplots (although species-area curves may not always fit the semilog form) (Palmer et al. 1991; Pielou 1977).

There were also several general lessons learned by testing the two sampling designs. It is important to consider the attributes of a fair and comprehensive comparison of sampling designs for plant diversity:

- Unbiased sample site selection. In this case, all sites were selected randomly and the outside boarders of the plot designs were identical. Unbiased site selection allows for the generalization of results to similar habitats in the surrounding landscape. In this way the data gained by the comparison of techniques can be used to broaden the knowledge about the ecosystems under study.
- Random order of techniques. Applying the techniques in random order reduces bias and allows for training in the local taxonomy as two or more techniques are being tested.
- Adequate replication. Too few replicates may not detect significant differences among techniques. In this example, significant differences in species richness for the various-size subplots were easily found with as few as seven replicate plots in a vegetation type (Stohlgren et al. 1995b).
- Tests in multiple vegetation types and biomes. It may be necessary to test alternative field designs in various vegetation types and biomes

to determine the general utility of the methods. In this case, similar results were found in grasslands and forests. Additional tests in other vegetation types and biomes may also be necessary, but there is at least some confidence in the results in sites that varied from 10 to 69 plant species per 0.1 ha, in two states.

- Analysis of multiple parameters. Tests of alternative techniques are made stronger by analyzing multiple parameters. In this case, species richness is compared at multiple spatial scales (Figure 6.2). Also, the mean cover and height by species were different because of the different environmental gradients covered by the 1 m² subplots. Testing the different techniques relative to the theoretical species-area relationships is important.
- Designs for broad applications. In this case, either study design has potentially broad applications in comparing various land use practices and successional patterns, and to investigate species-environment relationships. Nested designs can be shrunk (proportionately) for smaller-scale studies (e.g., thin riparian zones, tundra vegetation, lichen surveys, etc.) or expanded for larger-scale habitats, such as widely scattered trees, to assess plant species patterns at multiple spatial scales.

There are several limitations of design comparison studies. Cost is a primary consideration. Given a certain amount of funding to complete a given study, many investigators might be reluctant to use a portion of the funds to test alternative techniques. (This is a shame since we have much to learn about measuring plant diversity and other aspects of vegetation science). And because of the general requirements for adequate replication, multiple parameters, and tests in various habitats and biomes, only a few different designs can be tested at one time. Given the modest budgets for most ecological studies, even sampling in several different habitat types and biomes may be cost prohibitive. Another limitation is that, to be successful, new designs must be flexible to meet multiple and general objectives. Maintaining high comparability with previous datasets can constrain new designs. There may be trade-offs in meeting multiple objectives such as complete independence of subplots and long-term monitoring in confined relocatable areas. Often, some compromises must be made to accommodate the specific needs of land managers. In this study, the land managers wanted to limit trampling within the boundaries of the plot to increase the value of the plots for long-term monitoring. The systematic placement of the subplots along the plot boarders (Figure 6.1, bottom) makes the design easy to use in the field for relocation of long-term study plots. However, random placement of the subplots may produce less-biased species richness data and less of an affect of spatial autocorrelation among subplots. But random placement of the nonoverlapping subplots (i.e., x, y coordinates, angle, etc.) would be more difficult for field crews, and permanent marking of the subplots would be necessary if trend analysis is a study objective. The last limitation of sampling design comparison studies is simply the reluctance of scientists, land managers, students, and others to shift from old to new sampling techniques. The inertia of plant ecologists, agencies, and others should not be underestimated; it is often easier to grab an off-the-shelf protocol or an historic methods paper than to question or test long-established techniques.

Technique comparison studies are very valuable. We are, unfortunately, far from standardizing field techniques for assessing plant species diversity at landscape, regional, and national scales, and in vegetation types around the globe. Decisions must be made about the number and placement of plots needed to describe the plant species diversity of large study areas and landscapes, and in disturbed and restored landscapes (Stohlgren 1994, 1995b, 1998c). The timing of sampling, species' rareness, and the patterns of landscape features must also be considered. Determining the efficiency of sampling in vascular plant species richness projects will depend on species discovery rates and the rate of observed changes (Heltshe and Forrester 1983; Miller and Wiegert 1989). Other sampling designs may be more appropriate in these and other vegetation types or for other objectives. Data comparability is becoming an increasingly important consideration (e.g., Baker 1990; Nevah and Whittaker 1979; Rice and Westoby 1983), but an experimental approach to sampling designs is absolutely necessary to improve the science of measuring plant diversity. It is healthy to put your favorite designs to the test!

7

Case Study

Comparing Rangeland Vegetation Sampling Techniques

The Issue

As we learned from chapter 6, studies that compare older sampling techniques to newer ones sometimes are necessary to advance the science of plant ecology. In this chapter I present another comparison of techniques to challenge very commonly used rangeland sampling techniques. In terms of measuring and monitoring native and nonnative plant species richness in grassland habitats, never has such a study been so urgently needed. The inertia of rangeland sampling techniques is extremely apparent in the literature.

Between 40 and 50 years ago, various rangeland sampling techniques were developed, primarily to assess the cover of dominant plant species. Other objectives commonly included monitoring changes in the cover of "preferred forage species," percent bare ground (as a measure of erosion potential), and maintaining an elusive quality termed "rangeland health" (National Research Council 1994). In the past few decades, many of the same techniques have been used to continue assessments of grazing effects on native species richness and diversity, and the spread of invasive exotic species. The objectives changed, but the sampling designs did not (Stohlgren et al. 1998c).

Commonly used rangeland sampling methods in the past half-century included Parker transects (Parker 1951) and Daubenmire transects (Daubenmire 1959). The Parker transects were 30.5 m (100 ft.) long and placed perpendicular to the environmental gradient, with a primary objective to assess the frequency (as a surrogate for foliar cover/dominance) of plant species in an area (Parker 1951). Readings of plant species or bare ground were made through a small, 1.9 cm (3/4 in.) diameter ring at 30.5 cm (1 ft.) intervals along the transect (Figure 7.1). The three Parker transects per sample site were placed 25 m apart and produced 300 points per sample site and 1200 points per prairie type.

Daubenmire (1959) preferred quadrat measurements to point measurements of plant cover. He found that about forty 20 cm × 50 cm quadrats along a transect were adequate to capture more than 70% of the plant species in an area. Perhaps noting that the mean cover of dominant species changed little after the first 20 quadrats (Daubenmire 1959, p. 49), the U.S. Forest Service (1996; and prior documents no longer available) suggested a modification of Daubenmire's sampling technique with 20 quadrats at 1.5 m (5 ft.) intervals along a 30.5 m (100 ft.) transect, with a minimum of two transects per vegetation type. At issue is whether these commonly used transect techniques adequately describe local plant species richness and diversity, the cover of dominant species, or newly invading exotic species in, say, a 0.1 ha area. How well do these techniques compare to alternative sampling designs such as the modified Whittaker nested scale design (see Figure 6.2, bottom)? Again, an experimental approach is needed to critically compare various sampling techniques. Field techniques must be tested simultaneously, as typically employed by range ecologists, with several replicate trails in several rangeland types.

Background and Sampling Considerations

There are distinct advantages to using published sampling designs in rangeland inventories and monitoring. The inertia of rangeland sampling meth-

Figure 7.1.   The Parker Loop. Photo by author.

ods may have arose from a practical need to bring some level of standardization, cost efficiency, and precision to rangeland vegetation sampling (Daubenmire 1968). When legislation in the 1970s (e.g., Resources Planning Act of 1974, Soil and Water Resources Conservation Act 1977) mandated rangeland inventories and monitoring on federal lands, the Parker and Daubenmire publications provided readily available, accepted methodologies. Federal agencies were quick to adopt the Parker and Daubenmire methods, often modifying them somewhat for various reasons, and they are still in use today in many areas (Brady et al. 1991; Coughenour et al. 1991; U.S. Forest Service 1996). However, there are three reasons why it was time to reevaluate rangeland sampling methods: (1) ecological paradigms had changed; (2) rangeland conservation objectives had changed; and (3) the need for rangeland inventory and monitoring had increased in the face of decreasing funding to monitor rangelands (i.e., sampling techniques must be more cost efficient and information rich than in the past) (Stohlgren et al. 1998c).

Despite Gleason's (1926) "individualistic concept," many plant ecologists remained heavily influenced by Braun-Blanquet, Clements, and Daubenmire in "The matter of selecting a unit of maximum homogeneity in sampling natural vegetation" (Daubenmire 1959, p. 47). Perhaps it is not surprising that until 1985, the U.S. Forest Service *Range Analysis and Management Handbook* stated that permanent transect sites must be entirely within a single vegetation type and/or plant association, condition class, with similar aspect, location, and soil type (U.S. Forest Service 1985). Reinforced in textbooks by Daubenmire (1968) and others (e.g., Mueller-Dombois and Ellenberg 1974; Tansley and Chipp 1926), the emphasis placed on sampling homogeneous areas may have resulted in the development of vegetation sampling methods that employ smaller-size quadrats and fewer replicate quadrats or transects than might be needed in more heterogeneous habitats. This, in turn, may have limited our ability to extrapolate information to larger, unsampled areas that contain heterogeneous and rare habitat types (Stohlgren et al. 1998a,c).

Another paradigm is that reducing rather than understanding variance is an important goal of vegetation sampling (Daubenmire 1968; Mueller-Dombois and Ellenberg 1974; Tansley and Chipp 1926).While this is certainly important in determining a precise mean value for, say, aboveground biomass, this approach may lead to methods that reduce variance artificially when measuring the foliar cover of locally rare species. Assessing biomass with circular plots will produce a lower variance than square or rectangular plots due to the smaller perimeter:surface area ratio of the circular plot. However, rectangular plots are more appropriate for estimates of foliar cover and species richness because they cover more heterogeneous areas (see Figure 3.2) (Daubenmire 1959; Stohlgren 1994), despite the fact that the variation may be increased relative to circular plots. Transect designs may reduce the variation in biomass, foliar cover, and species richness measurements due to spatial autocorrelation effects and the limited environmental gradients covered. In heterogeneous areas, circular plots and linear transects will likely miss more species than in homogeneous areas.

Spatial autocorrelation also affects cost efficiency. One aspect of cost efficiency is determining the additional cost required to obtain new information. Cost efficiency was an obvious concern of early range conservationists (Daubenmire 1959; Heady 1957; Johnston 1957; Parker 1951), but it was not always properly evaluated with information gained by alternative techniques. Quadrats restricted to locations along transects may miss locally rare or patchy habitats in the sample area while oversampling similar areas, thereby reducing one aspect of cost efficiency: more information might be gained by using a different pattern of sampling (Fortin et al. 1989; Stohlgren 1994).

The primary concern of rangeland conservation generally has been to maintain forage productivity. In most environments, most of the above-ground biomass comes from a few dominant species (Barbour et al. 1987; Daubenmire 1974). Historically rangeland evaluations concentrated on these dominant species or were based on qualitative assessments (U.S. General Accounting Office 1991). Over the past 20 years there has been growing concern over invasive weeds and the protection of native plant diversity (Joern and Keeler 1995; Mack 1981; National Research Council 1994; Randall 1996). We know from the well-known example of cheatgrass (*Bromus tectorum* L.) in the western United States (Mack 1981), that early detection of invasive species is important because these species can quickly spread and dominate regional flora. The scale of rangeland conservation has also changed from site-by-site assessments to the evaluation of landscape and regional land use patterns to sustain productivity and biodiversity (National Research Council 1994). It has become increasingly important to extrapolate transect or plot-level information to landscape and regional scales (Stohlgren et al. 1997b). To meet these new objectives, new sampling techniques must make full use of mathematical modeling, geographic information system technologies, and multiscale vegetation sampling designs (Kalkhan et al. 1995). Rangeland handbooks in the United States offered little in the way of suggestions on how to inventory and monitor plant species diversity and provided options for transect layouts without a rationale for each alternative (e.g., Stohlgren et al. 1998c; U.S. Soil Conservation Service 1976).

Our philosophy was that we were still in the experimental phase of developing rangeland sampling techniques for plant diversity and cover given the changing ecological paradigms and conservation objectives (Stohlgren et al. 1998c). Our objective was to compare commonly used or proposed rangeland vegetation sampling techniques using the same metrics (e.g., the numbers of native and nonnative plants species, foliar cover, percent bare ground) with adequate replication in shortgrass steppe, mixed grass prairie, northern mixed grassland, and tallgrass prairie in the Central Grasslands. We included the original modified Whittaker sampling design (see Figure 6.2, bottom) (Stohlgren et al. 1995b) as described in the previous chapter.

Study Areas and Methods

To adequately evaluate vegetation sampling techniques, comparisons must be made in several vegetation and soil types, and under a wide variety of grazing and disturbance regimes. Often, cost limits ambitious study designs. Still, we had four study locations (Figure 7.2): shortgrass steppe at the Central Plains Experimental Range (Pawnee National Grassland, Nunn, Colorado), mixed grass prairie at the High Plains Experiment Station (Cheyenne, Wyoming), northern mixed prairie in Wind Cave National Park (Hot Springs, South Dakota), and tallgrass prairie in Pipestone National Monument (Pipestone, Minnesota) (Table 7.1).

At each sampling site we superimposed the four rangeland sampling techniques in random order and as typically applied in the field, as described below.

The Parker transects were placed perpendicular to the environmental gradient, with a primary objective to assess the frequency as a surrogate for foliar cover/dominance of plant species in an area (Parker 1951). One Daubenmire transect bisected each sampling site parallel to the environmental gradient (Figure 7.3). The primary objective of the Daubenmire transect is to accurately quantify the foliar cover of most of the plant species at a site. We established four modified Daubenmire transects (henceforth called Daubenmire transects) in each prairie type, resulting in 80 quadrat readings. Daubenmire (1959) and the U.S. Forest Service recommend the use of foliar cover classes (e.g., 0–5%, 5–25%, 25–50%, etc.) to reduce the variance among observers. For spatial and temporal trend analysis, these cover classes may be too broad (particularly if most species are less than 1% cover). To be consistent with the quadrat and point (Parker) methods, the foliar cover of

Figure 7.2.   Map of study locations adapted from Stohlgren et al. 1998c.

Table 7.1.   Study locations, sample sites, vegetation type, grazing regime, and elevation.

Study Location	Sample site	Vegetation type	Grazing regime	Elevation (m)
Cheyenne	1	Mixed grass	1. light, season-long	1977
High Plains	2	prairie	grazing.	
Experiment	3		2. Exclosed from grazing.	1933
Station,			3. Heavy, season-long	1918
Cheyenne,	4		grazing.	
Wyoming			4. Eight paddocks heavily	1947
			stocked, rotational grazing.	
Central Plains	1	Short grass	All plots not exclosed are	1666
Experimental	2	steppe	subject to grazing by cattle,	1627
Range,	3		prairie dogs, and antelope.	1711
Nunn,	4			
Colorado				1573
Wind Cave	1	Northern	1. Lightly grazed, native	1379
National Park,	2	mixed	ungulates.	
Hot Springs,	3	grass	2. Heavily grazed, native	1165
South Dakota	4		ungulates.	
			3. Heavily grazed, native	1161
			ungulates.	
			4. Lightly grazed, native	1317
			ungulates.	
Pipestone	1	Tallgrass	All sites are not grazed,	
National	2	prairie	but they are prescribe	530
Monument,	3		burned at 3- to 4-year	468
Pipestone,	4		cycles, and site 4 was	520
Minnesota			seeded withnative grasses	472
			in 1993, 1994, and1995.	

each species and bare ground were recorded to the nearest percent in each quadrat.

A variety of rangeland ecological condition sampling methods for vegetation and soils are being considered by the U.S. Agricultural Resource Service (Herrick et al. 1996). One proposed transect and quadrat technique (which we termed the TAQ) was designed for grass/herbaceous vegetation (details provided by W. Whitford, Environmental Protection Agency, personal communication, June 1996). The method uses a larger quadrat (0.5 m × 1.0 m) than the Daubenmire quadrat (20 cm × 50 cm). Similar to the Daubenmire transect, the primary objective of the TAQ transect is to accurately quantify the foliar cover of most of the plant species at a site. One 100 m long TAQ transect was located in the center of each sample site parallel to the environmental gradient (Figure 7.3). Five 0.5 m × 1.0 m quadrats were placed randomly along alternate sides of the transect. The quadrat contained

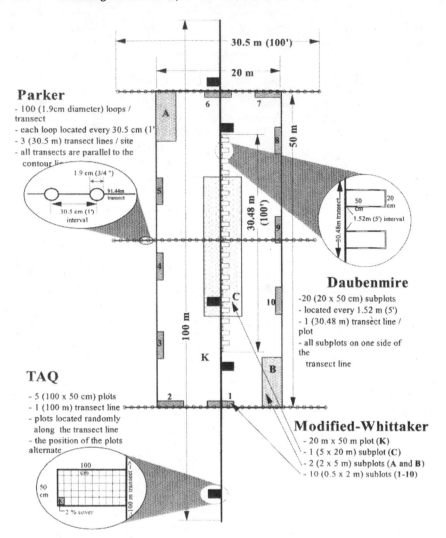

Parker
- 100 (1.9cm diameter) loops / transect
- each loop located every 30.5 cm (1')
- 3 (30.5 m) transect lines / site
- all transects are parallel to the contour li...

Daubenmire
-20 (20 x 50 cm) subplots
- located every 1.52 m (5')
- 1 (30.48 m) transect line / plot
- all subplots on one side of the transect line

TAQ
- 5 (100 x 50 cm) plots
- 1 (100 m) transect line
- plots located randomly along the transect line
- the position of the plots alternate

Modified-Whittaker
- 20 m x 50 m plot (**K**)
- 1 (5 x 20 m) subplot (**C**)
- 2 (2 x 5 m) subplots (**A** and **B**)
- 10 (0.5 x 2 m) sublots (**1-10**)

Figure 7.3. Layout of plots, quadrats, and transects at each sample site, with four sample sites per prairie type.

a decimeter grid of nylon string, and foliar cover for each species and percent bare ground were estimated to the nearest percent. Four TAQ transects per prairie type resulted in 20 quadrat readings.

One multiscale modified Whittaker plot (see chapter 6) was established at each sample site (Figure 7.3). Foliar cover for each species and percent bare ground were estimated to the nearest percent in the ten 1 m² subplots and cumulative plant species are noted in the 10 m² subplots, the 100 m² subplots, and the 1000 m² plot. Ancillary data recorded for each plot included the universal transverse mercator (UTM) location and elevation from a global positioning system, slope, and aspect (Stohlgren et al. 1998c).

The field ecologists were trained in the use of the four sampling techniques at the Cathy Fromme Prairie, Fort Collins, Colorado. At each sample site in each prairie type we staggered the order of the four methods and recorded the time required to collect data using each technique to assess cost efficiency relative to the amount of information gained. We sampled each site as close to the phenological maximum (peak biomass) as possible (Stohlgren et al. 1998c). About 150 of the 1700 plant specimens encountered (9.1% of the total specimens) could not be identified to species due to phenological stage or missing flower parts, so the numbers of native plus nonnative species do not always equal the total number of species.

For each prairie type we used one-way ANOVA (Systat, version 6.0; SPSS Inc., Chicago, Illinois) to test for method effects for total species richness, the number of native and nonnative species, the number of species with less than 1% cover, total foliar cover, total bare ground, and the time required for each sampling method with four replicates per prairie type. We used the total number of species found in the 20 m × 50 m modified Whittaker plot to assess the accuracy (completeness) of the various transect methods in describing plant species in a local (0.1 ha) area or vegetation type. For comparisons of percent foliar cover and bare ground, only the 1 m² subplots in the modified Whittaker plots were used in the comparison to transect techniques. Data from the four locations were pooled and two-way ANOVA was used to test for method effects, sites effects, method × site interactions for vegetation characteristics, and the time required for each sampling method. Whenever ANOVA results showed a significant effect ($P < 0.05$; with no significant interaction), Tukey's honestly significant difference (HSD) test was used to determine significant differences among means (Day and Quinn 1989).

Results, Discussion, and Lessons Learned

Choice of Sampling Design Greatly
Affects Measurements of Plant Diversity:
The Plot Thickens

One of the primary concerns of point transect and quadrat and transect techniques in plant diversity studies is that they may miss some locally rare species or patchily distributed species that may not grow along a straight transect. That was precisely what happened (Figure 7.4). The transect techniques missed about half of the native species and half to two-thirds of the nonnative species in each study area.

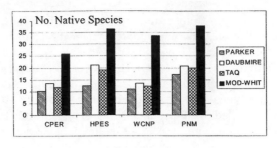

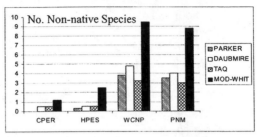

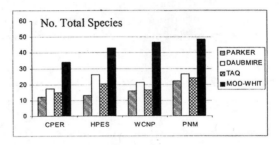

Figure 7.4.   Mean number of native plant species (top), non-native plant species (middle), and total plant species (bottom) captures with the various sampling designs at the Central Plains Experiment Research site (CPER), High Plains Experiment Station (HPES), Wind Cave National Park (WCNP), and Pipestone National Monument (PNM). Adapted from Stohlgren et al. 1998c.

Be Prepared for Site-Specific Variability in Results

Some field measurements are less affected by various sampling designs than others. Despite the differences in native and nonnative species richness among sampling techniques, mean foliar and soil cover did not significantly differ among sampling techniques. The modified Whittaker method took about 40–50 minutes longer to complete than the transect methods in that vegetation type. Combined species richness information from four replicates of each of the methods in the shortgrass steppe type resulted in 41, 56, 57, and 109 plant species from the Parker, Daubenmire, TAQ, and modified

Whittaker methods, respectively. Thus, even with replicates of each technique, about half the plant species found in the four modified Whittaker plots were found by the transect methods. Two nonnative species [Cypress spurge (*Euphorbia cyparissias*) and Russian thistle (*Salsola iberica*)] detected by searching the four 0.1 ha modified Whittaker plots were not picked up by replicate sampling with the transect methods. At site 1, the Daubenmire transect detected only 8 of 26 species at the site.

Plant species measurements are affected by both sampling design and site-specific variation. At the mixed grass prairie area in Wyoming, the transect methods missed two nonnative species at each sample site. Just as in the shortgrass steppe, mean foliar and soil cover did not significantly differ among sampling techniques, even though fewer species were captured by the transect methods. The modified Whittaker method took almost 3 hours per site because half the plant species had less than 1% foliar cover each, and several of them were short, well-camouflaged, and difficult to find in knee-high vegetation in the 1000 m² plot. The Parker transects took less time to complete, but they also recorded the fewest plant species (Figure 7.4).

Having four replicates of the transect techniques did not improve the measurement of species richness in the mixed grass prairie. Combined, the methods resulted in 51, 87, 78, and 155 plant species from the Parker, Daubenmire, TAQ, and modified Whittaker methods, respectively. At site 1, the Parker transect method recorded only 8 of 37 plant species present in the 0.1 ha site.

At the northern mixed prairie area in South Dakota, the transect methods missed about four to five nonnative species per site. Species richness, foliar cover, and soil cover were patchy due to localized prairie dog (*Cynomys* sp.) disturbance and microtopographic variability (often missed by the linear transects). The modified Whittaker method tallied 171 plant species in the four 0.1 ha plots. The six nonnative species present in the modified Whittaker plots were missed by the other methods.

At the tallgrass prairie area in Minnesota, the transect methods missed about four to five exotic species per site. Foliar and soil cover did not differ significantly among methods. The modified Whittaker method took almost 4 hours per site and the Daubenmire transects averaged 2 hours 11 minutes per site. The four replicate Parker, TAQ, and Daubenmire transects detected only 83, 91, and 84 plant species, respectively, while the four modified Whittaker plots tallied 187 plant species in the tallgrass prairie study area.

Summarizing Data Across Sites Can
Show General Results

Assessing broad-scale patterns when comparing various field methods can be tricky. In this case, ANOVA results showed significant methods effects and site effects for all species richness characteristics, but no significant method × site interactions except for the number exotic species. Thus the methods behaved similarly in each prairie type. If there were significant method × site interactions, the interpretation of results would be ambiguous. Here it was obvious that the transect methods (Parker, Daubenmire, and TAQ) significantly underestimated the total species richness, the number of native species, and the number of species with less than 1% cover in each prairie type. For all prairie types combined, the modified Whittaker plot captured an average of 42.9 plant species per plot compared to 15.9, 18.9, and 22.8 plant species per plot for the Parker, TAQ, and Daubenmire methods, respectively.

The four methods produced similar results for total foliar cover and soil cover. The detection of exotic plants, especially noxious weeds, was greatly enhanced by searching the 0.1 ha plot in the multiscale modified Whittaker sampling design. Even with replicate transects, the transect methods usually captured less than 50% of the total species richness in the local area.

We expected the transect methods to capture about 70% of the plant species at a given sample site, and perhaps up to 80% of the plant species in a study area with four replicate transects. Daubenmire (1959) suggested that 40 quadrats captured, on average, 22.5 of 31.5 taxa (71.4 % of the taxa) in 31 areas that he sampled. Even with 80 quadrats (four transects) per prairie type, we found, on average, that the Daubenmire transects recorded only between 46% (shortgrass steppe) and 52% (tallgrass) of the species present in a prairie type (48.8% overall). At some sites the Daubenmire transects captured only 31% (shortgrass steppe, site 1) or 39% (tallgrass prairie, site 3) of the plant species present. The TAQ and Parker transects captured only 46.9% and 38.1%, respectively, of the plant species recorded in this study, which certainly does not provide an accurate description of local plant species diversity.

There are two primary reasons why transect methods fail to capture much of the plant diversity at a site: (1) the small total sample area, and (2) the high degree of spatial autocorrelation due to the linearity of the methodology and clustering of plots in more localized areas. The Daubenmire transect method, for example, sample only 2 m² (20 cm × 50 cm × 20 quadrats) at each site for information on species richness and foliar cover, and only 8 m² after four replicate transects are established. The TAQ method covers slightly more area at 2.5 m² per site and 10 m² after four transects. Even with twice as many TAQ and Daubenmire quadrats per site (i.e., similar sampling time compared to the modified Whittaker method) (Table 7.2), the total area sampled would be 50% and 40%, respectively, of that sampled using the modified Whittaker method. The Parker transect's 1.9 cm diameter hoop covers just 0.09 m² per site (300 points) and 0.3 m² (1200 points) after four replicates. The ten 1 m² subplots in the modified Whittaker plot collected foliar cover data from 10 m² per site and 40 m² after four replicates (five times the area of the Daubenmire quadrats), and the four scales of nested subplots collect species richness information from a 1000 m² area at each site and 4000 m² after four replicates. This larger sample area is significantly

Table 7.2. Mean and (standard error) of characteristics for the various sampling techniques in all locations.

Characteristics	Sampling design				F_{site} (P)	F_{method} (P)	$F_{site \& method}$ (P)
	Parker	TAQ	Daubenmire	Modified Whittaker			
Total no. of species	15.9[a] (1.3)	18.9[a] (1.2)	22.8[a] (1.6)	42.9[b] (2.4)	9.3 (0.001)	72.4 (0.001)	0.88 (0.548)
No. of native species	12.8[a] (1.1)	15.7[a] (1.3)	17.2[a] (1.4)	33.5[b] (1.9)	11.0 (0.001)	59.6 (0.001)	0.65 (0.747)
No. of exotic species	1.9 (0.5)	1.7 (0.4)	2.4 (0.5)	5.5 (1.0)	72.5 (0.001)	35.5 (0.001)	3.77 (0.001)
No. of species with < 1% cover	7.4[a] (0.9)	8.8[a] (1.0)	13.6[b] (1.4)	18.5[c] (1.6)	3.6 (0.019)	18.7 (0.001)	0.90 (0.533)
Percent foliar cover	53.4 (4.5)	62.0 (6.6)	50.4 (3.6)	66.1 (6.8)	4.6 (0.006)	2.0 (0.130)	0.691 (0.713)
Percent soil cover	19.0 (3.9)	21.0 (4.2)	26.0 (3.7)	22.3 (3.4)	0.7 (0.576)	0.5 (0.66)	0.32 (0.963)
Time minutes	65.5[a] (16.2)	58.4[a] (10.1)	92.9[a] (13.5)	176.2[b] (20.4)	6.2 (0.001)	15.0 (0.001)	0.38 (0.592)

AOVA F and P, df = 3, 60; row means with dissimilar superscripts are significantly different ($P < 0.05$) using Tukey's test.

Adapted from Stohlgren et al. (1998c).

more effective at describing plant species richness in heterogeneous and patchy environments.

Both Parker (1951) and Daubenmire (1959) suggested that a larger area around their transects be searched. Daubenmire (1959, p. 48) mentioned in a footnote that the presence of rare species should be noted, but he didn't specify the size or shape of the search area, somewhat contrary to his original objective of standardizing a methodology. Parker (1951) suggested searching a 45.7 m × 30.5 m (150 ft. × 100 ft.) plot to "assure a record of rare but important indicator species which may not otherwise be encountered on the transect." It is unclear what was meant by "rare but important indicator species," though a complete taxonomic listing was not suggested. In any case, perhaps due to the increased time and taxonomic expertise needed to search the larger plot, we know of no examples where the "searching" part of either protocol has been adopted and consistently used (Francis et al. 1972; Hutchings and Holmgren 1959; Johnston 1957).

The second problem with transect methods is the high degree of spatial autocorrelation. Because two points close together are more similar than two points that are farther apart, quadrat sampling or point sampling along the Parker, TAQ, and Daubenmire transects tends to record redundant information among quadrats, while missing many species in the local area (Table 7.2). Corrections to frequency data collected with the Parker transects have been proposed (Brady et al. 1991; Cook et al. 1992), but we have seen no attempts to correct for spatial autocorrelation effects on species richness (the method's biggest problem). The ten 1 m² subplots of the modified Whittaker plot were arranged on the perimeter of the 20 m × 50 m plot and were less influenced by spatial autocorrelation, but there was also an obvious effect of sample area when comparing the results of the different methods. The larger and more widely spaced subplots averaged 28 (±2.4) species per site, producing foliar cover data on 23% more plant species than in the Daubenmire transects. The modified Whittaker subplots also provide cover and frequency data that are more spatially independent, which can be useful in evaluating the spread of exotic species (Stohlgren et al. 1995b). Furthermore, transects simply miss many nonlinear components of the landscape (e.g., animal disturbances, microtopographic features, and small-scale edaphic differences). Transect methods, designed under the Clementsian paradigm of landscapes dominated by homogeneous communities, may not have prepared us for sampling the heterogeneous landscapes of the Central Grasslands or other areas.

We could have compared the methods in three ways: (1) as they are typically applied; (2) equal sample time devoted to each method; or (3) equal area sampled by each method (Stohlgren et al. 1998c). We reasoned that the methods should be sampled as typically applied in a given vegetation type to assess what information each of the techniques generally records in a rangeland unit and to serve as a comparison to past sampling efforts. The equal time approach would have required twice as many TAQ and Daubenmire transects and quadrats and still would have resulted in (1) far less sample area than the modified Whittaker methods (as show above), and (2) the new transects likely would have missed many rare species and habitat patches (as shown in the data on total species per prairie type). An equal area approach would have taken five times longer for the Daubenmire method and four times longer for the TAQ method just for the cover and frequency data (i.e., without searching the 0.1 ha area for total species richness, which took more than an hour at many sites). An equal area approach with the tiny Parker loop is unrealistic. Adding a 0.1 ha search area around transect methods would take nearly as much time as the modified Whittaker method, but would not eliminate the problems associated with spatial autocorrelation, nonindependent frequency, and collecting cover data on fewer plant species. Comparing the techniques as typically applied may be one indication that previous rangeland monitoring may have gathered little information on most locally rare and nonnative plant species.

Our "replicate sites" suggest that heterogeneity is the rule rather than the exception in the Central Grasslands. Species overlap between sites averaged 42–49% for the four prairie types. Typically plant diversity in the Central Grasslands is patchy because of the underlying patchiness of nutrient rich and poor areas (Cambardella and Elliott 1993; Seastedt 1995) superimposed on variations in topography with broad-scale temperature and microtopographic moisture gradients (Coffin and Lauenroth 1989b,c; Collins and Glenn 1995), patchy disturbances (prairie dogs and other animals, ungulate soil disturbance) (Knight 1994; Polley and Collins 1984; Polley and Wallace 1986; Whicker and Detling 1988), seed bank variability and chance establishment (Coffin and Lauenroth 1989a; Joern and Keeler 1995), competition (Coffin and Lauenroth 1989b; Joern and Keeler 1995; Nasri and Doescher 1995), fire (Dix 1960; Hulbert 1969, 1988; Whisenant and Uresk 1990; White and Currie 1983), and heterogeneous herbivory (Parsons et al. 1991; Singer 1995). Temporal variation from succession, invasion, or climate change can be equally impressive (Bock and Bock 1995; Collins and Glenn 1995; Fargione and Tilman 2005; Mack 1981). Larger, long-term plots and greater sampling intensity are needed to accurately quantify spatial and temporal heterogeneity. It may be time to wean field ecologists from the fixation of sampling only homogeneous units.

Methods Can Also Affect Measurements of the Cover of Dominant and Subdominant Species

The most obvious way field techniques can affect the measurements of cover by species is to miss species entirely. Often the transect techniques missed native dominant and subdominant species at a site, and they frequently missed important exotic plant species. At site 4 in the Central Plains Experimental Range, the Parker transect recorded only 10 of 41 plant species and missed 1 of the 5 dominant species at the site, hairy grama (*Bouteloua hirsuta* (H.B.L.) Lag. (Table 7.3).

Table 7.3. Comparisons of average percent cover of the five most dominant species from representative sites.

Location (sample site)	Species	Native/ exotic	Parker	TAQ	Daubenmire	Modified Whittaker
				Sampling techniques		
Central Plains	Bouteloua gracilis	Native	25.4	11.8	16.0	22.5
Experimental	Opuntia polyacatha	Native	1.3	9.6	2.1	12.2
Range	Bouteloua hirsuta	Native	0	2.6	2.7	7.9
(plot 4)	Eriogonurn effusum	Native	2.3	3.6	17.2	7.3
	Aristida longiseta	Native	5.0	4.6	8.9	6.0
Cheyenne	Bouteloua gracilis	Native	30.9	15.2	20.6	15.9
(plot 1)	Koeleria pyramidata	Native	5.1	14.2	11.2	10.8
	Stipa comata	Native	2.9	0	2.1	7.0
	Agropyron smithii	Native	0	1.9	1.4	4.6
	Artemisia frigida	Native	0.7	1.9	0.8	4.2
Wind Cave	Bromus japonicus	Exotic	4.1	22.1	3.4	14.1
National	Psoralea tenuflora	Native	1.0	6.5	17.7	13.2
Park	Poa pratensis	Exotic	23.7	14.8	16.2	13.1
(plot 3)	Symphoricarpos occidentalis	Native	0	0	0	4.2
	Agropyron smithii	Native	0.7	2.9	2.3	2.5
Pipestone	Bromus inermis	Exotic	7.7	21.8	17.2	9.3
National	Aster oblongifolia	Native	0	0	0	9.0
Monument	Symphoricarpos occidentalis	Native	0.7	1.8	1.3	5.4
(plot 3)	Solidago missouriensis	Native	0	1.4	8.4	4.3
	Poa pratensis	Exotic	7.7	2.4	2.8	3.4

Five dominant species were determined from modified Whittaker method.

At site 2 in Cheyenne, the Daubenmire transect missed all five of the exotic species present, including some highly invasive species: crested wheatgrass (*Agropyron cristatum* L.), cheatgrass, Dalmatian toadflax (*Linaria dalmatica* L.), yellow sweet clover (*Melilotus officinalis* (L.) Lam.), and Kentucky bluegrass (*Poa pratensis* L.). The TAQ and Parker methods each missed one of the five dominant species at two sampling sites (sites 1 and 4) (Table 7.3). In Wind Cave, patches of snowberry (*Symphoricarpus occidentalis* Hook.) (42% cover in one 1 m² modified Whittaker subplot) at site 3 were not detected by the transect methods (Table 7.3). Clumps of another native species, *Rosa blanda* (5.6% cover in the 1 m² modified Whittaker subplots), at site 4 were not detected by the transect methods. The modified Whittaker method took almost 3.5 hours per site, more than twice as long as the transect techniques (Table 7.3). At Pipestone National Monument, at site 3, patches of the native herb *Aster oblongifolia* L. (mean of 9.0% cover in the 1 m² modified Whittaker subplots and present in 9 of 10 subplots) were not detected by the transect methods (Table 7.3). Seven exotic species at Pipestone National Monument were not detected by the transect methods: quack grass (*Agropyron repens* L.), asparagus (*Asparagus officinalis* L.), yellow toadflax (*Linaria vulgaris* Mill), yellow sweet clover, timothy (*Phleum pratense* L.), goatsbeard (*Tragopogon dubius* Scop.), and red clover (*Trifolium pratense* L.). It must be tougher to capture plant species in small quadrats than we realized!

Importance of Detecting Locally Rare Species

Why is it so important to detect and monitor locally rare species in rangeland sampling? First, locally rare species are a major component of plant diversity, and probably biodiversity. Almost half the plant species we encountered in each prairie type had less than 1% foliar cover (Table 7.2). Assuming that many soil organisms and terrestrial invertebrates may be mutualistically linked to many of these plant species, underestimating the number of locally rare plant species may directly translate to underestimating local and regional biodiversity (Stohlgren et al. 1997a,b).

The second reason that detecting locally rare species is important is purely economic. Early detection of exotic species and noxious weeds (i.e., while they are less than 1% cover) is important in range management: eradication and restoration efforts are less expensive for smaller affected areas (Peters et al. 1996). The use of cover classes such as 0–5%, 5–25%, etc. (Daubenmire 1959; U.S. Forest Service 1996) may be too broad to evaluate the detection and spread of exotic species. Searching the 0.1 ha modified Whittaker plots led to the early detection of Cypress spurge and Russian thistle in the shortgrass steppe and to the detection of crested wheatgrass, cheatgrass, Dalmatian toadflax, yellow sweet clover, and Kentucky bluegrass in the mixed grass prairie. The rapid spread of these weeds could reduce the productivity of preferred forage and decrease native plant diversity (Rosentreter 1994). In the northern mixed prairie of Wind Cave National Park, yellow sweet clover was barely present in the corner of the modified Whittaker plot at site 4, but it extended well into the surrounding landscape. This is significant because park managers consider yellow sweet clover to be one of their most invasive weeds. In the tallgrass prairie at Pipestone National Monument (site 3), only the large 20 m × 50 m plot was able to capture two native species that had been seeded by resource managers into a burned site: silver-leaved psoralea (*Psoralea argophylla* L.) and purple prairie clover (*Dalea purpurea* L.). Both species were growing in large, but widely spaced patches. Detecting these species could be an indication that seeding efforts are starting to work.

Finding new or rare species at a site is not a primary objective of rangeland sampling, but it is exciting for conservation biologists. In the reseeded tallgrass prairie site (site 4) we found a single unknown legume in the 0.1 ha plot (undetected by the transect methods). Because there was only one individual present, we were unable to collect it, but with so little of the tallgrass ecosystem left in the United States (Leach and Givnish 1996; Stubbendieck and Willson 1987; Swengel and Swengel 1995), this could be an exciting find. In any case, as rangeland conservation objectives are modified to protect native plant diversity, manage for rare species and habitats, and monitor the spread of nonnative species, it is clear that field sampling must result in a more complete picture of plant diversity than can be gained from the commonly used transect methods of the past.

Always Consider the Ability to Extrapolate Results from Plots to Landscapes to Regions

The key to evaluating local- to regional-scale range conditions and trends is the ability to extrapolate field data beyond the sample plots (i.e., to the surrounding unsampled landscape). The nested scale design of the modified Whittaker or similar techniques allows for the development of species-area curves (generally, species richness = $m \times \log$ area + b) based on the average number of species found in the 1 m², 10 m², 100 m², and 1000 m² subplots and plots (Stohlgren et al. 1997a). The species-area curves allow estimates of the number of native and exotic species in the larger, unsampled area. Species accumulation curves (or species effort curves) can be developed from transect data, but we know of no way to correct for bias due to spatial autocorrelation. Furthermore, species accumulation curves developed from transects aligned such that they miss species-rich patches, which frequently happened in this study, will greatly underestimate the number of species in a larger area.

The cover and height data from the modified Whittaker 1 m² subplots serve as ground truth data for classification and vegetation type mapping (Stohlgren et al. 1997a), productivity estimates (Heady 1957), and plant diversity analyses (Stohlgren et al. 1997a), and for mapping important species (weed invasions, rare plants) and habitats (riparian zones, wetlands) (Stohlgren et al. 1997a,c). Predictive geographic information system (GIS) models can be based on physical and botanical data, supplemented with a few vegetation plots to validate and improve the models to optimize cost-efficient field sampling (Stohlgren et al. 1997a).

Cost Efficiency Is Always a Concern

The old adage holds that "you get what you pay for." The large modified Whittaker method took longer to complete at a site than the transect methods (Table 7.2). However, cost efficiency is based on (1) information gained per unit of time, and (2) the total amount of information gained. The number of plant species recorded per minute was similar among techniques: 0.32, 0.25, 0.24, and 0.24 species per minute for the TAQ, Daubenmire, modified Whittaker, and

Parker methods, respectively. The TAQ method recorded more plant species per minute than the other methods because it took the shortest time and species accumulation curves rise quickly in the early stages of botanical surveys. Capturing more species becomes increasingly difficult, especially with transect methods, due to spatial autocorrelation effects. In fact, we recommend an alteration to the modified Whittaker design by staggering their placement along the exterior of the 100 m² subplot (Figure 7.5) to better disperse the 1 m² subplots throughout the plot while aiding their placement and relocation.

Even with the design used, the total amount of information gained from the larger modified Whittaker plot exceeded that of the transect methods as measured by total species richness, the number of native and exotic species, and the number of species with less than 1% cover (Table 7.3). The ability to create accurate species-area curves is also a cost-saving device, since fewer field plots are needed when information can be accurately extrapolated to larger, unsampled areas.

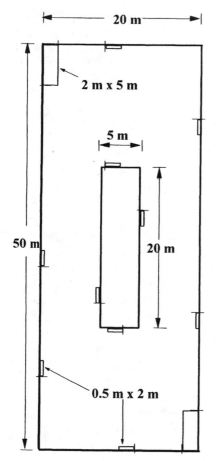

Figure 7.5. Revised sampling layout for the Modified-Whittaker sampling plot. Adapted from Stohlgren et al. 1998c.

Ease of use is an extremely important component of cost efficiency. Our field crews found that the modified Whittaker technique worked well in all vegetation types, but it was time consuming in species-rich habitats (Stohlgren et al. 1998c). The snap-together, plastic subplot frame (0.5 m × 2.0 m) was easy to assemble and light. The subplot size was able to accommodate shrubs and tall grasses. The 20 cm × 50 cm Daubenmire frame was difficult to use in shrublands, and very difficult to use in tallgrass prairie sites: with so many tall grasses and forbs overlapping the quadrat, it was difficult getting accurate cover estimates. The 0.5 m × 1.0 m TAQ quadrat was easy to use on shortgrass sites also, but was almost impossible to use in tallgrass or shrub patches. Plants had to be threaded through the nylon decimeter grid, and it was difficult to get accurate readings for litter, duff, and soil. All the techniques were easy to use on shortgrass prairie. The Parker transects were fast and easy, though they did require a great deal of stooping. The Parker methods frustrated field crews, who realized how many plant species the technique was missing at each sampling site (Stohlgren et al. 1998c).

Another consideration of cost efficiency is that the major expenditure of time is traveling to the site, so that spending more time at the site to gain more information may be time well spent. Better monitoring data on rare and exotic species, independent frequency data, and accurate species-area curves improve the ability to extrapolate to larger, unsampled areas. Thus despite the increased time at each site, the overall cost efficiency of sampling landscapes and regions is improved with multiscale sampling techniques.

As rangeland conservation objectives evolve and adapt, so must rangeland vegetation sampling techniques. Due to the threats of exotic plants to rangeland productivity, increasing concern about noxious weeds, and the desire to protect native plant diversity and biodiversity, sampling techniques will be expected to accurately quantify the status and trends of plant diversity at local, regional, and national scales. We found that the commonly used Parker and Daubenmire transects and the proposed TAQ transect methods significantly underestimated the total species richness, the number of native species, and the total number of species with less than 1% cover in each prairie type. The multiscale modified Whittaker method that included an exhaustive search for plant species in a 20 m × 50 m area captured twice as many plant species than replicated transect methods. The detection and measurement of exotic plants, including noxious weeds, was greatly enhanced by using a multiscale sampling design. Other nested designs based on large plots (e.g., Carrington and Keeley 1999; Keeley et al. 1995) would have worked equally well, as long as the subplot data were not confined to a linear or small portion of the plot. Even replicated transect methods were less effective than multiscale techniques because of small sample areas and spatial autocorrelation bias and because they tended to miss rare species and habitats. Multiscale sampling methods may better monitor the spread of exotic species and evaluate range conditions and trends at local, regional, and national scales compared to the commonly used transect methods. At the very least, sampling techniques should be carefully tested to be sure they are meeting the ever-changing objectives of land managers and the public.

PART III

SCALING TO LANDSCAPES

8

Case Study on Multiphase and Multiscale Sampling

The Issue

Only a small portion (usually less than 1%) of any landscape can be affordably sampled. Estimating the patterns of plant diversity in the much larger, unsampled landscape is important and challenging. For example, many national and state parks, national forests, wildlife refuges, and nature preserves require detailed information at multiple scales to evaluate the status and trends of native plant species, nonnative plant species invasions, and the effects of grazing and fire on plant diversity (Stohlgren et al. 1997b). Rocky Mountain National Park in Colorado provides a typical example. There is a growing effort to protect highly diverse or unique habitats such as riparian zones (Baker 1990) and aspen stands (Peet 1981; Suzuki et al. 1999). However, attempts to quantify the relative contributions of various plant communities in the total plant diversity of large areas have been hampered by single-scale, small-plot sampling designs and poor vegetation sampling methods. The park, like many landscapes, provides a great opportunity to test multiphase, multiscale methods to rapidly assess patterns of plant diversity in a complex landscape.

In this case study, the methods were based on (1) interpretation of satellite data or aerial photography to develop a vegetation type map to include homogenous and heterogeneous vegetation types, and small areas suspected of having unique species; (2) random selection of plot locations in each type; (3) the use of multiscale sampling design to assess key aspects of plant diversity (e.g., species richness at multiple scales, plant species cover, patterns of nonnative species invasions); (4) determination of species composition overlap within and among vegetation types; and (5) assessment of native and nonnative plant species patterns in the landscape.

Study Area and Testing Field Methods

Diverse landscapes are ideal for testing sampling designs. The Front Range of the Rocky Mountains in Colorado ranges in elevation from 1600 m to nearly 4350 m and contains a variety of vegetation communities from prairie to alpine tundra. Dominant types and species (generally from low to high elevation) include prairie vegetation dominated by short grasses (*Bouteloua gracilis, Buchloe dactyloides*) and sagebrush (*Artemisia tridentata*), ponderosa pine (*Pinus ponderosa*), Douglas-fir (*Pseudotsuga menziesii*), lodgepole pine (*Pinus contorta*), aspen (*Populus tremuloides*), limber pine (*Pinus flexilis*), spruce (*Picea engelmannii*), and subalpine fir (*Abies lasiocarpa*) (Peet 1988). The test site was a 754 ha area (2500–3000 m elevation) in the Beaver Meadows area of Rocky Mountain National Park (Figure 8.1).

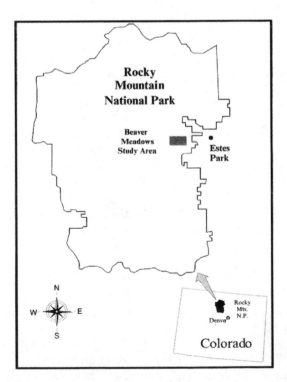

Figure 8.1.   Map of the test site in Rocky Mountain National Park, Colorado.

Testing a Multiphase and Multiscale
Sampling Design

Multiphase sampling provides a means to scale information from relatively small vegetation plots to large landscapes with remotely sensed data collected at various intermediate resolutions and scales. For example, Thematic Mapper satellite data are fairly coarse in resolution (i.e., 30 m grid cells, six reflectance bands), but they are relatively inexpensive and provide timely and frequent information at landscape to regional scales (Figure 8.2). Aerial photography can provide higher-resolution data at landscape and regional scales (e.g., 1:24,000 scale natural color or infrared photography), but it is relatively expensive to acquire and photointerpret. Because of the increased expense of aerial photography, it is often less frequent and timely. However, the higher resolution photographs may allow the delineation of small, but

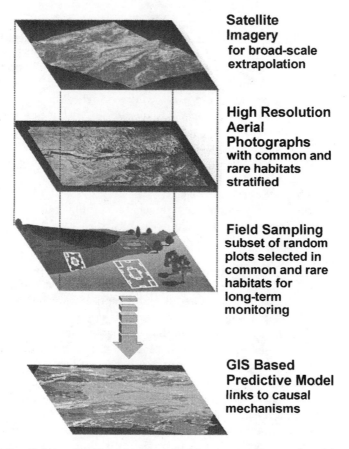

Satellite Imagery
for broad-scale extrapolation

High Resolution Aerial Photographs
with common and rare habitats stratified

Field Sampling
subset of random plots selected in common and rare habitats for long-term monitoring

GIS Based Predictive Model
links to causal mechanisms

Figure 8.2. Schematic diagram of multi-phase sampling and multi-scale sampling (field sampling. Adapted from Kalkhan and Stohlgren 1999.

rare and important habitat types that are less likely to be detected with coarser resolution imagery.

The power of multiphase sampling is in linking multiple layers of remotely sensed imagery with field data such that predictive models can interpolate patterns of plant diversity to larger spatial scales with known precision and accuracy. Since you cannot do detailed botanical studies from space, evaluating regional patterns in plant diversity from satellite data usually requires an intermediate stage of interpreting high-resolution aerial photographs. For example, Nusser et al. (1998) used multiphase sampling to investigate for natural resource inventories with two scales of sampling—3110 ha (2 miles × 12 miles) and 64 ha (0.5 mile × 0.5 mile), with three sample points in the smaller area. The example I present below is one of many possible approaches that can be used in multiphase sampling.

Aerial Photo Interpretation and the Minimum Mapping Unit

Aerial photographs are available for many land management units. Care must be taken to use the most recent and complete set of photographs that meet study needs. In this test example we used 1:15,840 natural color aerial photographs taken on September 28, 1987. We stratified the vegetation to include lodgepole pine, ponderosa pine, wet meadow (dominated by *Poa palustris, Deschampsia caespitosa,* and *Poa interior*), dry meadow (dominated by *Carex helianthus* and *A. tridentata*), and aspen communities (Stohlgren et al. 1997a,b). The ponderosa pine community was further stratified into burned (prescribed fire in September 1994) and unburned ponderosa pine. We had a minimum mapping unit of 0.02 ha.

In this example we overlaid a 48 unit × 32 unit grid (north to south) over the aerial photograph and randomly selected (computer random number generator) five to seven potential sample points in each of the six vegeta-

---

Reminder: A "minimum mapping unit" is the smallest area routinely delineated and classified on an aerial photograph when creating a vegetation type map. Delineating and classifying very small areas is more labor intensive and expensive than delineating larger areas and combining vegetation types. However, smaller minimum mapping units may help to identify locally rare but important habitats within larger areas or vegetation types. Generally, smaller minimum mapping units allow for a greater number of vegetation types to be recognized and displayed on a vegetation map. Many soil maps are created with 2–5 ha minimum mapping units, while many vegetation type maps are made with 0.5–100 ha minimum mapping units, though maps with 2 ha minimum mapping units are common.

tion types. We located the points in the field with the aid of the photographs, other maps, and a compass and later checked and mapped the locations with a global positioning system (GPS; Trimbal Pathfinder Professional, Trimble Navigation Limited, Sunnyvale, California). For each field site we calculated slope, aspect, and elevation (using digital elevation models and our GPS information). At each point we established a modified Whittaker nested sampling plot (see Figure 7.5) (Stohlgren et al. 1995b, 1998c). Other multiscale sampling designs (e.g., Keeley et al. 1995) could have been substituted for the modified Whittaker design used here.

Testing the Importance of Resolution

Scientists continue to struggle to develop strategies to quantify the biological diversity of landscapes and regions (e.g., Magurran 1988; Noss and Cooperrider 1994; Peters and Lovejoy 1992; Soulé and Kohm 1989; Wilson 1988), much like plant ecologists struggle to link vegetation analyses across scales (Franklin 1993; Peterson and Parker 1998; Short and Hestbeck 1995). This is because biological diversity is greatly affected by scale and resolution (Huston 1994). The three potential resolutions of investigation in this case study were coarse resolution (typically a 100 ha minimum mapping unit), moderate resolution (typically a 2–50 ha minimum mapping unit), and fine resolution (a less than 2 ha minimum mapping unit, such as 30 m × 30 m). There have been a disproportionate number of studies conducted at the two extremes. For example, there are active research programs to quantify patterns of biological diversity with coarse resolution with complete coverage (Austin and Heyligers 1991; Messer et al. 1991; Palmer et al. 1991; Scott et al. 1993; Stoms 1992), and many plant studies are conducted with fine resolution, but very incomplete coverage, by using collecting data from small plots in large areas (Kareiva and Anderson 1988).

Attempts to quantify the relative contributions of various plant communities to the total plant diversity of large areas remains hampered by poor multiscale vegetation sampling methods. The primary challenges for plant ecologists are to link data from plots to landscapes and to capture small but important patches of diversity. Linking data from plots to landscapes requires multiscale field techniques to assess plant diversity and mathematical models to estimate the number of species in larger areas. Since costs constrain the portion of a landscape that can be sampled, researchers often reduce the number of vegetation classes considered or increase the minimum mapping unit (decreasing the resolution). Ecologists rarely know how these decisions affect the accuracy and completeness of information on plant diversity maps.

The second challenge of identifying small but important patches of plant diversity, such as hotspots of high species richness, unique plant assemblages, and rare species, demands detailed spatially explicit data of plant species distributions from many cover types occupying patchily distributed small areas (Cherrill et al. 1995). In the case study area, aspen (*P. tremuloides*) occupies less than 4% of the forested landscape in the Front Range of the Colorado Rockies, often in small patches (Peet 1981, 1988; Suzuki et al. 1999). Vegetation maps created with minimum mapping units of 2 ha or larger may grossly underestimate aspen cover (Kaye et al. 2001). Not knowing the actual cover of aspen forests translates into a gap in the knowledge needed to manage this rare habitat that is valuable for many wildlife and plant species (DeByle and Winokur 1985; Mueggler 1985; Salt 1957) and for fire prevention (Jones and DeByle 1985).

For example, consider two vegetation maps of the same 750 ha area that might be used as a basis for a stratified random sampling scheme (Figure 8.3). The vegetation map created with a 0.02 ha minimum mapping unit from aerial photography (Figure 8.3, top) shows detailed information on vegetation type boundaries, including isolated patches of aspen and contiguous wet meadow (riparian type) vegetation. In contrast, the vegetation map created from 30 m × 30 m pixel data from a satellite (Figure 8.3, bottom) has sharp, boxy vegetation boundaries, discontinuous wet meadow vegetation, and blocks of aspen in slightly different locations than those identified by aerial photography (Figure 8.3, top).

The vegetation map based on satellite imagery is affected by "average" reflectance values from the 30 m × 30 m pixels, which tend to homogenize dominant vegetation types (e.g., lodgepole pine) by combining areas with small forest gaps and mixed vegetation types to the most dominant type nearby. The map created from aerial photography is more expensive to produce because it requires stereophotographic imaging by an expert remote sensing specialist. However, the cost may be justified if capturing rare, im-

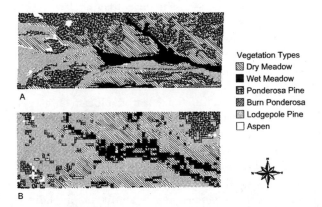

Vegetation Types
▨ Dry Meadow
■ Wet Meadow
▦ Ponderosa Pine
▨ Burn Ponderosa
▨ Lodgepole Pine
☐ Aspen

Figure 8.3.   Vegetation map created with a 0.02-ha minimum mapping unit from aerial photography (top), and a vegetation map created from 30 m x 30 m pixel data from a Thematic Mapper satellite (bottom). The scale and extent are the same in the two maps, but the resolution is much finer (or higher) in the top map.

portant vegetation types provides better information on patterns of plant diversity.

High-resolution vegetation type maps are an important starting point for evaluating patterns of plant diversity at landscape scales. Additional information is needed on species diversity within vegetation types, species-area relationships, and species diversity between types. This information can then be used to quantify patterns of plant species richness in an area with a minimal amount of field sampling.

Unbiased Site Selection

Sample sites are selected in an unbiased manner using remotely sensed information and a stratified random sampling design. High-resolution satellite imagery or aerial photography is obtained for the study area. Vegetation is stratified to include common homogeneous types (typically recognized in most vegetation mapping efforts), heterogeneous types of special interest (e.g., ecotones and ecoclines, mixed stands and communities), and rare vegetation types (Stohlgren et al. 1997a,c; Figure 8.4). The size of the minimum mapping unit must be selected to accommodate particular rare vegetation types, such as aspen in this test case. After a preliminary vegetation map is prepared, four or five ground truth plot locations per vegetation type are selected randomly using a randomizing function in a geographic information system or with a grid system on a plastic overlay atop the aerial photograph (Stohlgren et al. 1997b).

Figure 8.4. Lesson 14. Many readers will not be familiar with the commands and workings of geographic information system software. An immediate lesson is to befriend a GIS expert while becoming familiar with basic commands and the potential of the software to be able to ask and answer general and specific questions. (Photograph by author)

Multiscale Vegetation Sampling

In this case study, a modified Whittaker nested vegetation sampling plot was established (see Figure 7.5) at each ground truth sampling point. The dimensions of the plot and subplots were halved to fit the plots in the burned areas. Just as in the last case study, percent cover by species and average height by species were recorded in the 1 m² subplots. Cumulative species (additional species found in the subplots or plots) were recorded successively in the other subplots and plots. Three to five plots were established in each vegetation type. The unique-to-type species, which were likely to be the species of most conservation interest, were also noted. GPS was used to document the locations of the plots to incorporate the data directly into a geographic information system.

Determining the Appropriate Species-Area Curve Model

Mean species richness data from the 1 m², 10 m², and 100 m² subplots from each 1000 m² plot were fit to species-log(area) curves and log(species)-log(area) curves after initial tests on subsets of the data showed these models produced similar, high coefficients of determination (generally $r^2 > 0.95$), while direct species-area curves fit the data poorly. The semilog relationship has proven to be a robust species-area curve (Shmida 1984; Stohlgren et al. 1995b).

Evaluating Species Composition Overlap

Jaccard's coefficient was used to compare species composition overlap among plots and vegetation types (Krebs 1989, chapter 4). The mean Jaccard's coefficient for each vegetation type was calculated from all possible pairwise comparisons between random sets of three plots. Remember that this coefficient is sensitive to sample size, plot size, and spatial autocorrelation among plots.

Evaluating the Effects of the Minimum Mapping Unit

After digitizing the high-resolution (0.02 ha minimum mapping unit) vegetation map from the aerial photography, the ELIMINATE command (with KEEPEDGE option; Arc/Info version 6.0) was used to create vegetation maps with minimum mapping units of 2 ha, 50 ha, and 100 ha. For the 50 ha and 100 ha minimum mapping unit maps, the DISSOLVE command was used to remove polygons less than the minimum mapping unit and to maintain homogeneity within the combined polygons. For each map, the area and number of polygons were calculated in each of the recognized vegetation types (STATISTICS command).

Results, Discussion, and Lessons Learned

In this 754 ha study area, 330 plant species were found in the 25 sample plots (four 0.025 ha plots and twenty-one 0.1 ha plots). By stratifying the vegetation into both large, common types (e.g., lodgepole pine, ponderosa pine, and dry meadow) and potentially important small-area types (e.g., aspen, wet meadow, and burned ponderosa pine) (Stohlgren et al. 1997b), and using unbiased plot locations, we accounted for about one-third of Rocky Mountain National Park's plant species list in just 2.2 ha of sampling area.

Species-Area Curves Vary by Vegetation Type

Average species-area curves for the various vegetation types showed that the aspen type had both the steepest slope and the greatest intercept, while the burned ponderosa pine type was on the other extreme (Figure 8.5). These species-area relationships are only accurate over the scales of measurement (i.e., they are based on plot sizes of 1 m², 10 m², 100 m², and 1000 m²), but identical field methods make comparisons valid among vegetation types. Based on these species-area relationships, one can see the importance of the species-rich aspen and wet meadow types in the study landscape. However, these average species-area relationships cannot be extended to larger areas because they do not consider species composition overlap, the extent of environmental gradients, and the patchiness of species richness at landscape scales (Figure 8.5). Clearly additional information is needed on species composition overlap within and among vegetation types at multiple scales.

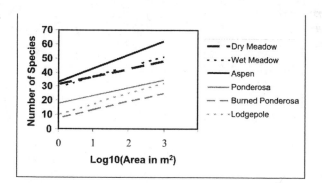

Figure 8.5.   Average species-area curves for sampled vegetation types in the Beaver Meadows area of Rocky Mountain National Park, Colorado. The curves are log-linear regressions of the mean number of species at 10-m², 100-m², and 1000-m² scales (four plots in each type). Adapted from Stohlgren et al. 1997c.

Species Composition Overlap Varies
Within Vegetation Types

Spatial variation in species composition was high in the study area. Species composition overlap between the replicate plots within a vegetation type ranged from 19.9% in the wet meadow and lodgepole types (i.e., the mean of pairwise comparisons of species lists in the 1000 m² plots) to 47.0% in the dry meadow type (Table 8.1). The low standard errors of the means suggest that species overlap was fairly consistent within all types. Still, knowing that there is 20% or 50% overlap among plots in a vegetation type helps to quantify heterogeneity within a vegetation type.

Several factors can affect the measurement of species composition overlap within a vegetation type. Plot size is an obvious factor. Small plots tend to capture common species and miss locally rare species. In fairly homogeneous habitats, increases in plot size may produce similar or increased values in species composition overlap. However, in heterogeneous types, or if greater environmental gradients are included in larger plots and additional rare microhabitats are incorporated in the larger plots, then mean species composition overlap can decrease with larger plots. Such was the case here (Table 8.1). In the dry meadow, aspen, burned ponderosa, and lodgepole pine types, the large plots contained more common, patchy, and widely distributed species, so species composition overlap increased among the larger plots. In contrast, the larger plots in the wet meadow and ponderosa pine types encountered more locally rare species, so species composition overlap decreased with increasing plot size (Table 8.1).

Table 8.1.   Mean species composition overlap (Jaccard's coefficient) within vegetation types based on four plots per type (smaller plots used in the burned ponderosa type).

	Jaccard's coefficient	
Vegetation type	100 m² plot data	1000 m² plot data
Dry meadow	43.5%	47.0%
	(3.1%)	(3.9%)
Wet meadow	20.2%	19.9%
	(4.0%)	(3.0%)
Aspen	22.8%	25.5%
	(2.1%)	(2.5%)
Ponderosa pine	33.1%	32.3%
	(2.1%)	(2.8%)
Burned ponderosa pine	27.5%	31.8%
	(2.5%)	(2.2%)
Lodgepole pine	19.1%	19.9%
	(1.6%)	(1.7%)

Standard errors are in parentheses.

Other factors can also influence species composition overlap, including the distance between plots (Stohlgren et al. 1999b), chance plot locations in unique or species-rich habitats, and unequal sampling within or among plots. Four plots in the 750 ha test area would likely have higher mean species composition overlap than four plots in a 75,000 ha area. Likewise, with a small number of sample plots, the chance placement of a plot in a species-rich patch or a microhabitat with a unique plant assemblage could greatly decrease mean species composition overlap in a vegetation type. There are a host of assumptions when comparing species composition overlap, including equal sampling effort among plots, equal detection of all plant species at each site, and similar average environmental gradients sampled within and among plots. As usual, caution should be used in interpreting these types of results.

Species Composition Overlap Also Varies Among Vegetation Types

Species overlap varied greatly among vegetation types (Table 8.2). The combined species lists from four plots showed that the composition of the wet meadow vegetation type overlapped 27.9% with the aspen community type, but less than 17% with the other vegetation types. The cross-comparisons of the ponderosa pine, burned ponderosa pine, and lodgepole pine types had between 28% and 31% overlap, which is fairly high species overlap among the pine types. Although the wet meadow and dry meadow vegetation types were close to each other on the landscape, their species composition overlapped only 16.3%.

Like species composition comparisons within vegetation types, comparisons of species composition overlap among vegetation types must be cautiously interpreted. Species composition among vegetation types is affected

Table 8.2.   Mean species composition overlap (Jaccard's coefficient) among vegetation types based on four plots per type (smaller plots used in the burned ponderosa type).

Vegetation type	Wet meadow	Aspen	Ponderosa pine	Burned ponderosa	Lodgepole pine
Dry meadow	16%	20%	37%	*21%*	23%
Wet meadow		28%	14%	*11%*	14%
Aspen			26%	*19%*	27%
Ponderosa pine				*31%*	28%
Burned ponderosa					*30%*

The burned ponderosa plots were 250 m², while all other plots were 1000 m², so their values are in italics—they cannot be directly compared to the other types, but may be an index of values if larger plots were used.

Adapted from Stohlgren et al. (1997b).

by all the factors that influence species overlap results within vegetation types. Small plots may exaggerate the differences between vegetation types. The physical and environmental distance between plots and groups of plots will affect these results. However, some general patterns can be quite revealing. The mesic and more rare vegetation types, aspen and wet meadow, are more similar to each other than each is to the more xeric pine types.

It may be satisfying to many "community ecologists" that species composition overlap among vegetation types is consistently lower than species composition overlap within vegetation types. This provides some support for the notion that "plant communities" might be loosely defined by groups of species with somewhat higher affinities for specific overstory tree species, a confined set of environmental conditions, or a subset of topographic, edaphic, or climatic conditions. However, it should be realized that the degree of species composition overlap might be strongly influenced by the completeness of sampling (i.e., plot size, sampling intensity, observer expertise, etc.). Additional plots in each type could easily lead to higher species composition overlap among types, since most plant species occur in more than one vegetation type, more than one state, and more than one country. Few plant species are very narrow endemic species. Likewise, the boundaries of most vegetation types cannot be clearly delineated; many are gradients and ecoclines, providing ample opportunity for a desegregation of communities in intermediate habitats.

Vegetation Patterns Depend on the Minimum Mapping Unit Selected

Describing vegetation patterns with large minimum mapping units may significantly underestimate plant community diversity and the number of polygons (e.g., habitat patches and landscape complexity) (Table 8.3). At the 0.02 ha resolution, six vegetation types in 117 polygons were recognized, including the aspen type, which was scattered in clumps throughout the landscape. With the 2 ha minimum mapping unit, five vegetation types were recognized for the study area. The aspen type was absent, the size of the burned ponderosa pine habitat was half of that recognized with the 0.02 ha minimum mapping unit, and the number of polygons delineated was three times greater than for the 0.02 ha minimum mapping unit. With the 50 ha minimum mapping unit, four vegetation types were recognized, while the 100 ha minimum mapping unit showed only three common vegetation types in three polygons (Table 8.3).

Determining How Various Vegetation Types Contribute to Plant Species Richness Patterns

By incorporating species richness and uniqueness data with the aerial coverage estimates of vegetation types, it is possible to evaluate the relative

Table 8.3.   Vegetation types recognized, area of each type (main table entries), and number of polygons recorded (in parentheses) with different minimum mapping units for the same 754 ha study area.

Vegetation type	Minimum mapping unit			
	100 ha	50 ha	2 ha	0.02 ha
Lodgepole pine	132 ha (1 polygon)	137 ha (2 polygons)	147 ha (9 polygons)	141 ha (16 polygons)
Ponderosa pine	281 ha (1 polygon)	281 ha (1 polygons)	266 ha (20 polygons)	270 ha (48 polygons)
Dry meadow	341 ha (1 polygon)	279 ha (2 polygons)	270 ha (7 polygons)	261 ha (31 polygons)
Wet meadow		56 ha (1 polygons)	65 ha (2 polygons)	63 ha (3 polygons)
Burned ponderosa			5 ha (1 polygons)	10 ha (7 polygons)
Aspen				9 ha (12 polygons)
All types combined	754 ha (3 polygons)	754 ha (6 polygons)	754 ha (39 polygons)	754 ha (117 polygons)

Adapted from Stohlgren et al. (1997a).

contribution of various vegetation types to plant species richness at landscape scales (Table 8.4). It is difficult to fully identify plant species that are truly unique to a specific vegetation type without a complete survey of all plant species from strictly distinct vegetation types. The unique-to-type species in this case (Table 8.4) is an index of "relative uniqueness" based on a given similar sample size, sampling effort, and pattern of sampling. In this case, about half the plant species recorded in the wet meadow type were not recorded in the other five vegetation types, suggesting the wet meadow type is an important contributor to plant diversity in the landscape.

The importance of various vegetation types to landscape-scale plant diversity also can be seen with area-weighted measures. In this example, the aspen and burned ponderosa vegetation types contained the highest number of unique species per hectare, with 25.2 and 10.2 species/ha, respectively, observed in the plots. Considering their large combined area, the pine community types contained relatively few plant species (Table 8.3). The most telling result of this study was that the habitats with the greatest number of unique species per hectare were vegetation types that are not likely to be recognized with larger minimum mapping units (Figure 8.6).

High-resolution maps may be needed to accurately depict patterns of species richness in complex landscapes. We know from the species-area curves (Figure 8.3) and data on species richness and uniqueness in the various vegetation types (Table 8.4) that, in this landscape, it was important to

Table 8.4.   Total number of plant species observed in each vegetation type, unique-to-type plant species, area of vegetation type, and number of unique-to-type species per hectare of habitat.

Vegetation type	No. of species observed	No. of unique-to-type species	Area of vegetation type (ha)	No. of unique species per hectare of habitat
Dry meadow	81	31	260.8	0.4
Wet meadow	148	76	63.1	3.3
Aspen	150	50	8.8	25.2
Ponderosa pine	88	12	269.9	0.6
Burned ponderosa pine	59	10	9.8	10.2
Lodgepole pine	88	18	141.1	1.6
Total duplicates removed)	330		753.8	

delineate and adequately sample locally rare habitats such as the aspen and burned ponderosa vegetation types. Even a map created with a 2 ha minimum mapping unit would not detect the critical aspen stands in the area (Figure 8.6, bottom). In addition, high-resolution maps (Figure 8.6, top) help to depict the contagion or patchiness of habitat.

Effectiveness of Multiscale Sampling Techniques

Multiscale techniques helped to quantify plant diversity patterns in the 754 ha study area in a few weeks of sampling. Because of the nested, large plots, the variation within and among vegetation types could be approximated with only three to five replicate plots per vegetation type (Table 8.3). Still, many of the metrics used in evaluating patterns of plant diversity at landscape scales are heavily influenced by plot size, sample size, habitat heterogeneity and patchiness, the complexity of environmental gradients underlying the patterns of habitat contagion and plant species distributions, and ecological processes such as herbivory, competition, and various disturbances.

Establishing large, multiscale vegetation plots can be expensive. One plot took two ecologists 2–3 hours to complete, and longer in species-rich habitats. However, the multiscale vegetation plots worked equally well in expansive vegetation types (e.g., lodgepole pine) and small habitats (e.g., aspen, burned areas). Changing the size of the plots from 20 m × 50 m in most vegetation types to 10 m × 25 m in the burned ponderosa pine type did not radically affect major results.

The plots were extremely efficient at capturing known plant species in Rocky Mountain National Park. Sampling occurred in a restricted elevation range of 2500–3000 m and did not encompass subalpine, alpine tundra, and lower elevation riparian zones. Still, the plots produced 330 plant species

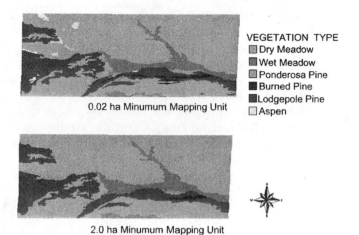

0.02 ha Minumum Mapping Unit

VEGETATION TYPE
▨ Dry Meadow
▥ Wet Meadow
▢ Ponderosa Pine
■ Burned Pine
■Lodgepole Pine
☐ Aspen

2.0 ha Minumum Mapping Unit

Figure 8.6.   The Beaver Meadows area of Rocky Mountain National Park, Colorado, with a 0.02 minimum mapping unit and a 2.0 minimum mapping unit.

(approximately one-third the number of plant species recorded in the park) in the 2.2 ha area within the plots in the 754 ha sampling area. This suggests that multiscale sampling techniques may be very efficient in conducting landscape-scale plant surveys in areas with poor existing data (i.e., most natural areas) (Stohlgren et al. 1995c) and plant species distribution patterns may be highly redundant within larger landscapes in the same ecoregion.

Capturing one-third the number of plant species in the 1074 km² Rocky Mountain National Park in just 25 plots (0.025–0.1 ha) suggests that existing vegetation maps in many national parks, wildlife refuges, and other natural areas could be improved substantially. Maps that usually contain information only on overstory types could be improved with a minimal amount of field work by conducting multiscale plant surveys to quantify understory species richness, cover and distribution of nonindigenous plants, and locations of habitats with high diversity or unique species assemblages. The obvious next step is to model species richness related to environmental variables (e.g., slope, aspect, and elevation) to develop a predictive plant diversity model (see chapter 14). Because plots were precisely located with a GPS and the data were collected within a geographic information system framework, they provided resource managers with a means to monitor long-term changes in plant diversity and weedy plant invasions and to evaluate the effects of various land use practices on plant diversity. They also provide an independent dataset to assess the accuracy of present and future vegetation maps (Kalkhan et al. 1995).

These or other multiscale sampling techniques could be used in any area with high success because they require a minimum amount of fieldwork. The efficiency is the result of (1) recognizing potentially important vegetation types before stratification (small aspen stands and wetlands in our case) (Figure 8.7); (2) unbiased selection of plot locations within vegetation types; and (3) the multiscale sampling methods and evaluations of species composition overlap within and among vegetation types. The techniques are adaptable to a wide range of vegetation types simply by adjusting the dimensions of the plot and subplots (Stohlgren et al. 1995b). For example, we used 0.025 ha plots in some small, burned forest types and in alpine tundra, but larger plots may be necessary in some systems. The most important requirements of the technique are to stratify common and rare vegetation types, select unbiased sampling sites, include four scales of sampling (e.g., the 1 m², 10 m², and 100 m² subplots from each 1000 m² plot for our area), and select an appropriate minimum mapping unit.

High-Resolution Mapping Helps in Evaluating Patterns of Plant Species Diversity

The size of the minimum mapping unit greatly influences our potential understanding of plant diversity patterns in three ways. First, if the minimum mapping unit is too large, some large-area vegetation types appear larger and more contiguous, some medium-area vegetation types appear reduced or increased in landscape cover, and some small-area vegetation types are entirely undetected (Table 8.3). We expected that a vegetation map created with a 100 ha minimum mapping unit (typically used in statewide vegetation maps) would report only a few cover types in our study area. However, even with the 2 ha minimum mapping unit (used in many states and National Park Service units), the species-rich aspen type would not be recognized (Figure 8.6 and Table 8.3).

The 2 ha minimum mapping unit recovered 86% of the plant species, but failed to detect the aspen vegetation type (Figure 8.6) and associated unique plant species (Table 8.4). The 2 ha minimum mapping unit produced better information on species diversity patterns than the 50 ha or 100 ha minimum mapping unit, but could have extreme repercussions in estimating wildlife diversity based on habitat availability. Peet (1981) showed that aspen stands were high in plant diversity, and we demonstrate that the aspen stands also have steeper species-area curves, greater numbers of unique species, and little species overlap with other communities (except with the wet meadow type, which also is rare on the landscape).

Second, increasing the minimum mapping unit size sharply decreased the number of polygons recognized (Table 8.3). If these polygons represent patches of important wildlife habitat, then our assumptions about habitat availability and connectivity may be heavily influenced by the resolution of vegetation maps. Too large a minimum mapping unit may suggest the presence of a large, contiguous habitat that does not really exist. Alternatively, important thin corridors or small patches of habitat (i.e., riparian

Figure 8.7.   Lesson 15. For all plant diversity studies, if you miss the rare habitats, hotspots of species richness, and areas with unique plant assemblages, then you miss much of the story. (Photograph of rare vegetation types, by Lisa D. Schell, used with permission)

zones, stands of aspen) may not be recognized on the vegetation map where they do exist in nature. For small mammals, amphibians, and patch-specific plants and invertebrates, these small sanctuaries could be the most important features for persistence, providing for the survival of populations and metapopulations throughout the landscape (Opdam et al. 1993).

The third major effect of large minimum mapping unit size is simply that finished, brightly colored maps and geographic information system themes may lead to complacency: land managers may assume that additional research, inventories, and monitoring are not a priority. Finer resolution vegetation surveys can aid in the detection and management of rare species by identifying distinctive ecosystems. The aspen, wet meadow, and burned ponderosa vegetation types covered very small portions of the landscape, but contained a high proportion of the unique plant species. Several rare species occurred in the wet meadow: the ladies' tresses orchid (*Spiranthes romanzoffiana*), wood lily (*Lilium philadelphicum*), and white bog orchid (*Limnorchis dilatata* ssp. *albiflora*). Although these species are not on any federal lists, they may be considered locally rare because their habitats are small and patchily distributed.

This case study teaches us that small minimum mapping units allow for the development of high-resolution maps. When combined with multiscale vegetation sampling of common and rare habitat types, this approach is very helpful for evaluating patterns of plant diversity.

9

Case Study

*Designing a Monitoring Program for Assessing Patterns
of Plant Diversity in Forests Nationwide*

The Issue

Consider the difficulties in designing a monitoring program for the condition and production of the nation's forests, including changes in understory plant diversity. The methods have to be flexible enough to work equally well in a variety of forest types, yet standardized enough to allow for highly comparable data across the nation. The U.S. Forest Service's Forest Health Monitoring Program accomplished this task. It is a national program that makes annual evaluations of the condition, changes, and trends in the health of forest ecosystems in the United States (Riitters et al. 1992). The monitoring program consists of a nationwide, uniform distribution of sample plots providing a large, unbiased sample of the nation's forests (1 plot/63,942 ha).

Since its inception and design tests in the early 1990s (Riitters et al. 1992), the monitoring program has focused on multiple aspects of forest ecosystems, including tree growth, regeneration, and mortality; visual crown symptoms; soil condition; contaminants in foliage; growth efficiency; vertical vegetation structure and diversity; and air pollution. From the outset, attention focused on sampling design efficiency, extensive training of field crews, quality control and assurance (Cline et al. 1995), and, combined with the Forest Inventory and Analysis Program, the forest monitoring program has been very successful in monitoring the timber resources of the nation's forests. Initially, vertical vegetation structure (i.e., the amount, arrangement, and composition of forest vegetation) was seen as one of the most promising indicators of biotic integrity and biodiversity (Riitters et al. 1992). There was early recognition that vegetation structure was relevant to plant species diversity (e.g., supporting the Endangered Species Act of 1973) and habitat diversity (e.g., supporting the Federal Policy and Management Act of 1976) as an early warning system to help identify the effects of environmental stress (Stapanian et al. 1993). However, due to changes in interagency support and decreased funding, primary emphasis was placed on tree resources, with lesser attention on plant species diversity and other ecological indicators.

In 1997, U.S. Forest Service officials recognized the incredible opportunity to enhance the timber aspects of the program with improved, systematic, long-term monitoring of understory plant diversity. Invasive nonnative plant species have been documented in several site-specific studies in forests (Stohlgren et al. 1999a, 2000b), posing problems in fire control, maintaining wildlife habitat, and protecting native plant species (Mack et al. 2000; Westbrooks 1998).

Background and Sampling Considerations

As initially measured in the Forest Health Monitoring Program, species richness and cover were recorded in 12 small (1 m²) quadrats per fully forested plot by vegetation height. Most field crews employed trained foresters rather than trained botanists, so recorders sometimes lumped species into simple groups such as grasses, forbs, and shrubs, or identified taxa only to the genus level. Some regions emphasized the importance of more complete and accurate data on understory plant species composition more than other regions. As a result, species richness by strata, species accumulation curves, and vegetation volume by strata could not always be meaningfully summarized at state, regional, and national scales.

This provided an opportunity to retrofit the original Forest Health Monitoring methods for evaluating understory plant diversity for two reasons: (1) the total 12 m² area (twelve 1 m² quadrats) sampled was probably too small to capture a significant portion of native plant diversity at each site, and (2) the single-scale approach could be easily modified to better evaluate plant diversity in patchy and heterogeneous habitats, which are typical of most habitats. Small quadrats miss species, especially those with low frequencies of occurrence (see chapter 7). Especially in forested vegetation types, the small 1 m² quadrats gathered only incomplete data on plant diversity and exotic species primarily because most plant species are locally rare (more than 50% of the species have less than 1% foliar cover) and because plants are never randomly distributed on a plot or landscape (Stohlgren et al. 1998d).

The previous methodology was modified in two important ways. First, multiscale techniques were incorporated that increased the area surveyed in each plot (Bull et al. 1998). Second, qualified botanists were added to the field teams. Highly qualified botanists were needed to field-identify most of the species at each site, including many locally rare species and invasive species from Europe, Asia, Africa, South America, and Australia. In this pilot study, a random subset of plots in various forest types in different regions added an "extended survey" of plant diversity on each of four subplots (Figure 9.1). For each plot, cover by species was recorded in the twelve 1 m² quadrats and the presence of all native and exotic plant species was recorded for each of the four 168 m² subplots.

In this chapter, the results of a pilot project are presented where colleagues and I tested the modified vegetation indicators (i.e., understory diversity and general vegetation structure) to provide better data on plant diversity in the nation's forests. This research was also designed to demonstrate the potential of multiscale plant diversity data (i.e., quadrat-, subplot-, and plot-level information) to evaluate species richness patterns (e.g., hot spots of native plant diversity, early detection of the spread of exotic plant species, and to assess the suitability of forest stands as habitat for wildlife).

Study Areas and Methods

Plant structure and diversity data were collected on 111 plots across Oregon ($n = 14$), Washington ($n = 12$), Colorado ($n = 33$), Michigan ($n = 37$), and Virginia ($n = 15$) in conjunction with annually scheduled field sampling in those states (Figure 9.1). For the 20 plots located on the eastern slope of Colorado's Rocky Mountains, ancillary information was collected on elevation, GPS location (x, y coordinates), community type (pinyon-juniper, ponderosa pine, Gambel oak, lodgepole pine, aspen, Douglas-fir, spruce), and primary land use (timberland, woodland, or rangeland). The presence of cattle grazing was also noted, though this is not part of the information

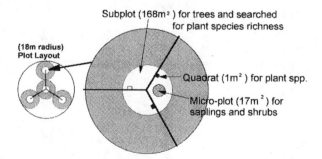

Figure 9.1.   The modified, multi-scale Forest Health Monitoring Program plot design. Foliar cover by species in two height strata was recorded by species in the three 1-m² quadrats in each 168-m² subplot, and the remaining area in each subplot was then surveyed for plant species not recorded in the quadrats. Four subplots comprise one plot.

normally collected on the systematically placed plots. All plots were located and established using the standard Forest Health Monitoring methods.

Plot locations were predetermined from an unbiased systematic sampling grid, which covers the United States (Stolte 1997). On each of the four subplots per plot, crews established three 1 m² quadrats at 4.6 m on 30°, 150°, and 270° azimuths from the subplot center (Figure 9.1). Crews repeated this process for each site.

Botanists on the field crews used a polyvinyl chloride (PVC) frame to delineate the 1 m² vegetation quadrat. Botanists then identified and estimated cover for all species in stratum 1 (0–0.61 m) and stratum 2 (0.61–1.83 m), with ground variables recorded in stratum 1. For each quadrat, three data elements were recorded: (1) species identification, (2) strata, and (3) plant canopy and ground variable cover. Species codes used for analysis were the standardized U.S. Natural Resource Conservation Service (NRCS) PLANTS database codes. The height stratum of the plant was recorded, with foliar cover estimated to the nearest percent. Cover was also recorded for the following ground variables: wood, water, rock, roots, duff/litter, lichen, moss, soil, trail/road, dung, and other (trash, bones, etc.) (Bull et al. 1998).

The 1 m² quadrat frame was calibrated (painted in 10 cm sections) to improve the accuracy and precision of cover estimates (Figure 9.2). Bota-

Estimating Cover in the 1 m₂ quadrat

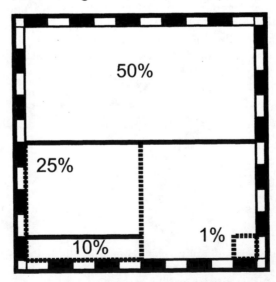

Figure 9.2.   Frame used to estimate cover by species in the 1-m² quadrat.

nists were carefully trained to estimate cover. Cover was estimated only on plants or portion of plants that were inside the quadrat frame.

After completing the three quadrats on a subplot, the botanist searched the entire subplot and recorded any new species not found in the quadrats. Cover classes (5% classes) were assigned to any new species encountered in the larger plots and general comments about the subplot were recorded. Botanists repeated this process for all subplots. Unknown plants found on the subplots were collected off-plot, pressed, and were later identified or mailed to the herbarium at the University of North Carolina–Chapel Hill Biota of North America Program for species identification. At a minimum, field crews obtained the number of unique species on each plot, even if identifications could not be made to the species level because of the stage of growth at the time of collection.

All statistical analyses were conducted with Systat (version 9.0; SPSS, Inc., Chicago, Illinois). Due to the small sample sizes in the pilot study, nonparametric Mann-Whitney U or Kruskal-Wallis tests were used to compare plant diversity characteristics among vegetation types, land use classes, and grazing types. Simple linear regression was used to correlate native and exotic species richness by state and for all plots combined. Kriging was used to map interpolations of exotic species richness in a portion of eastern Colorado.

Results, Discussion, and Lessons Learned

There was considerable variation found in plant diversity in the five pilot states. For example, native species richness per plot ranged from 6 species in a Colorado plot to 100 species in a Virginia plot. Exotic species richness per plot ranged from zero species in many plots to 21 species per plot in Oregon. The cover of exotic species ranged from 0% to 113% in one Oregon plot that contained 21 exotic species. For the 111 plots in five states, the mean cover of native species was 37.4% ($\pm2.9\%$; 1 S.E.), while the mean cover of exotic species was 3.7% ($\pm1.2\%$). These and the following results based on small sample sizes may not be representative of conditions throughout vegetation types, land use classes, states, or the nation, but they demonstrate the potential of the combined Forest Health Monitoring–Forest Inventory and Analysis data once the program becomes fully operational throughout the nation.

Improved Measurement of Plant Diversity
and Early Detection of Invasive Exotic Plants

One major feature of the revised vegetation sampling method included searching the four 168 m² subplots for additional plant species not previously recorded on the twelve 1 m² quadrats. We found that searching the larger plots aided in the early detection of exotic plant species. For all 111 plots, sampling only the twelve 1 m² quadrats resulted in capturing, on average, 33.5% (±3.3%) fewer exotic plant species per plot. The small quadrats missed 23.9% (±6.0%) of exotic plant species per plot in Michigan, and missed 53.1% (±10.1%) of exotic plant species per plot in Washington. When only one exotic plant species was found on a plot, the twelve 1 m² quadrats missed the species 66% of the time. On average, the quadrats also missed 32% (±1.4%) of the native plant species on each plot. These were primarily locally rare, patchily distributed plant species. Thus searching the larger subplots significantly improved measurements of native plant diversity and the early detection of exotic plant species.

Assessing Patterns of Plant Diversity Among States

Kruskal-Wallis test results identified significant differences among states in native and exotic plant species richness per plot—at least for this first set of test plots (Table 9.1). For example, Virginia had twice the native species per plot compared to Colorado and Michigan. The first set of plots in Washington had three times the exotic species richness of plots in Michigan. Again, these results are preliminary data from five states and sample sizes are small and may not be representative of a larger sample.

The first set of plots in the states also varied in the cover of native and exotic plant species (Table 9.1). The initial plots in Colorado had much lower foliar cover of native species compared to the other states. Exotic species cover was consistently different among states and highly variable locally, resulting in larger relative variances (coefficients of variation ranged from 37% in Virginia to 82% in Oregon).

Table 9.1.   Mean native and exotic plant species and foliar cover in the initial 111 plots in five pilot states [Kruskal-Wallis (KW) tests among states].

States	Native species	Exotic species	Total native cover (%)	Total nonnative cover (%)	No. of plots
Colorado	21.9 (1.5)	2.2 (0.4)	16.6 (2.1)	0.8 (0.4)	33
Michigan	23.2 (1.5)	1.5 (0.3)	41.2 (3.4)	4.7 (2.0)	37
Virginia	57.8 (7.1)	3.4 (0.8)	37.7 (7.6)	3.2 (1.2)	15
Oregon	35.6 (4.8)	4.7 (1.5)	52.2 (10.3)	9.7 (8.0)	14
Washington	37.5 (4.3)	5.2 (1.4)	65.0 (14.4)	2.6 (1.2)	12
Combined	30.6 (1.8)	2.8 (0.3)	37.4 (2.9)	3.7 (1.2)	111
KW	33.0	13.5	33.6	7.1	
P	< 0.001	< 0.009	< 0.001	< 0.129	

Standard in parentheses.

Assessing the Increased Area and Multiscale
Approach of the Design

For many plots, the number of native and exotic species was highly affected by the spatial scale (area) of the sample. For example, the exotic species richness might be fairly constant with increasing area (Oregon plot in Figure 9.3), or inconsistent with increasing area (Virginia plot in Figure 9.3).

This demonstrates an important benefit of a multiscale sampling design. The plot in Virginia suggests that invasive plant species are patchily distributed, with many coexisting invasive plant species at larger spatial scales compared to the Oregon plot. Based on the species-area relationship, we would expect to encounter several more exotic plant species in an area twice the size and adjacent to the selected plot in Virginia. Plant ecologists cannot assume consistent, linear increases in species richness with increasing area.

Relationship of Native
and Exotic Plant Diversity

The first year of data showed that areas of high native plant diversity were consistently more heavily invaded by exotic species than areas of low native diversity (Figure 9.4).

For the five pilot states combined, exotic species richness was significantly, positively correlated with native species richness ($r = 0.38, P < 0.001$). The plots in Colorado and Virginia had relatively strong relationships between native and exotic species richness. Plots in Oregon and Washington had weaker relationships, but still had positive slopes in the correlation between native and exotic species richness (Figure 9.3). Other factors such as distance to roads, riparian zones (Stohlgren et al. 1998d), and urban areas, etc., might improve these relationships.

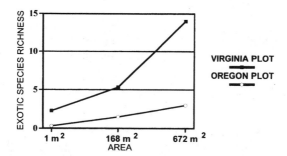

Figure 9.3. Example species-area relationship for exotic species in one plot in Oregon and one plot in Virginia. The 1 m² data points are means of the twelve 1 m² quadrats, followed by the mean of four subplots (168 m²), followed by the grand total of exotic species in the four subplots (672 m²).

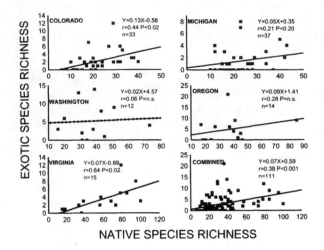

Figure 9.4. Relationship of native species richness to exotic species richness for test plots in five pilot states (and combined).

For the 111 plots in the five pilot states combined, plant species richness was significantly correlated to total foliar cover ($r = 0.46$, $P < 0.001$). Exotic species cover was significantly positively correlated to native species richness ($r = 0.24$, $P < 0.001$). The cover of exotic species was significantly, positively correlated to exotic species richness ($r = 0.64$, $P < 0.001$), indicating that the conditions favoring native species may also favor establishment of exotic species and their subsequent growth.

Native and exotic species cover varied significantly by land use type (Table 9.2).

Rangeland plots had low native species cover, while containing the greatest cover of exotic species. Woodland plots had the lowest native and exotic species cover compared to the other land use types (Table 9.2). Again, additional sampling is needed to confirm these patterns. Sites that were

Table 9.2. Example Kruskal-Wallis test results of mean species richness and foliar cover in stratum I in 240 quadrats 1 m^2 for land use types in Colorado.

Primary land use	Native species	Nonnative species	Native cover (%)	Nonnative cover (%)	No. of plots/ quadrats
Timber	3.4 (0.2)	0.2 (0.1)	11.7 (1.6)	0.9 (0.4)	10/120
Woodland	2.1 (0.2)	0.1 (0.0)	10.4 (2.1)	0.2 (0.1)	5/60
Rangeland	1.6 (0.4)	1.6 (0.4)	10.1 (4.7)	6.8 (2.5)	2/24
Timber/urban	2.4 (0.3)	0.6 (0.2)	15.8 (4.1)	2.4 (1.0)	3/36
KW	18.0	33.8	6.4	31.9	
P	< 0.001	< 0.001	< 0.093	< 0.001	

Standard errors in parentheses.

grazed and ungrazed by cattle could only be compared for 108 ponderosa pine/pinyon-juniper quadrats (only nine test plots) (Table 9.3) in Colorado. Ungrazed plots had significantly greater cover of exotic plant species compared to grazed plots in this small subsample. However, other environmental factors (e.g., light, water, nitrogen, soil disturbance, past land use) may have been different in the gazed and ungrazed sites, so the differences cannot be unequivocally associated only with grazing or lack of grazing.

Are these methods robust enough to monitor plant diversity in the nation's forests? The uniform sampling design in the five-state pilot program effectively detected important and alarming patterns that certainly justify continued monitoring. Exotic plant species have successfully invaded the most species-rich sites (Figure 9.3), the cover of exotic plant species was positively correlated with exotic species richness, and 69.4% of the plots contained at least one exotic plant species. These patterns have also been observed in local studies of Rocky Mountain National Park, Colorado, the Central Grasslands (Stohlgren et al. 1999a), and in selected riparian zones in the Central Grasslands (Stohlgren et al. 1998c,d) and the Pacific Northwest (DeFerrari and Naiman 1994). Still, it is only by systematic, unbiased monitoring at the national scale that national trends can be evaluated. The synthesis of results from small-scale experiments and local case studies would be insufficient in describing these broad-scale patterns.

With mandates and policies to protect native species diversity in the nation's forests, it may become more important to closely monitor hotspots of native and exotic plant species richness and cover. Only about 30% of the forested plots surveyed in the five states had no exotic plant species. Many of these plots may have had exotic species nearby in forest clearings or riparian zones, along roadways, or on rangelands. Often field crews did not record understory plant diversity in plots and subplots that were located in grasslands, shrublands, and other nonforested areas. Finding that species-rich areas are being more heavily invaded than species-poor areas (Figure 9.3) complicates exotic species control efforts in the nation's forests. More targeted chemical and biological control may be needed to battle exotic plant

Table 9.3. Mann-Whitney U-test results of mean species richness and foliar cover in stratum 1 in 108 quadrats (1 m^2) for grazed and ungrazed (by cattle) sites in ponderosa pine and pinyon-juniper vegetation types in Colorado.

Grazing status	Native species	Nonnative species	Native cover (%)	Nonnative cover (%)	No. of plots quadrats
Grazed	1.7 (0.2)	0.1 (0.0)	11.1 (2.9)	0.2 (0.1)	4/48
Ungrazed	2.5 (0.3)	0.3 (0.1)	12.1 (2.5)	1.3 (0.6)	5/60
U	1057	1286	1340	1318	
P	< 0.016	< 0.113	< 0.533	< 0.194	

Standard errors in parentheses.

invasions. In any case, the careful monitoring of vegetation types in the United States may allow for the rapid assessment of the vulnerability of various habitats to invasion by exotic species. This is especially needed to assess various land use practices and other factors such as ground disturbance by small mammals, proximity to disturbed roadsides and riparian zones, and proximity to infested urban sites, etc., as major contributors to the invasion process.

The real power in regional-scale analyses will come after several years of data collection and after some plots are resampled over several years to better evaluate spatial and temporal variation with larger sample sizes in multiple regions. This U.S. Forest Service monitoring program will continue to provide the first national-scale data on the plant diversity of the nation's forests. More importantly, the spatial attributes of the data allow for immediate use in controlling invasive plant species (Figure 9.5).

The spatial patterning of data can be immediately useful to land managers for the early detection of noxious weeds and the protection of rare/unique plant species and habitats, and to identify healthy forests versus those forests needing immediate management attention. Control efforts can take a

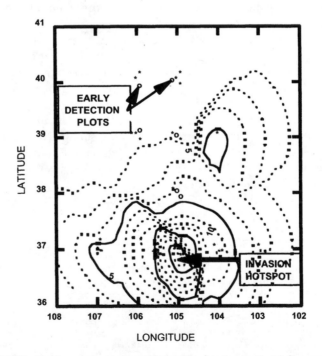

Figure 9.5. Example contour map of exotic plant species richness for plots in eastern Colorado. Kriging (SYSTAT 7.0) was used to interpolate between plots. Subplot locations are shown as asterisks except for subplots with only one exotic species, which are shown as a circle.

Figure 9.6.   Lesson 16. The hotspots of native plant diversity in the nation's forests are being invaded by non-native plant species. This story would have been missed if sampling continued to occur only in 1 m² quadrats that missed rare and patchily distributed non-native species in two-thirds of the plots. (Photograph by author)

two-pronged approach by focusing on small, newly established populations that are less expensive to treat, and on hotspots of invasion (i.e., large, dominating populations) where the threat of extirpation of native plant species is a concern (Figure 9.5).

This national-scale monitoring program is the first, statistically sound, quality assured, multiscale assessment of native and exotic plant diversity in the nation's forests. Greatly expanded datasets will document hotspots of native plant diversity and primary areas of invasion by exotic plant species, determine the effects of land use practices on plant diversity, and aid in evaluating the condition of the nation's forests. Such information is vital to adaptive management, prescribing control efforts for invasive species, and aiding in the preservation of native biodiversity. Another important benefit of the sampling design is increased comparability of data with other plant diversity monitoring programs. As a slight modification to this design, our field crews have been establishing one 168 m^2 subplot (including the three 1 m^2 quadrats) or pairs of subplots as a rapid assessment technique for native and nonnative plant diversity and distributions in small habitats (see http://www.NAWMA.org/documents/Mapping%20Standards/BEYOND%20NAWMA%20STANDARDS.pdf). These multiscale vegetation sampling methods are directly comparable to the multiscale vegetation sampling techniques used by other agencies.

The current sampling strategy with the uniform sampling grid should be augmented with stratified ground surveys in rare and important habitats and areas of high risk of invasion by exotic species (frequently missed by any systematic sampling grid), and by developing comparable datasets in nonforested areas. Based on this small case study, the number and cover of exotic species in this study is just 8–9% of total species richness, but some subplots had up to 100% cover of exotic species. Mapped displays of data (Figure 9.5) will become increasingly important in identifying and protecting areas of high native plant diversity (Figure 9.6), controlling invasive and noxious plant species, and synthesizing species-environment data for predictive modeling.

10

Case Study

Patterns of Plant Invasions in Forests and Grasslands

The Issue

Plant ecologists realize that invasive nonnative plant species threaten native biodiversity because they can poison livestock, clog waterways, compete with cash crops, and degrade rangelands (Westbrooks 1998). Managers of national parks, wildlife refuges, and other natural areas are equally concerned because of the potential negative effects of nonnative plant species on native plant diversity, wildlife habitat, native pollinators, fire regimes, and nutrient cycling (D'Antonio and Vitousek 1992; Stohlgren et al. 1999a; Vitousek 1990; Westbrooks 1998). Thus there is an urgent need to rapidly assess the vulnerability of natural landscapes and specific habitats to invasion (Loope and Mueller-Dombois 1989). Systematic surveys of where nonnative species have successfully invaded are needed to guide research, control, and restoration efforts. Again, since only a small portion of any large landscape or region can be affordably surveyed, modeled information on native and nonnative plant diversity, soil characteristics, topography, and climate may be needed to guide the management of invasive species in larger, unsampled areas (Chong et al. 2001; Stohlgren et al. 1997a). In this case study we carefully considered current theories, experimental evidence, and various sampling design strategies before initiating the field studies (Stohlgren 2002; Stohlgren et al. 2002).

Background and Sampling Considerations

How might plant ecologists develop hypotheses and evaluate gaps in existing knowledge in invasion ecology? Several mathematical models suggest that areas of high species diversity should be resistant to invasion by non-native species (Case 1990; Law and Morton 1996; Post and Pimm 1983; Rummel and Roughgarden 1983; Turelli 1981). The mathematical models generally claim that colonization by nonnative species should decline in the face of many strongly interacting species, which are thought to use resources more completely. However, it is difficult to measure competition or resource availability in the field, especially at landscape scales.

A few field studies and small-scale experiments have reported a negative relationship between native and nonnative species richness (e.g., Fox and Fox 1986; Tilman 1997; Figure 10.1). One small-scale experiment (Levine 2000) found a positive relationship between nonnative species success and native plant diversity in one riparian site in California. In a seed-addition experiment in mature oak savanna in Minnesota, Tilman (1997) found that invasibility correlated negatively with plant species richness ($n = 60$, 1 m^2 plots). Many of these observations, theories, and small-scale experiments could lead plant ecologists to believe that species-rich plant communities might somehow be less vulnerable to invasion by nonnative plants than species-poor communities because there might be no available

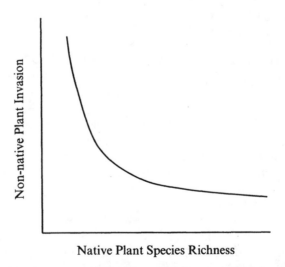

Native Plant Species Richness

Figure 10.1. Hypothetical relationship of native and non-native species richness in natural communities assuming similar environmental characteristics among sites and similar levels of immigration and turnover, and assuming competitive exclusion is a relatively strong force (i.e., available resources are more fully used in species-rich areas). Adapted from Tilman 1999.

resources or unused niches in species-rich areas (Grime 1973; Herben et al. 2004; MacArthur and Wilson 1967; McNaughton 1983, 1993; Tilman 1982, 1997).

However, a growing number of observational studies have demonstrated that locally not all species-rich vegetation types are immune to nonnative plant invasions. Robinson and Quinn (1988) and Robinson et al. (1995) showed that species-rich areas of annual grasslands in California were more easily invaded than species-poor areas. Timmins and Williams (1991) found that the number of weeds in New Zealand's forest and scrub reserves did not correlate with the number of native species. Recently our survey of five forest and meadow vegetation types in the Colorado Rockies and four prairie types in Colorado, Wyoming, South Dakota, and Minnesota reported more extensive nonnative plant invasions in species-rich vegetation types (Stohlgren et al. 1999a).

Our initial observations were limited to nine vegetation types (four 1000 m² study plots per type) (Figure 10.2), but they raise the possibility that at a local scale, some species-rich vegetation types could be invaded. To guide research, control, and restoration activities at landscape and regional scales, additional systematic surveys are badly needed to provide land managers

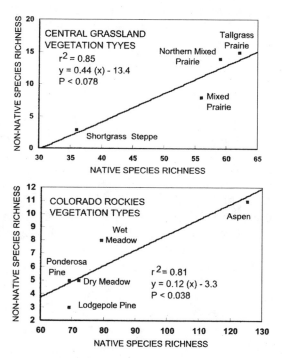

Figure 10.2.   Number of native and non-native species in sets of four 1000 m² plots in vegetation types in the Central Grasslands and Colorado Rocky Mountains. Adapted from Stohlgren et al. 1999a.

with information on the patterns and environmental factors associated with the successful invasion of nonnative plant species.

Studies of plant invasion may be affected by resolution (or grain), scale (or plot size), and extent (range of habitats studied, area of the region assessed). Small-scale, site-specific experiments have yielded contradictory results showing that species-rich areas can be either less invaded (Tilman 1999) or more invaded (Levine 2000), while global-extent studies suggest species-rich areas have been heavily invaded (Lonsdale 1999). The results of small-scale experiments and evaluations of regional floras have not been particularly useful to land managers who demand landscape-scale information on which habitats are (or may be) heavily invaded. It remains a top research priority of several land management agencies to conduct systematic surveys at multiple spatial scales (e.g., plot, landscape, and biome scales) for the early detection and management of nonnative plants.

The objective of this case study was to synthesize data from several studies to greatly expand the number and spatial distribution of sampling sites to assess patterns of nonnative plant invasions relative to vegetation type characteristics, topography, level of disturbance, and soil characteristics (e.g., soil texture, nitrogen, and carbon). We also develop explanatory models based on the data from the 22 vegetation types throughout the north-central United States from smaller datasets from four previous studies (Stohlgren et al. 1997b, 1998c,d, 1999b) that used the same multiscale vegetation and soil sampling methods.

Case Study Areas

Four different studies were conducted in nine areas (25 sets of four plots in 22 vegetation types) between 1995 and 1998 (Figure 10.3). At each site, four multiscale 20 m × 50 m vegetation plots were sampled as described below [see Stohlgren et al. (2002) for details].

Grassland Study Sites

Study locations included shortgrass steppe at the Central Plains Experimental Range (Pawnee National Grassland, Nunn, Colorado), mixed grass prairie at the High Plains Experiment Station (Cheyenne, Wyoming), northern mixed prairie in Wind Cave National Park (Hot Springs, South Dakota), and tallgrass prairie in Pipestone National Monument (Pipestone, Minnesota) (Stohlgren et al. 1998c).

Riparian Study Sites

There were four study locations: one in the shortgrass steppe at the Central Plains Experimental Range and three areas of northern mixed prairie in Wind

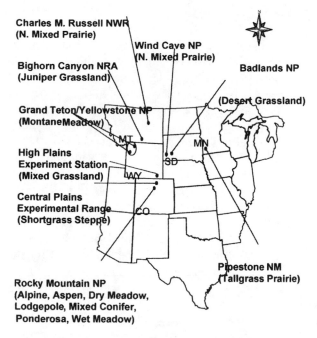

Figure 10.3.   Map of study sites from Stohlgren et al. (2002).

Cave National Park (South Dakota), Badlands National Park (South Dakota), and Bighorn Canyon National Recreation Area (Wyoming and Montana) (Stohlgren et al. 1998d).

Rocky Mountain National Park Sites

Seven vegetation types were characterized with three to five modified Whittaker plots in each type in Rocky Mountain National Park, Colorado (2500–3660 m elevation). These were lodgepole pine, aspen, ponderosa pine, wet meadow, dry meadow, mixed conifer, and alpine tundra (Stohlgren and Bachand 1997; Stohlgren et al. 1999a).

Grazing Study Sites

Modified Whittaker plots were placed in grazed and long-term ungrazed (more than 12 years of continued protection) at the Charles M. Russell National Wildlife Refuge, Montana; Yellowstone/Grand Teton National Parks, Wyoming; and Wind Cave National Park, South Dakota (Stohlgren et al. 1999b).

Case Study Methods

Modified Whittaker sampling plots were established at each site (see Figures 5.8 and 7.4) (Stohlgren et al. 1995b), placed with the long axis parallel to the environmental gradient. Foliar cover for each species in the understory and the percent bare ground were estimated to the nearest percent in the ten 1 m² subplots, and native and nonnative plant species were noted in each 1000 m² plot. Ancillary data recorded for each plot included UTM location from a GPS and elevation, slope, and aspect.

This case study included a variety of environmental data for each vegetation plot. Disturbance has been linked to plant invasions in many areas. So all plots were given a disturbance class rating as follows: 0, long-term exclosure; 1, almost no grazing by native ungulates or livestock; 2, light grazing; 3, moderate grazing; 4, heavy grazing or mowing; and 5, recent fire (past 3 years). Five soil samples were taken at each plot and analyzed for texture, carbon, and nitrogen [see Stohlgren et al. (2002)]. Climate data were gathered from the long-term weather station nearest to each site. Climate variables included mean October to June precipitation, July to September precipitation, maximum and minimum January temperatures, and maximum and minimum July temperatures [see Stohlgren et al. (2002)].

All statistical analyses were conducted with Systat (version 7.0; SPSS, Inc., Chicago, Illinois), and $P < 0.05$ was used to determine significance in all tests. Analyses were conducted at two scales. At the "plot scale" (0.1 ha scale), each of the 100 plots were entered into the regressions of nonnative species richness and cover as dependent variables and other plant, soil, and topographic variables were entered as independent variables. At the "vegetation type scale," the grand number of native and nonnative species (combined species lists from four 1000 m² plots in each vegetation type) and the mean soil, disturbance, and topographic characteristics were used in the multiple regressions.

Simple linear regressions were used at both scales to determine the relationships between nonnative and native species richness and cover, topographic variables, and soil characteristics and to predict nonnative species richness and cover as described in the previous case studies. In this case study we used path coefficient analysis (Dewey and Lu 1959) (see chapter 13 for more details) to evaluate the direct and indirect relationships of the environmental factors to nonnative species richness and cover. Again we relied on forward stepwise regression, the most widely used multiple regression model (Neter et al. 1990), to compare nonnative species richness and cover at multiple scales (plot scale and vegetation type scale) in a consistent manner. This regression model may not always result in the "best" regression model for all comparisons [see Neter et al. (1990, pp. 452–453)], but the reported relationships agreed with field observations. Path coefficient analysis simply displays the standardized partial regression coefficient (direct influence) of an environmental factor on the dependent variable, with significant ($P < 0.05$) simple correlation coefficients (indirect influences) shown among environmental variables. The residual factors from the stepwise linear regressions are not presented, but they are easily calculated as $R_{(X)} = \%(1 - R^2)$ (Dewey and Lu 1959; Stohlgren and Bachand 1997).

Results, Discussion, and Lessons Learned

As might be expected by broad-scale sampling, mean species richness varied considerably among plots and vegetation types (Table 10.1), a fact underscored by the range of values recorded. Native species richness ranged from 9 species/0.1 ha in an ungrazed northern prairie type in the Charles M. Russell National Wildlife Refuge in Montana to 71 species/0.1 ha in a grazed northern prairie type in Wind Cave National Park, South Dakota. Nonnative species richness ranged from 0 species/0.1 ha in the high elevation alpine type in Rocky Mountain National Park, Colorado to 18 species/ 0.1 ha in the riparian juniper-grassland type in Bighorn Canyon National Recreation Area, Wyoming. The greatest foliar cover of nonnative species was 50.5%, recorded in desert grassland in Badlands National Park in South Dakota, followed by 43.5% in tallgrass prairie at Pipestone National Monument, in Minnesota.

The wide variation in native and nonnative species richness can be combined with information on soil characteristics for a better understanding of the patterns of successful invasions. Soil characteristics also varied considerably among plots and vegetation types (Table 10.2), again highlighted by recorded ranges. Percent sand in the top 15 cm of soil, for example, ranged from only 2.4% in a northern prairie plot in the Charles M. Russell National Wildlife Refuge to 87.6% in a montane meadow plot in Montana (Yellowstone/Grand Teton National Parks). Conversely, percent clay ranged from 6.7% in an aspen plot in Rocky Mountain National Park to 89.3% in the Charles M. Russell National Wildlife Refuge. Percent total nitrogen in the soil ranged from 0.02% in the riparian juniper-grassland type to 0.8% in a wet meadow plot in Rocky Mountain National Park. Likewise, percent total carbon varied from 0.3% in the riparian juniper-grassland type to 14.2 % in a mixed conifer forest plot in Rocky Mountain National Park. How might soil characteristics affect nonnative plant invasions?

Multiple Factors are Correlated to Successful Invasions

There are several general patterns of invasion at the 1000 m² plot scale. There were significant positive relationships between nonnative species richness and native species richness, total soil nitrogen, and the silt + clay content in the soil (Table 10.3). Likewise, nonnative species cover was significantly positively correlated to those same variables at the plot scale (Table 10.3).

Table 10.1. Vegetation characteristics in one hundred 0.1 ha plots (25 sets of four plots in 22 vegetation types) in the northcentral United States.

Study area	Vegetation type	Native		Nonnative	
		Species richness	Percent cover	Species richness	Percent cover
Rocky Mountain National Park, Colorado	Alpine	25.0 (3.9)	74.4 (4.3)	0.0	0.0
	Aspen	50.2 (3.2)	38.7 (10.5)	5.5 (1.8)	6.4 (4.2)
	Wet meadow	35.0 (8.7)	79.0 (6.0)	6.5 (1.2)	5.8 (2.6)
	Lodgepole pine	29.0 (2.9)	8.0 (2.0)	1.5 (0.3)	0.3 (0.2)
	Dry meadow	41.5 (1.7)	44.3 (1.9)	2.0 (0.7)	0.6 (0.2)
	Ponderosa pine	34.8 (3.2)	15.6 (1.1)	2.0 (0.7)	0.2 (0.1)
	Mixed conifer	22.5 (6.1)	48.2 (15.8)	1.8 (1.2)	1.0 (1.0)
High Plains Experiment Center, Wyoming	Mixed grass prairie	35.5 (2.1)	45.9 (5.9)	2.5 (1.0)	0.04 (0.04)
Wind Cave National Park, South Dakota	Northern mixed prairie, grazed	41.6 (4.9)	31.9 (4.5)	6.8 (0.5)	20.8 (4.4)
	Northern mixed prairie, riparian	56.5 (5.4)	38.1 (3.4)	9.0 (0.8)	12.0 (4.8)
Charles M. Russell National Wildlife Refuge, Montana	Northern mixed prairie ungrazed	23.0 (5.9)	19.8 (4.9)	1.2 (0.5)	0.4 (0.2)
	Northern mixed praire, grazed	19.4 (2.8)	19.6 (3.2)	0.8 (0.3)	0.2 (0.1)

Location					
Yellowstone/Grand Teton National Park, Wyoming	Montane meadow, ungrazed	34.0 (1.7)	56.8 (6.4)	2.5 (1.4)	0.2 (0.1)
	Montane meadow, grazed	34.0 (3.4)	50.0 (5.8)	3.3 (0.9)	2.0 (1.0)
Central Plains Experimental Range, Colorado	Shortgrass upland (1996)	26.0 (1.8)	57.5 (4.1)	1.2 (0.4)	0.04 (0.01)
	Shortgrass upland (1997)	29.4 (1.9)	48.2 (5.3)	1.1 (0.4)	0.03 (0.01)
	Shortgrass riparian (1997)	44.5 (9.6)	31.6 (5.6)	3.0 (1.1)	0.16 (0.11)
Badlands National Park, South Dakota	Desert/mixed grass upland	26.0 (2.2)	22.7 (2.0)	8.8 (1.5)	21.6 (11.4)
	Desert/mixed grass riparian	33.0 (5.5)	37.8 (10.6)	9.5 (0.6)	2.4 (0.4)
Bighorn Canyon National Recreation Area, Wyoming, Montana	Mixed grass upland	24.3 (1.7)	20.6 (2.7)	2.8 (1.9)	1.2 (1.0)
	Mixed grass riparian	29.0 (4.8)	38.0 (6.3)	10.0 (2.9)	20.9 (10.7)
Pipestone National Monument, Minnesota	Tallgrass prairie	37.5 (4.1)	57.6 (12.1)	8.8 (1.0)	20.4 (8.3)

Mean of four to eight plots (standard error)

Aapted from Stohlgrcn et al. (2002).

Table 10.2. Soil characteristics in one hundred 0.1 ha plots (25 sets of four plots in 22 vegetation types) in the northcentral United States.

Study area	Vegetation types	C (%)	N (%)	Sand (%)	Silt (%)	Clay (%)	Elevation (m)	Disturbance (class)
Rocky Mountain National Park, Colorado	Alpine	2.8 (0.5)	0.1 (0.02)	65.8 (1.9)	14.0 (2.3)	20.2 (0.9)	3007 (63)	1.0 (0.0)
	Aspen	3.5 (0.8)	0.1 (0.01)	75.6 (1.4)	14.3 (2.2)	10.2 (1.5)	2704 (35)	1.0 (0.0)
	Wet meadow	5.6 (1.9)	0.4 (0.2)	59.7 (11.4)	18.9 (4.4)	21.4 (7.2)	2545 (4)	1.0 (0.0)
	Lodgepole pine	4.2 (0.6)	0.2 (0.1)	72.8 (0.8)	14.5 (1.4)	12.7 (2.1)	2665 (33)	1.0 (0.0)
	Dry meadow	1.5 (0.5)	0.1 (0.03)	72.2 (1.1)	15.7 (0.5)	12.1 (1.3)	2612 (42)	1.0 (0.0)
	Ponderosa pine	3.0 (0.8)	0.1 (0.04)	77.3 (3.4)	11.8 (1.5)	10.9 (2.1)	2625 (44)	1.0 (0.0)
	Mixed conifer	7.0 (2.5)	0.3 (0.1)	66.1 (3.5)	15.8 (2.0)	18.1 (3.0)	2723 (113)	1.0 (0.0)
High Plains Experimental Center, Wyoming	Mixed grass prairie	1.8 (0.1)	0.2 (0.01)	63.0 (1.8)	11.0 (1.6)	26.0 (2.9)	1944 (13)	2.5 (1.0)
Wind Cave National Park, South Dakota	Northern mixed prairie, grazed	3.1 (0.5)	0.3 (0.03)	23.1 (2.1)	31.9 (2.4)	45.0 (0.9)	1250 (30)	2.5 (0.5)
	Northern mixed prairie, riparian	2.8 (0.1)	0.2 (0.03)	61.4 (2.9)	15.1 (2.9)	23.5 (4.4)	1187 (26)	2.5 (0.5)

Location		Col1	Col2	Col3	Col4	Col5	Col6	Col7
Charles M. Russell National Wildlife Refuge, Montana	Northern mixed prairie, ungrazed	1.8 (0.2)	0.2 (0.01)	17.1 (6.7)	25.8 (1.3)	57.2 (7.4)	865 (27)	2.0 (0.0)
	Northern mixed prairie, grazed	1.4 (0.1)	0.2 (0.01)	19.3 (6.4)	23.1 (2.7)	57.7 (7.9)	863 (19)	3.6 (0.2)
Yellowstone/Grand Teton National Park, Wyoming	Montane meadow ungrazed	2.7 (0.5)	0.2 (0.05)	53.7 (4.7)	25.1 (2.9)	21.2 (3.2)	2077 (63)	0.0 (0.0)
	Montane meadow grazed	4.3 (0.7)	0.3 (0.06)	58.6 (4.7)	20.4 (3.5)	21.0 (2.4)	2063 (37)	2.5 (0.2)
Central Plains Experimental Range, Colorado (shortgrass steppe)	Upland (1996)	0.9 (0.1)	0.1 (0.01)	74.3 (2.8)	11.1 (1.5)	14.7 (1.5)	1644 (29)	3.0 (0.4)
	Upland (1997)	0.9 (0.1)	0.1 (0.01)	72.7 (3.4)	6.1 (1.4)	21.3 (2.2)	1645 (16)	3.1 (0.2)
	Riparian (1997)	0.9 (0.2)	0.1 (0.02)	56.7 (4.2)	14.6 (1.7)	28.7 (4.8)	1620 (9)	2.3 (0.3)
Badlands National Park, South Dakota (desert mixed grass)	Upland	1.5 (0.2)	0.1 (0.01)	24.2 (5.0)	30.4 (5.2)	45.3 (8.2)	795 (27)	1.8 (0.3)
	Riparian	1.5 (0.05)	0.1 (0.02)	28.0 (8.3)	26.0 (2.5)	46.5 (8.6)	783 (16)	2.3 (0.5)
Bighorn Canyon National Recreation Area, Wyoming (mixed grass)	Upland	2.2 (0.6)	0.1 (0.05)	53.5 (4.7)	9.9 (6.9)	36.6 (4.7)	1378 (58)	2.0 (0.0)
	Riparian	0.8 (0.2)	0.1 (0.05)	62.1 (8.6)	13.4 (6.2)	24.5 (2.7)	1161 (65)	3.5 (0.6)
Pipestone National Monument, Minnesota	Tallgrass prairie	4.4 (0.2)	0.4 (0.02)	11.7 (1.2)	54.1 (0.8)	34.3 (1.7)	498 (16)	2.8 (1.0)

Mean of four to eight plots (standard error).

Adapted from Stohlgren et al. (2002).

Table 10.3. Simple linear regressions of vegetation and soil characteristics as predictors of nonnative species richness and cover for the one hundred 1000 m^2 plots and the 25 sets of four plots in 22 vegetation types used in the study.

Dependent variable/predictors	Coefficient	t	P	F	r
1000 m^2 plot scale (n = 100 plots):					
Log$_{10}$ No. nonnative spp.					
No. of native spp.	0.013	5.08	0.001	26	0.46
Log$_{10}$ soil % N	1.69	2.12	0.036	4.5	0.21
Soil (% sand + % clay)	0.003	2.00	0.048	4.0	0.21
Log$_{10}$ cover nonnative spp.					
No. of native spp.	0.014	3.60	0.001	13	0.34
Log$_{10}$ No. nonnative spp.	1.12	11.9	0.001	142.7	0.77
Soil (% sand + % clay)	0.005	2.38	0.019	5.7	0.24
Log$_{10}$ soil % N	2.92	2.60	0.011	6.8	0.25
Vegetation type scale (n = 25 sets of four plots):					
Log$_{10}$ No. nonnative spp.					
No. of native spp.	0.125	2.63	0.015	6.9	0.48
Soil (% sand + % clay)	0.096	1.76	0.092	3.1	0.12
Log$_{10}$ cover nonnative spp.					
Log$_{10}$ No. nonnative spp.	0.08	6.80	0.001	47	0.82
Log$_{10}$ soil % N	4.7	1.74	0.095	3.0	0.12

In addition to soil characteristics, topographic, climatic, and geographic factors were correlated to invasion success. For the 100 plots throughout the study region, there was a significant negative relationship between elevation and nonnative species richness (log$_{10}$ nonnative species richness; $r = -0.32$, $F = 11.5$, $P < 0.001$). Only the high-elevation alpine type in Rocky Mountain National Park had no nonnative species in the sample plots.

At the plot scale, 50% of the variation in nonnative species richness was explained by native plant species richness, longitude, latitude, soil total nitrogen percentage, and mean maximum July temperature (Figure 10.4a). For the 100 plots, 60% of the variation in nonnative species cover was explained by nonnative plant species richness, elevation, and total soil nitrogen (Figure 10.4b).

There was also evidence that climate and soils influenced total species richness at the 1000 m^2 plot scale. About 25% of the variation in total species richness is explained by mean July maximum temperature, January minimum temperature, percent clay content in the soil (a possible surrogate for water-holding capacity), October to June precipitation, and July to September precipitation ($F = 6.8$, $P < 0.001$, df = 5 and 86). Thus vegetation plots encompassed a very broad range of vegetation types, soils, and climatic systems in the synthesis that follows.

0.1-ha PLOT SCALE

a. Estimating Non-Native Species Richness

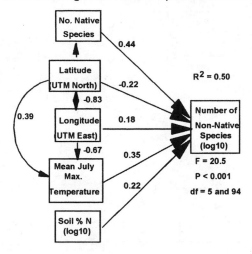

b. Estimating Non-Native Species Cover

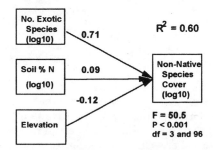

Figure 10.4.   Path coefficient diagram of environmental factors influencing non-native species richness (a) and cover (b) for one hundred 1000-m² plots in the north-central U.S. Direct arrows to non-native species richness or cover include standardized partial regression coefficient values, while arrows between environmental variables are simple correlation coefficients. R^2 is the adjusted coefficient of multiple determination. Adapted from Stohlgren et al. 2002.

Also, at the vegetation type scale (25 sets of four plots), nonnative species richness was significantly positively correlated to total native species richness and total soil nitrogen (Table 10.3). Compared to the plot-scale results, the relationship was stronger for native species richness and weaker for soil total nitrogen. Also at the vegetation type scale, 64% of the variation in nonnative species richness was explained by native plant species richness, elevation, and winter (October to June) precipitation (Figure 10.5a).

VEGETATION TYPE-SCALE

a. Estimating Non-Native Species Richness

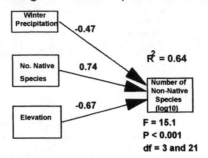

b. Estimating Non-Native Species Cover

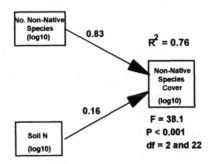

Figure 10.5. Path coefficient diagram of environmental factors influencing non-native species richness (a) and cover (b) for the 25 sites (vegetation-type scale) in the north-central U.S. Direct arrows to non-native species richness or cover include standardized partial regression coefficient values, while arrows between environmental variables are simple correlation coefficients. Adapted from Stohlgren et al. 2002.

One possible interpretation is that once several nonnative species become established, a habitat may be particularly vulnerable to greatly increased cover of invasive species or eventual invasion by a highly productive species (Figures 10.5b and 10.6). This explanation is supported by the strongly positive correlation between the nonnative species richness and foliar cover at the plot scale ($r = 0.77$, $P < 0.001$) and vegetation type scale ($r = 0.83$, $P < 0.001$) (Figure 10.4).

Regardless of the specific mechanisms involved in the invasion process, the patterns above (Figures 10.4 and 10.5, and Table 10.3) indicate that several biotic and abiotic factors may be interacting to affect the successful invasion of nonnative species. Single factors explain much less variation in nonnative species richness and cover (Table 10.3) than multiple factors (Fig-

Figure 10.6.   Relationship of non-native species richness and cover for the 25 vegetation types (sets of four 1000-m² plots) in the north-central U.S.

ures 10.4 and 10.5). It is also clear that different processes may be occurring at different spatial scales (e.g., the plot scale and the vegetation type scale). There are several factors that were not measured directly, such as plant biomass, seed bank or propagule pressure, resource availability for individual species, and plant competition for specific resources. It is possible that some of the factors that were directly measured are surrogates for some of the unmeasured variables. Still, more than half of the variability in nonnative species richness and cover can be obtained with relatively few biotic and abiotic factors in the study areas.

Reexamining the Relationship of Native Species Richness and Plant Invasion

In both grassland and montane biomes, species-rich sites have been heavily invaded at multiple spatial scales (Tables 10.1 and 10.3, and Figures 10.4a and 10.5a) (Stohlgren et al. 1999a). This alarming pattern may be more widespread than previously thought. A recent landscape-scale survey in arid, southwestern Utah found similar patterns, with heavy invasions in areas high in native species richness, rare habitats, and fertile soils (Stohlgren et al. 2001). Locally, species-rich riparian zones were more heavily invaded than species-poor upland sites nearby (Table 10.1) (Planty-Tabacchi et al. 1996; Stohlgren et al. 1998d). Of greatest concern, however, was a stronger correlation between native and nonnative species richness at the vegetation type scale compared to the plot scale. Species-rich vegetation types in the north-central United States appear to be highly vulnerable to invasion by nonnative species (Table 10.2). Conversely, there was little support for theories that areas of high species diversity might resist invasion by nonnative species

(Case 1990; Law and Morton 1996; MacArthur and Wilson 1967; Post and Pimm 1983; Rummel and Roughgarden 1983; Turelli 1981).

We didn't measure seed banks or seed dispersal, which surely influence patterns of plant diversity. Propagule pressure by native or nonnative species may be disproportionately higher in certain areas, but it is difficult and impractical to measure, monitor, regulate, or manage propagule pressure at landscape scales, especially for seeds that are ubiquitously distributed by wind, large and small mammals, and insects (Stohlgren 2002). The origin, autecology, and genetic variation of the nonnative invaders may also be important (Mack et al. 2000). In most of our study sites, disturbances such as grazing by ungulates and domestic animals had minimal effects on non-native species richness. Intensive grazing by cattle for more than 100 years on the relatively species-poor shortgrass steppe of Colorado has resulted in little invasion by nonnative species (Stohlgren et al. 1999b). There were no significant differences in native and nonnative species richness and cover between long-term grazed and ungrazed plots at the Charles M. Russell National Wildlife Refuge or in Yellowstone and Grand Teton National Parks (Table 10.1) (Stohlgren et al. 1999b). Nonnative species appear to be invading and thriving in both grazed and long-term ungrazed sites in our study areas (Stohlgren et al. 1999b). These sites tended to have long evolutionary histories of grazing (Milchunas and Lauenroth 1993).

Native species richness in an area is likely the result of habitat heterogeneity and available resources (Lonsdale 1999), seed supply (Coffin and Lauenroth 1989c; Tilman 1997), and many other factors such as disturbance history, land use, species migration and turnover, herbivory, competition, diseases, and pathogens (Stohlgren et al. 1999a). The attribute of "native species richness" may have little or no direct effect on invasion potential (Levine and D'Antonio 1999; Lonsdale 1999; Rejmánek 1996; Stohlgren et al. 1999a). However, this does not diminish the importance of native species richness as an indicator or predictor of habitat vulnerability to invasion (Table 10.3 and Figures 10.2 and 10.3). The simplest explanation may be that nonnative plant species thrive on the same resources (high light, nitrogen, and water) as native plant species (Stohlgren et al. 1997b, 1998d, 1999a, 2000b). While we do not fully understand the mechanisms and processes that create patterns of plant diversity, this case study points to an urgent and practical need to carefully measure native species richness at multiple scales.

Assessing the Role of Abiotic Factors in Plant Invasion

It is difficult to isolate and quantify potential causal factors from observational studies. Still, the relative effects of factors might be deduced by correlations of individual factors, such as soil nitrogen or disturbance, to response variables such as the number or cover of nonnative species. Disturbance, as my colleagues and I classified it, showed little statistical relationship to non-native species richness (Stohlgren et al. 2002). For the one hundred 0.1 ha plots, the number of plots from low (class 0) to high disturbance (class 5) were 5, 35, 26, 17, 15, and 2 plots. Despite this fairly broad range, there were no significant correlations between disturbance class and nonnative species richness at either the plot scale ($P = 0.56$, $F = 0.33$) or the vegetation type scale ($P = 0.33$, $F = 0.98$). This is not to say that disturbance is unimportant in the invasion process, it may merely draw attention to successful invasion of nonnative plant species in disturbed and undisturbed habitats (Stohlgren et al. 1999b), high variation in the biotic and abiotic factors in disturbed and undisturbed sites, or that disturbances at smaller or larger scales may influence plant diversity in different ways than we recorded in this case study (Stohlgren et al. 1999b, 2002). A simple explanation may be that many native and nonnative plant species may be well adapted to typical grassland and forest disturbances with which they evolved, such as grazing, fire, flooding, and small ground-dwelling mammals.

It may be that superficial disturbances, such as removal of aboveground biomass by herbivores, are less devastating than a plow, road grader, or excavations by small mammals. Traveling to our study sites we observed invasive plants along nearly all of the roadways and edges of agricultural lands. We detected no invasive plants at high-elevation alpine sites in Rocky Mountain National Park, Colorado, probably because most Mediterranean weeds in the area cannot tolerate cold temperatures (Stohlgren et al. 2000b). However, the common dandelion (*Taraxacum officinale*) can be found along high-elevation road cuts and trails. Other disturbances such as large-scale fire, insect and disease outbreaks, and flooding were not assessed in this case study, and they deserve more attention in future research (Hobbs and Huenneke 1992). Likewise, activities that increase available nitrogen on a site (e.g., fire, air pollution, fertilization) may promote invasion, especially if the site is near or connected to an already infested site.

Several abiotic factors are correlated to native species richness, such as soil nitrogen, elevation, and precipitation (Stohlgren et al. 2002). These factors are relatively easy to document over large areas and may greatly improve the precision and accuracy of spatial models of invasive species (Chong et al. 2001). That is, the nonnative species pool may include species that favor fertile sites. Escape from natural enemies may add to the success of nonnative species (Mack et al. 2000), but pathogens are more difficult to measure and monitor. Habitat characteristics are unquestionably important predictors of successful invasions (Table 10.3 and Figures 10.2 and 10.3), and they are relatively easier and less expensive to measure and monitor (Stohlgren et al. 1998d, 1999a, 2000b). Isolating the causes of the patterns reported here are beyond the scope of this observational study. Instead, we draw the plant ecologist's attention to the locations, habitats, and physical factors associated with the current patterns of successful invasion to aid in future control and restoration efforts (Stohlgren et al. 2000b).

Speculating on Theory, the Role of Species Turnover, and Plant Invasion

Community succession theory (Clements 1916) and competitive exclusion theory (Grime 1973) might assert that plant communities with high frequencies of co-occurring or "shared species" in confined areas may use resources more completely via niche differentiation and resource partitioning among the species. One would hypothesize that plant communities with few frequently occurring species would be more vulnerable to newly invading species, while communities with many frequently co-occurring species would sequester resources more completely, insuring against newly invading species. The data from the 25 vegetation types showed no correlation between the number of plant species shared among plots in a vegetation type and the number of nonnative species in the four plots (Figure 10.7a).

The theory of competitive exclusion might assert that areas of high native plant diversity would more completely occupy available niches and thus more resources, making few resources available for newly invading species. One would hypothesize a strong negative relationship between species richness and the number of nonnative species in a vegetation type. In direct contrast, the data from the 25 vegetation types showed a significant positive correlation between the number of native plant species (in four 1000 m^2 plots) and the number of nonnative plant species that have successfully invaded (Figure 10.7b). Twenty-four percent of the variation in nonnative species richness could be explained by a positive correlation with native species richness.

May (1973) argued that highly diverse communities are intrinsically unstable, with some species routinely dropping in and out. One would hypothesize that communities with many locally rare (or infrequent) plant species, and presumably high turnover rates, would be more susceptible to invasion than communities with few locally rare species. In this case study, as in many other field studies, my colleagues and I observed that most species in a plot had less than 1% foliar cover (Stohlgren et al. 2002). The data from the 25 vegetation types showed a strongly significant positive correlation between the number of native locally rare plant species (in four plots) and the number of nonnative plant species that have successfully invaded (Figure 10.7c). In fact, 35% of the variation in invading species richness could be explained by the richness of low-cover and locally sparse plant species. We also observed many species in cotyledon, seedling, and mature stages within and among plots. These small, young, individual plants and scattered subpopulations may be vulnerable to high turnover of individuals and local species composition. It is easy to imagine some native species dropping out and nonnative species replacing them. Theoretically many species can coexist as a result of biogenic small-scale heterogeneity and interactions among organisms for spatially and temporally variable resources (Huston and DeAngelis 1994), but species replacements also may occur in areas of

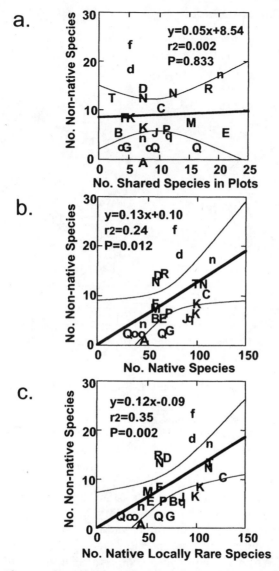

Figure 10.7. The number of non-native species in the 25 vegetation types (sets of four 1000-m2 plots) in the north-central U.S., relative to the number of shared species in each set of four plots, the number of native plant species, and the number of locally rare plant species (i.e., averaging <1% cover in the four plots). Codes are A = alpine, C = aspen, D = upland desert grassland d = riparian desert grassland, E = dry montane meadow, F = upland juniper grassland, f = riparian juniper grassland, G = lodge-pole pine (Pinus contorta), H = mixed conifer, J = ungrazed montane meadow, K = grazed montane meadow, M = mixed grassland, N = grazed upland northern mixed prairie, n = ungrazed riparian northern mixed prairie, o = grazed northern mixed prairie, P = ponderosa pine (Pinus ponderosa) Q = upland shortgrass steppe, q = riparian shortgrass steppe, R = tallgrass prairie, T = wet montane meadow.

Figure 10.8. Lesson 17. Non-native plant species have successfully invaded riparian zones (e.g., as in the *Poa pratensis* in Wind Cave National Park, South Dakota) and other hotspots of native plant diversity. High light, nitrogen, and water, the factors associated with high native plant diversity, are also associated with high non-native diversity and cover. The rich get richer! (Photograph by author)

high turnover (Palmer and Rusch 2001, Stohlgren et al. 1999a,b, 2000b; van der Maarel and Sykes 1993).

Plant ecologists and land managers may not be able to seek solace in the prevailing theories (and hope) that colonization by nonnative species in natural landscapes should decline in the face of many strongly interacting native species (Figure 10.8). Experiments should be designed to directly address observed patterns of native diversity and invasions in natural habitats. It may be important to conduct invasion experiments in habitats with high native species richness and soil fertility. Many questions must be resolved. Why were low-elevation areas more heavily invaded than high-elevation sites, and species-rich riparian zones more invaded than adjacent upland sites? Why are canopy gaps in coniferous forests, aspen stands, and productive montane meadows more heavily invaded than closed-canopy forests with sparse understory vegetation (Table 10.1) (Stohlgren et al. 1999a, 2002). Many questions are left for experimental plant ecologists.

11

Case Study

Evaluating the Effects of Grazing and Soil Characteristics on Plant Diversity

The Issue

Paired-plot designs are commonly used to compare the effects of fire, grazing, or other disturbances. Plots are placed in treated (or disturbed) and untreated (or control) sites and measured differences are meant to infer the direct effects of the treatments. As this next case study shows, such simple approaches are never as simple and straightforward as planned, but they do provide insightful information.

For more than half a century, plant ecologists have used exclosures (i.e., areas fenced to exclude wildlife or domestic livestock; Figure 11.1) as a means to evaluate the effects of grazing (Daubenmire 1940a; Heady 1968; Reardon 1996; Smith 1960; Stohlgren et al. 1999b). Such studies are vitally important to rangeland conservationists because of increased concerns about protecting native plant diversity (Bock et al. 1993; Stohlgren and Chong 1997) and preventing exotic species invasion and the spread of noxious weeds (Ellison 1960; Mack 1981; Stohlgren et al. 1998d). Comparisons of vascular plant diversity of grazed and ungrazed sites can yield important theoretical insights on the role of herbivory and competition in structuring plant communities (Belsky 1986; Harper 1969; McNaughton 1983). Commonly small plots or transects are placed on either side of a fence in a paired-plot or two-sample comparison of vegetation or soil (Bock et al. 1984; Chew 1982; Dormaar et al. 1994; Facelli et al. 1989; Heady 1968; Milchunas and Lauenroth 1993; Reardon 1996; Smith 1960; West et al. 1979). The two underlying assumptions of these types of studies are that (1) paired grazed and

Figure 11.1. Photograph by the author of the 30-year-old grazing exclosure (right) and adjacent heavily grazed winter range, Beaver Meadows, Rocky Mountain National Park, Colorado, USA. Lesson 18. Paired sample studies are never as simple or straightforward as planned, and the results are greatly affected by the sampling design and natural spatial variation.

ungrazed sites were identical prior to the advent of grazing, and (2) any measured difference between the paired grazed and ungrazed sites is due solely or primarily to the excluded grazers (Stohlgren et al. 1999b). Plant ecologists can indirectly test these assumptions and minimize the erroneous interpretation of data from exclosure studies (and most paired-plot study designs) by carefully designing field studies (Stohlgren et al. 1999b), but such assumptions are impossible to prove beyond a doubt.

Background and Sampling Considerations

Exclosure studies have produced inconsistent and questionable results because most exclosures are small (less than 1 ha), long-term grazing rates in adjacent rangeland are usually unknown, and studies have quantified grazing effects inconsistently (Daubenmire 1940a; Fisser 1970; Fleischner 1994; Woodward et al. 1994). The "long-term ungrazed" condition may be atypical for landscapes that evolved with grazing (Milchunas and Lauenroth 1993; Milchunas et al. 1988). Exclosures sometimes attract grazing animals so that grazing effects are accentuated adjacent to the exclosures. Exclosures may also differentially concentrate small herbivores and granivores (McNaughton 1983).

There are several other issues important to statistical inferences in grazing exclosure studies (Stohlgren et al. 1999b). First, most studies are poorly replicated (Webster 1992). Exclosures are expensive to construct, so they tend to be small and few, and thus are compromised by edge effects, unique local conditions, or poor representation of larger-scale processes (Daubenmire 1940a; Hughes 1996; Woodward et al. 1994). Large exclosures cover wider environmental conditions and incorporate larger-scale processes (e.g., patchy small mammal disturbance and seed dispersal effects), but their expense prohibits extensive replication. We are left with a few, small exclosures in different management units that were usually established to evaluate changes in forage production rather than plant diversity.

Most published studies using large exclosures involved pseudoreplication, where many small plots or transects were used on either side of a fence and statistical inferences were made about the "grazing effect" on the larger, unsampled landscape (Stohlgren et al. 1999b; Webster 1992). In our literature review of 28 exclosure studies, we found that 17 studies had a true sample size of $n = 1$ exclosure (Chew 1982; Dormaar et al. 1994; Frank et al. 1995; Lavado et al. 1996; Orr and Evenson 1991; Robertson 1971; Smeins et al. 1976; Tiedemann and Berndt 1972), while 20 studies relied on fewer than 5 exclosures (Chaneton and Lavado 1996; Cid et al. 1991; Collins and Adams 1983; Coughenour 1991; Evanko and Peterson 1955; Schulz and Leininger 1990; West et al. 1979). Sample sizes of $n = 1$ are obviously problematic except for very local inferences. Small sample sizes raise the possibility of atypical, site-specific results. Other problems of pseudoreplication have been reviewed by Hurlbert (1984) and Webster (1992).

Most exclosure studies used poor sampling techniques, particularly with respect to plant diversity. The actual area measured was usually quite small. For example, an evaluation of exclosures and grazed sites in Yellowstone National Park relied on actual measurements on only 3.5 m² per site (four 81.28 cm × 81.28 cm chart quadrats per site) on only five sites (Reardon 1996). Evanko and Peterson (1955) measured 3.0 m² per treatment (thirty 20 cm × 50 cm plots per treatment along a transect). An evaluation of grazed and ungrazed sites in Alberta, Canada (Dormaar et al. 1994) relied on just 1.5 m² per treatment (ten 0.1 m² plots, 10 m apart on a transect). As was shown in

Reminder: Pseudoreplication can occur when treatments are poorly replicated and samples are limited and nonindependent, and when inferences are made to the broader landscape or grazing regimes or other treatments that are not directly studied. My colleagues and I drew attention to three concerns limiting statistical inferences about potential grazing effects on plant diversity that have received little attention: subjective location of exclosures, poor sampling techniques, and inadequate assessment of spatial heterogeneity in the larger landscape (Stohlgren et al. 1999b). Most exclosures are subjectively located near roads and in flat terrain. This is what Krebs (1989) calls "accessibility sampling." None of the 28 studies we reviewed detailed the sample population (i.e., all potential sites from which the exclosure sites were selected). Questions arise whether local measurements of plant diversity are really typical of landscape-scale measurements of plant diversity (Brown and Allen 1989; Collins and Glenn 1997; Stohlgren et al. 1997b). Local species richness, for example, may be poorly correlated with landscape-scale species richness because species overlap and species richness are highly scale dependent (Collins and Glenn 1997; Stohlgren et al. 1997b, 1998a). Thus it is difficult to extrapolate information from subjectively selected sites to the broader landscape (Krebs 1989; Stohlgren et al. 1999b). Many study design flaws are preventable. Some randomized plots are needed.

chapter 4, it is unlikely that such small sample units tell a complete story. We did not find any exclosure studies that used well-replicated, large-plot designs in several vegetation types.

Lastly, nearly all exclosure studies confined sampling to a small study area within a much larger area of concern (Stohlgren et al. 1999b). The most extreme example we found was a series of ten 2 ha exclosures established in eastern Oregon. Only one 6.1 m × 6.1 m detailed chart plot was measured within each exclosure that had different external grazing regimes (Sneva et al. 1984). The use of tiny sampling areas is compounded when investigators attempt to evaluate different "treatment" effects without replicate exclosures (Webster 1992). Again, the major assumption is that the main effect detected by plant diversity or soil studies in paired plots or transects is due solely to the different grazing regimes on either side of a fence line [e.g., Bock et al. (1984), Dormaar et al. (1994), Hughes (1990, 1996) and nearly all exclosure studies we reviewed]. Do small plots capture all the spatial variation in plant composition at the sites or do they exaggerate differences between sites? Can the main effect of grazing by large mammals be isolated from natural spatial variation? Might other factors such as geography and soil quality also affect patterns of plant diversity, confounding the interpretation of grazing exclosure results?

Recognizing these problems, my colleagues and I designed a study that used standardized multiscale sampling techniques (Stohlgren et al. 1995b, 1998c) with improved replication (multiple exclosures at several areas in four states). Furthermore, we randomly selected plot locations inside and adjacent to the exclosures, with a third plot randomly located in a grazed area in the same vegetation type at each site. The third plot provided a means to assess spatial variation in grazed landscapes. Our objectives were to (1) examine several aspects of plant assemblages at multiple spatial scales in long-term grazed and ungrazed sites in several management areas; (2) determine the relative roles of grazing, soil characteristics, and climate in determining patterns of species richness; and (3) develop broad generalizations about the effects of grazing and cessation of grazing on plant diversity in typical grasslands in the Rocky Mountains (Stohlgren et al. 1999b). We hypothesized that native species richness would be lower in exclosures compared to grazed sites due to competitive exclusion in the absence of grazing (Grime 1973; Harper 1977; Figure 11.2).

We also hypothesized that grazed sites would have higher native and exotic (nonnative) species richness compared to ungrazed sites (Stohlgren et al. 1999b). First, grazing might reduce plant biomass and competition, and increase available nutrient and water resources, resulting in greater establishment of various plant species relative to long-term ungrazed sites (i.e., the intermediate disturbance hypothesis; Figure 11.2) (Connell 1978; Fox 1979; Grime 1973). Second, there is the long-standing paradigm, supported by studies in other regions, that grazing accelerates weed invasion

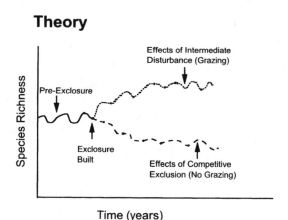

Figure 11.2.   Hypothetical relationship of mean species richness per plot over time and the potential effects of a grazing exclosure. Continued intermediate disturbance of grazers might increase species richness by adding early successional native species and non-native plant species. Inside the exclosure, some species would benefit more than others and competitively exclude smaller, locally rare species of plants.

(D'Antonio and Vitousek 1992; Daubenmire 1940b; Ellison 1960; Fleischner 1994; Hobbs and Huenneke 1992; Mack 1981). If grazed and ungrazed sites differed significantly in native or exotic species richness, cover, diversity, etc., and if they differed consistently among management areas, then grazing or cessation from grazing might be considered a major determinant of plant diversity. The comparison of long-term exclosures to long-grazed sites from highly productive to less-productive habitats (Huston 1979) should accentuate differences in plant diversity and recovery from grazing in Rocky Mountain grasslands.

Study Areas and Sampling Design

My colleagues and I surveyed federal land management agencies in Colorado, Wyoming, Montana, and South Dakota for potential study sites (Figure 11.3). Criteria for selection included more than 12 years continued protection from grazing [after McNaughton (1983)], exclosures larger than 0.4 ha in size, and moderate to moderately heavy grazing outside the exclosures in typical Rocky Mountain grasslands. We sampled 20 exclosures that met these criteria. The grazed areas included a variety of ungulates: Yellowstone (bison, elk, moose) and Grand Teton National Parks (bison, elk, moose, cattle) and Bighorn Basin Resource Area in Wyoming (cattle); Rocky Mountain National Park (elk, deer) and Uncompahgre (sheep) and Gunnison Resource Areas (cattle) in Colorado; Charles M. Russell National Wildlife Refuge (cattle), Custer National Forest/Pryor Mountain Wild Horse Range in Montana (wild horse); and Wind Cave National Park in South Dakota (bison, elk) (Figure 11.3).

Three large exclosures at Gunnison, Uncompahgre, and Rocky Mountain National Park contained multiple vegetation types, so additional plots were established in and adjacent to them. This created 26 exclosure sites in all. The sites were primarily montane grasslands, upper plains grasslands, and shrublands ranging in elevation from 776 m to 2675 m. The exclosures averaged 31.2 ± 2.5 years old (range 7–60 years) (Stohlgren et al. 1999b). The study sites likely had been at least moderately grazed in historic and presettlement times (Buchholtz 1983; Coughenour and Singer 1996). Although exact grazing intensity could not be determined for the sites, discussions with resource managers and field observations of plant cover and height confirm the sites to be at least moderately grazed annually.

To assess climate differences at the sites, precipitation and temperature data were gathered from the nearest long-term (more than 30 years) weather station for each management unit. Variables included long-term mean minimum and maximum temperature in January and July, winter/spring precipitation (October to June), and summer/fall precipitation (July to September).

Three multiscale vegetation plots were placed at each exclosure site. Modified Whittaker plots (see Figure 6.1) were randomly placed inside and

Figure 11.3.   Map of the study areas and associated vegetation types.
Adapted from Stohlgren et al. 1999b.

adjacent to each exclosure. Where the topography varied considerably inside or outside exclosures, plots were paired on similar aspects, slopes, and
elevations. Edges of fences were avoided by at least 2 m. A third plot for
each exclosure pair was randomly located within 1.5 km of the exclosure in
the similarly grazed landscape in the same vegetation type and with a similar slope, aspect, and elevation. The paired plots on either side of the fenced
exclosures averaged 100 m (±10 m; 1 SE) apart, while the two grazed plots
averaged 940 m (±230 m) apart.

Many multiscale designs could have been effectively used in this study,
but we used the original modified Whittaker plot (see Figure 6.1). We sampled
each site as close to the phenological maximum (peak biomass) as possible
(Stohlgren et al. 1999b). Plant species that could not be identified in the field
were collected and identified at the herbarium at Colorado State University
(Department of Biology). Fewer than 5% of the specimens encountered could
not be identified to the species level because of the phenological stage or
missing flower parts. In these cases, plants were identified to the genus level

and treated as individual species. Ancillary data recorded for each plot included UTM location and elevation from a GPS, slope, and aspect.

For each management area, my colleagues and I used ANOVA to test for differences in native and exotic species richness and cover, and exotic species frequency between exclosed sites, adjacent grazed sites, and randomly located grazed sites in the same vegetation types (Stohlgren et al. 1999b). In this abbreviated version of the study, I focus on the combined data from all management units for a summary ANOVA of ungrazed, grazed adjacent, and grazed distant plots. All statistical analyses were conducted with Systat (version 6.0; SPSS, Inc., Chicago, Illinois), and $P < 0.05$ was used to determine significance in all tests. Tukey's test was used as a means comparison test when the F-test was significant. We analyzed 1 m² subplot data and 1000 m² plot data separately to assess the scale dependency of the results (Stohlgren et al. 1999b). The same ANOVA approach was used to compare differences in species composition, diversity, evenness, and cover by life-form group, as described below. Exotic species richness and cover data, and distances between plots were not normally distributed, so those values were transformed ($\log_{10}$) prior to analysis.

Species Composition, Diversity, and Evenness

Jaccard's coefficient (Krebs 1989) was used to compare species overlap (1000 m² plot data) between exclosed plots and adjacent plots, adjacent plots and the randomly located grazed plots in the same vegetation types, and exclosed plots and the randomly located plots (see chapter 4). We selected Jaccard's coefficient over other similarity indexes because all species are equally important and we have found that more than 50% of all species sampled have less than 1% cover and few species have more than 5% cover (Stohlgren et al. 1997b, 1998c). We have found that in vegetation types with high even-

---

Reminder: Biological data often are not normally distributed in the statistical sense. Various transformations can be used to improve the normality of data prior to analysis (Zar 1974). In this case study, there were many "zeros" for the number of nonnative plant species richness and cover in 1 m² subplots and 1000 m² plots, so the appropriate transformation may be $\log_{10}(x + 1)$, where x is nonnative species richness or cover in a subplot or plot. In other cases, an arcsine or square-root transformation may be appropriate (Zar 1974). The resulting transformed data may not entirely conform to a normal distribution, but many statistical tests are fairly robust despite skewed distributions. Most statistical software packages can test the deviation from normality, and this is highly recommended.

ness, Jaccard's coefficient is an appropriate measure of similarity (Stohlgren et al. 1997a,b).

We used two diversity indexes and one evenness index recommended by Ludwig and Reynolds (1988). N1 is an index of the number of abundant species, while N2 measures the number of very abundant species. We measured foliar cover as a measure of abundance, so the diversity indexes measure the number of high-cover species and very high-cover species, respectively. These indexes complement the comparisons of total species richness or Jaccard's coefficients between grazed and ungrazed sites by down-weighting locally rare species. Thus higher N1 or N2 values would indicate higher dominant species diversity. Following Ludwig and Reynolds (1988), N1 was calculated as

$$N1 = e^{H'}$$

where H' (Shannon's index) for a sample is defined as

$$\hat{H}' = -\sum_{i=1}^{S}\left[\left(\frac{n_i}{n}\right)1n\left(\frac{n_i}{n}\right)\right],$$

where n_i was the cover of the ith species of S species in the sample and n is the total cover of all species in the sample. N2 is calculated as

$$N2 = 1/\lambda,$$

where λ (Simpson's index) for a sample is defined as

$$\hat{\lambda} = \sum_{i=1}^{S}\frac{n_i(n_i-1)}{n(n-1)},$$

where n_i and S were defined as above.

We used the modified Hill's ratio (E5) (Ludwig and Reynolds 1988) as an index of evenness where

$$E5 = \frac{(1/\lambda)-1}{e^{H'}-1} = \frac{N2-1}{N1-1}$$

and where E5 approaches zero as one species becomes increasingly dominant in foliar cover, thus higher E5 values indicate greater evenness in foliar cover among species.

Assessing Soil Quality in Grazed and Ungrazed Sites

Soil texture sampling and analysis can be used to assess the physical similarities between treatments (i.e., grazed and ungrazed sites), since soil texture may be slow to change due to grazing. In this case study, five soil samples were taken in each modified Whittaker plot (in the corners and the plot center) and combined. The surface litter, if present, was removed and the top 15 cm of soil was sampled with a 2.5 cm soil core. Particle size analysis was based on the standard hydrometer method (Gee and Bauder 1986). Samples were air dried for 48 hours, sieved with a standard #10 (2 mm pore size) sieve, ground in a standard three-ball grinder, and then oven dried at 55°C for 24 hours. Samples were analyzed for particle size (i.e., percent sand, silt, and clay), percent total carbon, and percent total nitrogen [see Stohlgren et al. (1999b) for details].

Multiple regression was used to determine the relationship of the number and cover of native or exotic plant species to soil characteristics for all 1000 m² plots in the study region. We tested the significance of each predictor with t-values (i.e., against the null hypothesis that the slope equals zero). We used stepwise forward multiple regressions to assess the ability of native species richness and cover, soil characteristics, elevation, and climate variables to predict exotic species richness. Soil characteristics included total nitrogen, total carbon, and percent sand, silt, and clay (only two soil texture variables were added into each model to reduce multicolinearity).

The forward linear regression models included only variables meeting the $P < 0.15$ criterion (Neter et al. 1990). Data distributions that were strongly skewed were transformed prior to analysis. Log_{10} transformations were used on percent nitrogen and exotic species cover to improve normality.

Results, Discussion, and Sampling Lessons Learned

My colleagues and I hypothesized that native species richness would be lower in the exclosures compared to grazed sites due to competitive exclusion in the absence of grazing (Figure 11.2). We also hypothesized that grazed sites would have higher native and exotic species richness compared to ungrazed areas due to disturbance (i.e., the intermediate disturbance hypothesis) and the conventional wisdom that grazing may accelerate weed invasion. Both hypotheses were soundly rejected at large spatial scales. Native species richness in 1 m² subplots was significantly higher ($P < 0.05$) on grazed sites, but only at the 1 m² scale (Table 11.1). However, there were no significant differences in the exotic species richness or the cover of native or exotic species in grazed and ungrazed sites (Stohlgren et al. 1999b).

Reminder: If two or more complementary variables such as percent sand, silt, and clay generally add up to 100%, at least one of the variables should be excluded in multiple regression analysis.

Table 11.1.   Mean number and cover of native and nonnative species in 260 subplots (1 m²) each in ungrazed exclosures, grazed adjacent plots, and grazed distant plots in nine study sites.

Grazing regime/ vegetation characteristic	Ungrazed	Grazed adjacent	Grazed distant
No. of native species	7.1[a]	8.3[a]	8.5[a]
	(0.3)	(0.3)	(0.3)
No. of nonnative species	0.9	1.1	0.9
	(0.1)	(0.1)	(0.1)
Cover of native species	40.1	38.6	37.1
	(2.1)	(1.6)	(1.8)
Cover of exotic species	8.4	7.5	5.4
	(1.0)	(1.0)	(0.7)

[a]Native species richness was significantly different between grazed and ungrazed sites ($F = 6.3$; df = 2 and 777; $P < 0.002$) at $\alpha = 0.05$ with Tukey's test.
Standard errors are in parentheses.

Adapted from Stohlgren et al. (1999b).

Plant species richness, cover, and frequency varied at multiple spatial scales throughout the study region. At the 1 m² scale, native species richness ranged from 1.9 ± 0.2 species/m² (mean ± 1 SE) in the ungrazed sites of the Uncompahgre area to 18.8 ± 1.2 species/m² in the grazed sites of the Wild Horse area (Stohlgren et al. 1999b). However, there were only a few consistent differences found between grazed and ungrazed sites in the number and cover of native and exotic species in any of the management areas. For example, native species richness in Grand Teton National Park averaged 8.8 native species/m² (±0.6 species/m²) in exclosures, while adjacent-grazed plots averaged 11.8 (±0.8) native species/m² and distant-grazed plots averaged 10.1 (±0.9) native species/m². There also was consistently greater richness and cover of exotic species in grazed sites in Grand Teton National Park compared to the ungrazed exclosures. The cover of native species was greater in the Yellowstone exclosures compared to grazed sites, but the reverse was true of the Wild Horse area (Stohlgren et al. 1999b).

On average, exotic species represented about 10–12% of the species per square meter and 13% (grazed distant plots) to 17% (ungrazed exclosure plots) of the foliar cover. The cover of exotic species in grazed distant sites was almost significantly lower than ungrazed sites (Table 11.1; $F = 2.3, P < 0.07$). This relationship was heavily influenced by the 24.8% mean cover of exotic species in exclosures in Rocky Mountain National Park.

At the 1000 m² plot scale, we found nearly identical native and nonnative (exotic) species richness in grazed and ungrazed sites (Table 11.2). We also found nearly identical frequencies of invasive species in grazed and ungrazed sites. The frequency of nonnative species was calculated as the mean number of 1 m² subplots/plot (minimum = 0; maximum = 10). Thus about half

Table 11.2.　Mean number of native and exotic species and frequency of
nonnative species (in ten 1 m^2 subplots) in twenty-six 1000 m^2 plots each in
ungrazed exclosures, grazed adjacent, and grazed distant plots in nine study sites.

	Ungrazed	Grazed adjacent	Grazed distant
No. of native species	31.5	32.6	31.6
	(2.5)	(2.8)	(2.9)
No. of nonnative species	3.1	3.0	3.2
	(0.5)	(0.6)	(0.5)
Frequency of exotic species	5.0	5.0	4.8
	(0.8)	(0.8)	(0.8)

Standard errors are in parentheses.

Adapted from Stohlgren et al. (1999b).

the subplots in grazed and ungrazed sites contained at least one exotic
species.

Mean native species richness in 1000 m^2 plots ranged from 9.0 ($\pm$1.2)
species/plot in the grazed sites of the Uncompahgre to 50.0 ($\pm$9.0) species/
plot in the grazed sites of the Wild Horse area (Stohlgren et al. 1999b). Mean
exotic species richness in 1000 m^2 plots ranged from 0.5 ($\pm$0.5) species/plot
in grazed and ungrazed sites of the Wild Horse area to 7.5 ($\pm$0.5) species/
plot in the grazed sites of Wind Cave. The frequency of exotic species (out
of 10 subplots in each plot) ranged from less than 10% in ungrazed plots at
the Charles Russell area, to 100% in grazed plots in the Wind Cave, Uncompahgre, and Bighorn Basin areas (Stohlgren et al. 1999b).

We also found no significant differences in species diversity (Hill's diversity indices N1 and N2) or evenness (Hill's ratio of evenness, E5) in grazed
and ungrazed sites. For all management areas combined, there were no significant differences between grazed and ungrazed plots in diversity or evenness, although mean diversity values were typically greater by about one
dominant species per plot in grazed sites (Table 11.3).

Table 11.3.　Mean Hill's diversity index values (N1 and N2 species equivalents)
and mean modified Hill's ratio for evenness (E5) in ungrazed, grazed adjacent, and
grazed distant 1000 m^2 plots.

Grazing Regime	Ungrazed	Grazed adjacent	Grazed distant
Number of abundant	9.0	10.1	10.2
species (N1)	(0.9)	(1.1)	(1.3)
Number of very abundant	7.8	8.7	9.3
species (N2)	(1.1)	(1.1)	(1.4)
Evenness (E5)	0.83	0.82	0.90
	(0.08)	(0.05)	(0.11)

Standard errors are in parentheses.

Adapted from Stohlgren et al. (1999b).

Mean N1 diversity index values ranged from 2.8 (±0.1) in ungrazed plots in the Uncompahgre area to 23.5 (±5.1) for grazed plots at Wild Horse. Mean N2 diversity index values were consistently lower than N1 values for all areas except the Charles Russell and the grazed distant plots at Wild Horse. Evenness values ranged from 0.5 (±0.01) at the Gunnison area to 1.2 (±0.4) for ungrazed plots at the Charles Russell area. Four management units (Grand Teton, Wind Cave, Rocky Mountain, and Uncompahgre) had slightly higher plant diversity in grazed sites, while three management areas (Charles Russell, Gunnison, and Bighorn Basin (by N1) had slightly lower plant diversity in grazed sites. Only two management areas (Wind Cave and Rocky Mountain) had consistently higher evenness in grazed sites (Stohlgren et al. 1999b). My colleagues and I hypothesized that forb and grass cover would be lower in grazed areas, while the cover of shrubs and bare ground would be higher in grazed areas. We found variable, insignificant differences in life-form composition between grazed and ungrazed plots in each management area and for all sites combined (Table 11.4).

For example, the cover of forbs was consistently higher in grazed compared to ungrazed plots in four of nine management areas, lower in two management areas, and inconsistent in three management areas. For percent bare ground, five of nine management areas had similar or inconsistent values for grazed and ungrazed plots. The frequency of forbs in 1 m^2 subplots was consistently higher in grazed plots (adjacent, 15.1 ± 1.6%; distant, 13.8 ± 1.7%) versus ungrazed plots (12.9 ± 1.5%). The frequency of grasses in 1 m^2 subplots also was consistently higher in grazed plots (adjacent, 6.6 ± 0.5%; distant, 6.8 ± 0.5%) compared to ungrazed plots (6.3 ± 0.5%). In contrast, the frequency of shrubs in 1 m^2 subplots was consistently lower in grazed plots (adjacent, 2.3 ± 0.3%; distant, 2.1 ± 0.4%) compared to ungrazed plots (2.7 ± 0.4%). However, none of the frequency means were significantly different between grazed and ungrazed plots.

Table 11.4.   Mean life form cover (%) and mean bare ground (%) by management area and grazing regime.

Grazing regime/life form	Ungrazed	Grazed adjacent	Grazed distant
Forb cover	14.8	16.6	13.9
	(3.8)	(3.1)	(2.8)
Grass cover	21.9	21.2	19.1
	(2.8)	(3.0)	(2.7)
Shrub cover	14.6	8.9	9.5
	(2.9)	(1.9)	(1.8)
Bare ground	53.7	52.8	57.3
	(5.8)	(6.3)	(6.1)

Standard errors are in parentheses.

Adapted from Stohlgren et al. (1999b).

Table 11.5.　Soil characteristics (top 15 cm) by grazing regime for the management areas.

Grazing regime/soil characteristic	Ungrazed	Grazed adjacent	Grazed distant
Percent sand	44.4	43.1	48.1
	(3.9)	(3.9)	(3.7)
Percent silt	20.8	21.2	17.2
	(1.8)	(1.6)	(2.3)
Percent clay	34.7	35.7	34.9
	(3.4)	(3.8)	(3.2)
Percent nitrogen	0.19	0.23	0.18
	(0.02)	(0.03)	(0.02)
Percent carbon	2.24	2.65	2.41
	(0.20)	(0.38)	(0.28)

Standard errors are in parentheses.

Adapted from Stohlgren et al. (1999b).

Some management areas had more consistent patterns of life-form changes than others. Grazed plots in Grand Teton National Park, for example, had consistently higher cover and frequency of forbs and grasses, more bare ground, and lower cover and frequency of shrubs compared to ungrazed plots. Grazed plots in Rocky Mountain National Park had consistently lower cover of forbs, grasses, and shrubs, but higher frequency of forbs and grasses than ungrazed plots. In the Yellowstone, Grand Teton, Rocky Mountain, and Bighorn Basin areas, and for all management areas combined, forb and grass frequency increased as shrub frequency decreased in grazed plots. Still, none of the comparisons of life-form composition was significantly different between grazed and ungrazed plots, and similarities and inconsistencies dominate most comparisons (Table 11.4).

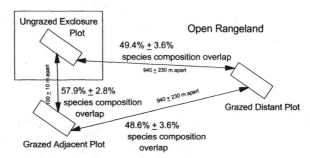

Figure 11.4.　Schematic diagram and mean species composition overlap (±1 SE) among ungrazed exclosures, grazed-adjacent sites, and grazed-distant sites. Adapted from Stohlgren et al. 1999b.

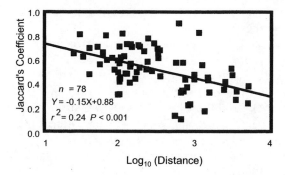

Figure 11.5.   Relationship of species composition overlap (J) and distance the plots are apart (within a management area). Adapted from Stohlgren et al. 1999b.

We found no significant soil characteristic differences between grazed and ungrazed sites (Table 11.5). The similarity in soil textures between grazed and ungrazed areas provides support for the underlying assumption that the exclosure sites were constructed in areas fairly similar in physical properties compared to the long-term grazed areas.

Species Composition Differences

We anticipated extreme differences in plant species composition between grazed and ungrazed sites, and virtually no differences between pairs of grazed sites (Stohlgren et al. 1999b). The species lists of the long-ungrazed and adjacent-grazed plots overlapped just $57.9 \pm 2.8\%$ (Figure 11.4). This difference in species composition is commonly attributed solely to the difference in grazing regimes. However, the species lists between pairs of grazed plots (adjacent and distant 1000 m^2 plots) in the same vegetation type overlapped just $48.6 \pm 3.6\%$ and the ungrazed plots and distant grazed plots overlapped $49.4 \pm 3.6\%$.

The distance between plots within each sampling area had a significant effect on species composition overlap. For all pairs of plots for each exclosure, species composition overlap significantly decreased with increasing distance apart (Figure 11.5). Thus the species lists of any two 1000 m^2 plots only 100 m apart would be expected to overlap just 58% on average, regardless of the grazing regime.

Rating Other Factors Relative to a Grazing Effect

Differences in vegetation and soils between grazed and ungrazed sites were minimal in most cases, but soil characteristics and elevation were strongly correlated to native and exotic plant diversity in the study region. For the seventy-eight 1000 m^2 plots, 59.4% of the variation in total species richness was explained by silt percent (coefficient = 0.65, $t = 5.11$, $P < 0.001$), elevation (coefficient = 0.01, t = 5.08, P < 0.001), and total foliar cover (coefficient = 0.11, t = 2.10, P < 0.039).

Native species cover and exotic species richness and frequency were also significantly positively correlated to soil percent nitrogen at the 1000 m² plot scale. Exotic species richness was positively correlated to native species richness at the P = 0.1 level of significance, but little of the variation was explained (r^2 = 0.04). Exotic species frequency was strongly correlated with exotic species richness (r = 0.70, P < 0.0001), suggesting that more invasive species meant wider distributions of the species.

How important were soil differences? Soil texture (percent sand, silt, and clay) and soil fertility (percent nitrogen and carbon) did not vary significantly between grazed and ungrazed plots. However, soil characteristics varied greatly among management areas. For example, mean percent sand varied from less than 30% in the Charles Russell area to greater than 65% in Rocky Mountain National Park. Mean soil nitrogen ranged from 9% to 11% in Uncompahgre to greater than 25% in Wind Cave National Park. Mean soil percent carbon also varied considerably within the study region.

For the nine management units in four states, species richness and cover were strongly correlated to climate. Mean exotic species cover was positively correlated with mean maximum January temperature (r = 0.62, P = 0.07, n = 9). Total plant cover was significantly negatively correlated with mean maximum July temperature (r = −0.50, P = 0.019), and total species richness was significantly positively correlated with total plant cover (r = 0.78, P = 0.013, n = 9). Factors other than grazing seemed to have greater effects on plant diversity than did long-term grazing.

Species-Specific Responses to Grazing Can Be Inconsistent

Few plant species showed consistent, directional responses to grazing or cessation from grazing. Examples of dominant species cover from Yellowstone, Bighorn Basin, and Wind Cave show highly inconsistent responses to grazing. For example, *Festuca idahoensis* was the dominant species inside the Blacktail and Junction exclosures in Yellowstone, with 9.5% and 24.6% cover, respectively. The species had much lower foliar cover in grazed plots near those exclosures. However, for the Lamar site, cover of *F. idahoensis* was higher in grazed plots relative to ungrazed plots. Typical of the response by most species in all management units, the cover of *Agropyron spicatum* varied greatly at the three exclosure sites at Yellowstone, with inconsistent patterns in grazed or ungrazed plots.

The cover of *A. spicatum* at Bighorn Basin was also variable. It averaged 3.8–11.4% cover inside exclosures, less than 3.5% to 4.5% in adjacent plots, and 5.7–8.7% at the random plot locations. The cover of the native species *Grindelia squarrosa* (curlycup gumweed) was low in ungrazed plots and grazed distant plots, but high in grazed adjacent plots (Figure 11.6).

Figure 11.6.   Photograph by Lisa D. Schell, used with permission, of *Grindelia squarrosa* (curly-cup gum weed) outside the grazing exclosure at Bighorn Basin, Wyoming.

The cover of another exotic grass, *Bromus tectorum* (cheatgrass), like many plant species, was also highly variable in the Bighorn Basin area. Due to patchy distributions, it had developed either high or low cover in grazed and ungrazed sites.

The exotic, sod-forming Kentucky bluegrass (*Poa pratensis*) was one of the few dominant species with a consistent pattern: it had higher cover in ungrazed plots at Bighorn Basin and Wind Cave. In contrast, the foliar cover of the exotic annual grass *Bromus japonicus* was as much as three times greater in sites grazed by bison, elk, and deer compared to ungrazed plots. The frequency (1 m² subplots) of *B. japonicus* was nearly twice as great in grazed versus ungrazed plots.

For all management areas, the four dominant species at the exclosure sites were different for each exclosure, adjacent plot, and randomly selected plot. Where plots had species in common, they often switched the order of dominance and were variable in cover. Patchy shrub species (e.g., *Artemisia tridentata*) and locally rare species had even more inconsistent patterns between grazed and ungrazed sites. Species seemed to respond individualistically to grazing.

Interpreting Observational Data

Observational studies of plant diversity involve carefully interpreting complex results with other studies. The section that follows is an example from our original journal article (Stohlgren et al. 1999b) as one approach in constructing a comparative discussion.

My colleagues and I acknowledged several assumptions and caveats of this case study and other paired-plot designs. This was an observational study, and the "effects of grazing" were not directly measured on vegetation or soils. Remember, the primary assumptions of exclosure studies are that (1) the vegetation and soils were initially similar on the grazed and ungrazed plots, and (2) vegetation and soil differences measured in subsequent years are presumed to be primarily caused by grazing in the grazed plots and by cessation of grazing in exclosed sites. It was reassuring to find similar soil textures, total percent nitrogen, and total percent carbon in grazed and ungrazed plots (Table 11.5) as a more consistent basis to assess differences in vegetation among treatments. However, we realize the possibility, albeit slim, that the grazed and ungrazed sites were initially different in mean soil quality, only to become more similar with either grazing or cessation of grazing. The data strongly confirm that no two sites are botanically identical (Figures 11.4 and 11.5), grazing is inherently heterogeneous, and current vegetation patterns represent a complex response to site-specific environmental factors, historic land uses, and species-specific responses to many natural processes. Still, it may be reasonable to assume because soil characteristics among treatments were very similar that the long-term exclosures should have shown the differences in plant diversity between ungrazed and grazed plots due to cessation of grazing and continued grazing, respectively.

We also have to acknowledge that vegetation and soils were measured at each site only once (Stohlgren et al. 1999b). Most of our grassland sites peak in biomass and plant species richness simultaneously in the early summer. However, we may have missed some early or late-season plant species in a few of the dry sites (e.g., Uncompahgre, Charles Russell) and plant species that occur in atypically wet or dry years. In any case, the sets of three plots in grazed and ungrazed sites at each exclosure were measured within 2 days of each other, providing a valid comparison of the main grazing effect and spatial variation. We are confident that many of the problems of previous exclosure studies were overcome with consistent vegetation and soil sampling methods, larger sampling areas, additional randomly placed plots in grazed sites, and increased replication. The broad generalizations that follow stem from the overwhelming similarities we found rather than the exaggerated differences we expected.

Cautious Generalizations Based on Available Data

The scope of the generalizations cannot exceed the spatial bounds of the study plots, the vegetation types sampled, and the grazing regimes in terms of intensity, frequency, timing, and animals involved in this case study. The results cannot be extrapolated beyond the vegetation types, soil characteristics, and the climate conditions represented in the sampling. Still, this case study represents the largest plot-based sampling of long-term exclosures in four states, and the only study that effectively quantified the effects of natural spatial variation, climate, and soil quality as other primary determinants of plant diversity patterns. So some cautions generalizations can be made based on these case study results and in the context of previous studies.

1. Grazing Probably Has Little Effect on Native Species Richness at Landscape Scales in These Rocky Mountain Grasslands

We hypothesized that native species richness would be lower in exclosures compared to grazed plots due to competitive exclusion in the absence of grazing. The average age of 31.2 (±2.5) years for the exclosures should have allowed ample time for succession and the sequestering of dominance by a few highly competitive species (e.g., Grime 1973; Harper 1977; Smith et al. 2004). At the 1 m^2 scale, there was significantly lower richness of native species in ungrazed exclosures compared to grazed sites (Table 11.1), about 1.5 fewer species/m^2 (a less than 20% difference). However, as the spatial scale increased to 1000 m^2 plots, the differences between grazed and ungrazed sites were unceremoniously muted. At the 1000 m^2 scale, mean native species richness between long-ungrazed and grazed plots differed by only 0.3–3.5% (Table 11.2). Thus higher native species richness in grazed sites at the 1 m^2 scale can be attributed to higher species packing (or species density), with locally common native plant species filling open sites throughout the 1000 m^2 plots in grazed sites. Since species richness was nearly identical at the larger plot scales (1000 m^2 plots), species richness may be largely unaffected by grazing at landscape scales in these vegetation types.

Some studies in other habitats have shown decreases in species richness on grazed sites [Reynolds and Trost (1980), sagebrush desert; Rummell (1951), pine habitat]. Some studies in other habitats have shown increases in species richness following a cessation of grazing [Chew (1982), desert grassland vegetation; Winegar (1977), riparian habitat), while some have

Reminder: Plant ecologists often compare their results to other studies to help corroborate findings to seek more general patterns. However, comparisons carry with them an obligation of the plant ecologist to evaluate several aspects of each study, including design limitations, specific environmental limits of the study plots, and the limits of extrapolation of results. Many of the grazing studies we investigated used small plots with limited replication of exclosures within and among vegetation types, so comparing results was difficult, whether the specific results agreed or disagreed with ours. Plant ecologists would do better to cite well-designed studies with contradictory results than to cite poorly designed studies with similar findings.

shown no major differences [Cid et al. (1991), similar habitats to our sites; Evanko and Peterson (1955), similar habitats to our sites; Hughes (1996), drier sites in Arizona; Smeins et al. (1976), drier sites in Texas]. Some studies have shown decreases in species richness at small scales, but little change at larger scales [Collins and Adams (1983), prairie sites in Kansas; Facelli et al. (1989), wetter sites in Argentina; this study, Tables 11.1 and 11.2]. Because different techniques were used in the same or different habitats, it is difficult to isolate the effects of study design from the effects of grazing and other environmental factors.

In this case study we found many more similarities than differences between grazed and ungrazed sites (Tables 11.1–11.5) using the same methods in a wide variety of vegetation types and grazing regimes. So despite hypotheses based on ecological theories of intermediate disturbance and competitive exclusion (Figure 11.2), in reality we found few major, consistent changes between grazed and long-ungrazed sites (Figure 11.7).

There are several possible mechanisms that may maintain similar species richness in grazed and ungrazed sites. It may be that 31 years, on average, is not long enough for competitive exclusion to be demonstrated. This seems highly unlikely given that 12 years was long enough to show drastic reductions in species richness in the Serengeti (McNaughton 1983) and 13 years was long enough to show major changes in life-form composition in old-field succession in tallgrass prairie in Oklahoma (Collins and Adams 1983). Still, the moderate to high diversity of species in many management units should have allowed for some highly competitive species to dominate exclosure sites (Connell 1978; Fox 1979; Horn 1975). In most management areas we also observed an equally rich variety of growth forms (tall perennial shrubs to short

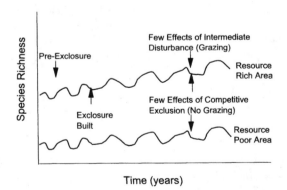

Figure 11.7.   (Compare with Figure 11.2). Hypothetical relationship of species richness over time in ungrazed then grazed habitats in resource-rich areas and resource-poor areas in Rocky Mountain grasslands. Natural variation may be due to climate variation and species metapopulations dynamics, while slight increases over time are the result of non-native species invasions. Resource rich areas might gain more invasive species than resource poor areas.

annual herbs), functional groups (C3 and C4 grasses), and physiological types (rhizomatous types, obligate seeders). It may be that grazing at current and past levels is not as strong a regulator of species richness in Rocky Mountain grasslands as it is in other areas (e.g., the Serengeti) (McNaughton 1979). These grasslands may lack dominant competitors with or without grazing, or they may be controlled more by belowground competition than by aboveground interactions (see Milchunas et al. 1988). Another possibility is that local extinction and immigration rates are similar in grazed and ungrazed sites in Rocky Mountain grasslands. Gibson and Brown (1991) showed that sheep grazing increased species colonization rates, but local extinction rates were similar on grazed and ungrazed sites of British limestone grassland. Glenn and Collins (1992) found that grazing had little impact on immigration and extinction rates of plant species in tallgrass prairies in Kansas, and the same may be true in Rocky Mountain grasslands. In any case, any or all of these "possible mechanisms" cannot be proven with observational studies alone; controlled experiments will be needed.

2. Grazing Probably Has Little Effect on the Accelerated Spread of Most Nonnative Plant Species at Landscape Scales

We also hypothesized that grazed plots would have higher exotic species richness and cover compared to ungrazed sites due to disturbances associated with grazing (i.e., more bare ground and the intermediate disturbance hypothesis) and the typical claim that grazing may accelerate weed invasion. Again we assumed that the vegetation and soils were initially similar on the grazed and ungrazed plots prior to the construction of the exclosures. The measured similarities in soil characteristics (Table 11.5) and nonnative species richness, cover, and frequency (Tables 11.1 and 11.2) between pairs of grazed plots in this study support this assumption. At the 1 m² scale, the number and cover of exotic species were similar in grazed and ungrazed sites (Table 11.1). The ungrazed plots consistently had slightly higher cover of exotic species, but this was due primarily to Rocky Mountain National Park, where control efforts may have reduced the cover of exotic species in some of the grazed sites (J. Connor, National Park Service, personal communication, September 1997). At the 1000 m² scale, mean exotic species richness between long-ungrazed and grazed plots differed by less than 3.5% (Table 11.2). Of the nine study areas we examined, only Wind Cave National Park had consistently higher exotic species richness in grazed sites compared to ungrazed sites. Likewise, the frequency of exotic species was extremely similar between grazed and ungrazed plots (Stohlgren et al. 1999b). Thus, assuming similar pre-exclosure conditions in the study sites, there was very little evidence that either continuous grazing at current levels or cessation from grazing radically altered exotic species richness, cover, or frequency in these vegetation types.

Other studies in various habitats have shown that exotic plant species invade sites with or without grazing. In mixed prairie in North Dakota, the cover of *P. pratensis* inside an exclosure increased from 0% in 1916 to 56% in 1994 (Frank et al. 1995). The cover of *P. pratensis* in grazed sites was 29% in moderately grazed sites and 0% in heavily grazed sites, but it may have been undersampled with ten 1 m² plots per treatment. Schulz and Leininger (1990) reported that *P. pratensis* cover was greater in grazed riparian sites, while *Poa palustris,* another exotic grass, was greater in long-ungrazed riparian sites. Diffuse knapweed (*Centaurea diffusa*), which occupies more than 1.2 million ha of the western United States (Lacey 1989), has been shown to invade pristine, ungrazed native plant communities (Lacey et al. 1990) and long-ungrazed sites (Sheley et al. 1997). Smith and Schmutz (1975) reported rapid increases in Lehmann lovegrass (*Eragrostis lehmanniana*), an exotic perennial grass, in long-ungrazed desert grasslands in Arizona. In contrast, Mack (1981) strongly suggested that overgrazing and disturbance were key factors in the spread of *B. tectorum* and other weeds in the historically lightly grazed perennial grasslands in Washington, Idaho, Oregon, Nevada, Utah, and British Columbia. Robertson (1971) found that *B. tectorum* could increase in sites protected from grazing, but the area had been previously heavily grazed. We also found *B. tectorum* in grazed and ungrazed plots, with higher cover generally in grazed plots (Stohlgren et al. 1999b). One study reported a decrease in the cover of exotic species in the absence of grazing, but that occurred as light levels were reduced in a more forested area (Woodward et al. 1994). In short, many exotic species invade grazed and ungrazed sites, and we found little evidence to suggest that grazing at current levels accelerates the spread of most species of weeds in these vegetation types (Tables 11.1 and 11.2) (Stohlgren et al. 1999b).

3. Grazing Affects Local Plant Species and Life-Form Composition and Cover, but Spatial Variation Is Considerable

Similarities in species richness (Table 11.2), similarities in diversity and evenness (Table 11.3), and differences in species composition between grazed and ungrazed sites, and between grazed adjacent and grazed distant sites (Figure 11.3) suggest that these vegetation types may have a "free substitution rule" for many species; that is, local extinctions are balanced by local immigration. Because species composition overlap generally increases with spatial scale within a vegetation type (Stohlgren et al. 1998c), the substitute species are likely part of the same landscape-level species pool.

One fairly consistent pattern was that shrub cover and frequency tended to decrease slightly in grazed plots relative to ungrazed plots (Table 11.4). However, greater shrub cover inside exclosures did not necessarily translate into less forb and grass cover (Table 11.4). Other studies in similar habitats (Coughenour 1991; Schulz and Leininger 1990; Singer 1996) and elsewhere

(Bock et al. 1984; Collins and Adams 1983; Kindschy 1987; Smith 1960; Smith and Schmutz 1975; Sneva et al. 1984; Tiedemann and Berndt 1972) have shown that cessation of grazing can increase the cover and frequency of shrubs. Still, there are some exceptions. Hughes (1980, 1983) found higher shrub frequencies on grazed sites in desert shrub communities in Arizona, and Smeins et al. (1976) found no significant increase in shrub cover in Texas after 25 years of protection from grazing. Three of nine management areas in our study region had inconsistent results among grazed and ungrazed sites, and the Charles Russell area had greater shrub cover in grazed plots (Table 11.4). While forb and grass cover is often lower on grazed sites (Table 11.4) (Bock et al. 1984; Cid et al. 1991; Hughes 1983), as would be expected with herbivory, a higher frequency of forbs and grasses in grazed sites may allow greater resilience and recovery when grazing pressure is reduced.

The cover of forbs and grasses can be highly variable, both spatially and temporally, in grazed and ungrazed sites. Woodward et al. (1994) reported a decrease in the cover of forbs and grasses in the absence of grazing in Olympic National Park, Washington, presumably due to decreased light levels in exclosures resulting from succession. In Yellowstone National Park, Coughenour (1991) found that the cover of grasses was increased in some exclosures and decreased in others. Reardon (1996) reported increased forb cover on grazed areas at some sites in Yellowstone National Park, but few major differences between grazed and ungrazed sites overall.

Our sampling techniques of foliar cover, which covered about five times the commonly sampled area of previous studies, also found no consistent differences in the cover of forbs, grasses, shrubs, or bare ground in Yellowstone National Park and elsewhere (Stohlgren et al. 1999b). We suspect that small quadrat sample areas and sample sizes exaggerated differences reported in many grazing studies. For example, the significant differences in native species richness between grazed and ungrazed sites at the 1 m^2 scale were insignificant at the 1000 m^2 scale. Given the many vegetation studies that recognized high spatial variability (e.g., Belsky 1983; Brown and Allen 1989; Collins and Adams 1983; Evanko and Peterson 1955; Frank and McNaughton 1993; Young 1943), it is difficult to understand why so many studies evaluating plant diversity change rely on paired-plot studies using small quadrats without many replicates.

Plant distributions are influenced by complex spatial and temporal parameters such as environmental gradients, seed dispersal, site occupancy, lag effects, patchy nutrient and water resources, competition, disturbance at multiple scales, selective herbivory, disease and pathogens, and species-specific demography. The Collins and Barber (1985) description of grassland communities seems equally true of the vegetation types we studied: biotic factors include large-scale effects of grazing superimposed on small-scale effects of burrowing, excavation, or wallowing. Pacala and Crawley (1992) theorized that spatial variability in herbivory could create ephemeral, local refuges for each plant species if there is not a negative correlation between a plant's palatability and its competitive ability. This case study suggests that at current levels of grazing, high spatial variability (i.e., well-dispersed populations and seeds) may be all that is necessary to maintain native plant diversity at landscape scales (Stohlgren et al. 1999b).

Even within so-called homogeneous vegetation types, plant species are commonly distributed in patches. In this study and others (Stohlgren et al. 1995c, 1997b, 1998b,c) we found that (1) about 50% of the vascular plant species had less than 1% foliar cover; (2) only a few dominant species were shared among plots within a vegetation type; and (3) plant frequency (i.e., the number of times a plant species occurred in 1 m² subplots) was extremely variable, suggesting that plant species distributions were patchy at 1000 m² scales [also see Collins and Barber (1985); McNaughton (1983)]. The consequence of low species overlap (Figure 11.4), besides the obvious claim that no two plots are alike, is that investigators are obliged to evaluate how spatial heterogeneity influences study results. It is unlikely that sweeping generalizations about treatment effects can be made by surveying a few square meters on either side of a fence line, as has commonly been the case in exclosure studies.

Hubbell (2001) assumed that the abundance of new immigrants must be offset by a decline in abundances of the cumulative resident species (i.e., that there are no empty niches in plant communities). This may not be the case in broadly disturbed ecosystems. There is evidence that the abundance and packing of species may be increased (or at least altered) by grazing, and there is no direct support for the "zero-sum" equilibrium theory described by Hubbell (2001) at all spatial scales. The cover and abundance of nonnative species in grazed and ungrazed areas may be additive rather than displacing native species richness or cover (Tables 11.1–11.4) (Stohlgren et al. 1999b). Patterns of species cover and abundance are complex at multiple scales.

4. Soil Fertility, Climate, and Other Factors Have a Greater Effect on Plant Species Diversity Than Does Grazing

This case study demonstrates that current levels of grazing may have little effect (approximately ±10% at the 1000 m² scale) on species richness (Table 11.2), foliar and life-form cover (Tables 11.1 and 11.4), plant diversity (Table 11.3), and selected soil characteristics (Table 11.5) in these vegetation types in the Rocky Mountains. Yet the species richness ranged from about 12 species/1000 m² plot (Uncompahgre) to more than 50 species/1000 m² plot (Wild Horse), greater than a fourfold difference. Obviously other factors have more of a controlling influence on plant diversity than livestock and wild ungulate herbivory at these regional scales.

At the regional scale (i.e., for these rangeland types in the four-state study region), soil characteristics and elevation play a major role in determining the richness and cover of native and exotic plant species. Grazed sites high in soil nitrogen and carbon in the Rocky Mountains tend to have higher native and exotic species richness and cover than sites with low soil fertility (Stohlgren et al. 1998d, 1999b). Soil fertility and water availability are known to overshadow the effects of grazing in many areas (Hongo et al. 1995).

Often, exclosure studies have shown that increased or decreased precipitation can have greater effects than herbivory in altering plant productivity and species richness (Chew 1982; Coughenour 1991; Milchunas et al. 1988; Orr and Evenson 1991; Sneva et al. 1984). For the nine management areas in four states, we found that mean exotic species cover was strongly positively correlated with mean maximum January temperature—with less cover in high-elevation cooler habitats (Stohlgren et al. 1999b). Meanwhile, total cover was significantly negatively correlated with mean maximum July temperatures—with less cover in lower-elevation, hot, dry habitats. And since total plant cover and total species are positively correlated in these sites at all special scales, species richness at landscape and regional scales may be largely controlled by climate and soils rather than by grazing.

It may be that important resources such as light, nitrogen, and water remain at relatively similar levels in grazed and ungrazed sites in most of the management units, prohibiting rapid growth by some species in the absence of grazing or extirpation of species in grazed sites. Plant diversity at landscape scales may not be controlled by ungulate herbivory (Crawley 1983; McNaughton et al. 1989) or competitive exclusion (Grime 1973; Harper 1977) as much as it is by other factors.

Disturbances such as rodent activity (Cid et al. 1991; Hulme 1996; Whicker and Detling 1988), insect outbreaks (Sneva et al. 1984), intermittent fire (Collins and Barber 1985; Hart and Hart 1997; Walker and Peet 1984), and occasional flooding (DeFerrari and Naiman 1994) likely play a significant role in maintaining plant diversity. We observed slightly more rodent activity inside exclosures. Evaluating other disturbances and factors was beyond the scope of this study. Milchunas et al. (1990) found that plant diversity at one site in the shortgrass steppe in Colorado increased with increased levels of perturbation, while a study by Collins and Barber (1985) in tallgrass prairie supported the intermediate disturbance hypothesis. More research is needed on these other disturbances in montane meadows.

5. Few Plant Species Show Consistent, Directional Responses to Grazing and Cessation of Grazing

In this case study we found that the vast majority of species showed inconsistent responses to grazing and protection from grazing due to high spatial variability (Stohlgren et al. 1999b). We question whether the terms "increaser" and "decreaser" (Ellison 1960; Weaver and Hansen 1941) are useful concepts for even a few plant species at landscape and regional scales. *B. tectorum* (Mack 1981) and *B. japonicus* may indeed spread faster in grazed and disturbed sites, but Daubenmire (1940b) classified *F. idahoensis* and *A. spicatum* as decreasers under grazing in southeastern Washington, while we found that the cover of *F. idahoensis* was higher in grazed plots relative to ungrazed plots, and the cover of *A. spicatum* varied greatly in and around the three exclosures in Yellowstone. Our use of a third plot randomly located in the same vegetation type strongly suggests that the natural patchiness of vegetation, spatially heterogeneous grazing and selective grazing, and inconsistent responses to grazing make it very difficult to classify plant species in simplistic ways so as to have meaning for landscapes and regions.

Consistent, directional changes in species cover and frequency in grazed plots compared to adjacent exclosed sites is usually considered evidence of the "grazing effect." Our study shows that it is difficult to attribute this effect to grazing alone. And since the responses of most species differ by site (Evanko and Peterson 1955), it is difficult to isolate the effects of grazing on plant diversity from differences due to other biotic and environmental factors.

This case study led to five broad generalizations about current levels of grazing on these vegetation types in the Rocky Mountains: (1) grazing probably has little effect on native species richness at landscape scales; (2) grazing probably has little effect on the accelerated spread of most exotic plant species at landscape scales; (3) grazing affects local plant species and life-form composition and cover, but spatial variation is considerable; (4) soil fertility, climate, and other factors may have a greater effect on plant species diversity than grazing; and (5) few plant species show consistent, directional responses to grazing and cessation of grazing. Land managers may be somewhat relieved that plant species diversity in these vegetation types in Rocky Mountain grasslands is fairly resilient to grazing and cessation of grazing. Nature's abilities to increase plant frequencies when foliar cover decreases due to grazing, and to freely substitute many species within landscapes under a wide variety of grazing regimes, may add to the redundancy of species distribution patterns (Stohlgren et al. 1997b) and a hedge against extirpation and extinction. We found no evidence that current levels of grazing have led to a loss in plant species richness and diversity at landscape scales in the vegetation types and management areas we surveyed. At local scales, exotic species may be additive in many ecosystems for many years before replacing native plant species. However, noxious weed species may call for immediate control measures.

Other biomes with different evolutionary histories may have very different responses to grazing (Mack and Thompson 1982; Milchunas and Lauenroth 1993). In most cases in our study region, dead tissues are removed by grazing because the plants are grazed when they are dormant (Coughenour 1991), and few measurable effects on diversity and productivity have been shown in Rocky Mountain grasslands. However, since few exclosures exist in riparian zones, wetlands, and rare habitats, additional research is needed in these habitats. Other effects of grazing such as mechanical damage to soils and trees, soil erosion, and gullying may be more pronounced in rare habitat types and riparian zones (Belsky and Blumenthal 1997; Fleischner 1994), and these should be regionally studied. A new system of large exclosures as suggested by Bock et al. (1993) is needed to fully evaluate grazing effects in rare habitat types and riparian zones.

Increased replication helped to evaluate spatial variation in grazed sites in each management area, without which broad generalizations about the effect of grazing on plant diversity could be erroneous. It is highly unlikely that the error term in unreplicated exclosure studies is represented by the sampling error as stated in Dormaar et al. (1994, p. 29). Even with increased replication, simple paired-site designs on either side of a fence line need to evaluate the effects of spatial variation.

The existing grazing exclosures have provided important insights into the effects of grazing on plant diversity patterns. A comparison of only grazed sites would not have yielded these insights. Additional exclosures are needed to make inferences about grazing in rare habitats and riparian zones. New exclosure sites should be randomly selected and carefully surveyed for pre-treatment data. Multiscale vegetation sampling and the use of large plots may help isolate the effects of spatial scale from the effects of grazing and cessation of grazing.

12

Case Study

Assessments of Plant Diversity in Arid Landscapes

The Issue

Thus far our case studies have focused on patterns of plant diversity, soil characteristics, and geographic/topographic factors in fairly mesic environments. At the plant neighborhood scale (approximately 1 m² scale or less), we have highlighted the coexistence of native and nonnative plant species relative to bare ground, soil texture, and soil nutrients. In arid environments, the patterns of native and nonnative plant diversity may be affected by a third constituent, cryptobiotic crusts.

Cryptobiotic (or microbiotic) soil crusts are filamentous webs of cyanobacteria, lichens, green algae, mosses, and fungi that play a critical role in the sustainability of desert ecosystems (Figure 12.1). The crusts mechanically stabilize soils and increase soil fertility by fixing atmospheric nitrogen and preventing other nutrients from being leached from surface soils (Garcia-Pichel and Belnap 1996). In addition, crusts influence vascular plant seed germination and water infiltration rates (Belnap and Gillette 1998; Belnap and Harper 1995). However, the fragile organisms that make up the crusts may take hundreds of years to recover from disturbances such as trampling by livestock and people and the use of off-road vehicles (Belnap 1995, 1998). Assessing damage to the crusts at landscape scales is vital to land managers in arid environments, who must protect vegetation and soils while providing for multiple uses such as recreation, livestock grazing, and mineral exploration.

Figure 12.1.   Cryptobiotic crusts in Grand Staircase-Escalante National Monument, Utah. Lesson 19. Soil crusts are common in many parts of the world, and they are very important in stabilizing soil, nutrient cycling, and providing resources important to protecting native plant diversity.

In all environments, biological conservation efforts are increasingly moving toward an ecosystem and landscape approach, recognizing the prohibitive cost and difficulty of a species-by-species approach (Agee and Johnson 1988; LaRoe 1993; Noss 1983; Stohlgren et al. 1997b,c). A key ingredient of our approach is a careful analysis of hot spots of plant diversity and rare/ unique habitats to identify critical habitats for preservation. Although rare habitats are small in total area, they often are heavily used by wildlife (McNaughton 1993; Simonson 1998). As in most arid landscapes, riparian zones and small wetlands in Grand Staircase-Escalante National Monument, Utah, are expected to be hot spots of biodiversity. It is important to know if these hot spots of native plant diversity and rare/unique habitats are vulnerable to invasion by exotic plant species (D'Antonio and Vitousek 1992).

Our understanding of the role of cryptobiotic crusts in maintaining and stabilizing soils and protecting native plant diversity can be greatly improved with systematic surveys combined with controlled experiments (Belnap and Gillette 1998; Belnap and Harper 1995). Cryptobiotic crusts are more prevalent in some areas of the monument than in others. Crusts have different levels of resistance and resilience to disturbance. Crusts on coarse-textured soils are much less able to handle disturbance and recover much more slowly than crusts on fine-textured soils. Also, some areas receive more trampling by livestock and people, off-road vehicle use, and other disturbances. Patterns of native plants may depend heavily on landscape-scale patterns of crust cover, level of crust development, soil characteristics, and invasion by exotic plant species.

The objectives in this case study were to (1) quantify patterns of native and nonnative plant species, cryptobiotic crust habitats, rare/unique habitats, and soil characteristics at landscape scales; and (2) determine which habitats in the southeast portion of the monument were more invaded by exotic plant species. This would provide an opportunity to evaluate the interactions between various components of vascular plant diversity and nonvascular plants (crusts) in complex arid environments.

Study Site and Methods

The recently proclaimed Grand Staircase-Escalante National Monument was established to preserve many important ecological features, including extremely high endemism in a floristically rich region of the southwestern United States. The monument is home to 50% of Utah's rare plant species, 11 of which are found nowhere else on Earth, and 84% of the state's flora (Shultz 1998). However, less is known about the patterns of native and exotic plant diversity in the broader landscape, creating several sampling challenges for plant ecologists. First, the monument is huge and floristically complex. The more than 850,000 ha (2.1 million acres) monument ranges in elevation from 1372 m (4500 ft.) to 2530 m (8300 ft.), from low desert shrub, steppe, sage (*Artemisia* ssp.), and pinyon-juniper (*Pinus edulis-Juniperus osteosperma*) woodlands to forests of ponderosa pine (*Pinus ponderosa*) (Welsh and Atwood 1998). Second, land use practices such as grazing, recreation, mining, and oil and gas exploration and extraction threaten the native plants, cryptobiotic crusts, and soils in the monument.

A 100,000 ha area in the southeast corner of the monument was selected for intensive study (Figure 12.2). The area contains a high plateau and low canyon land forms. Random sample sites were selected in vegetation types recognized on the monument's vegetation map. Additional sites were randomly located in rare types of special interest, such as washes and relict plant habitats (e.g., wetlands, riparian zones, washes, relict plant habitats) (Stohlgren et al. 1997b,c, 1998b), as they were encountered in the field. Eleven vegetation types (identified by dominant species) included aspen, blackbrush (*Coleogyne ramosissima*), desert mixed shrub (*Gutierrezia sarothrae* and many others), juniper (*Juniperus* ssp.), lowland riparian (*Salix* ssp., *Tamarix* ssp.), pinyon (*P. edulis*), pinyon-juniper (*P. edulis-Juniperus* ssp.), ponderosa pine, rabbitbrush (*Chrysothamnus* ssp.), sagebrush (*Artemisia tridentata, Artemisia bigelovii*), and wet meadow (*Agrostis stolonifera, Juncus arcticus*).

At each sample site, a multiscale plot was established for vegetation, cryptobiotic crust, and soil sampling (Figure 12.3). The modified Whittaker plot was used in this study (see Figure 7.5) (Stohlgren et al. 1998c), but other multiscale designs could have been substituted. Cover by species and the developmental stage of cryptobiotic crusts was recorded in the ten 1 m^2 subplots in eight classes from 1 (weakly developed) to 20 (fully developed) (see Belnap 1995, 1996).

Soil samples were collected and analyzed as in chapter 11, except that phosphorus, organic and inorganic carbon, and major cations were included (see Bashkin et al. 2003).

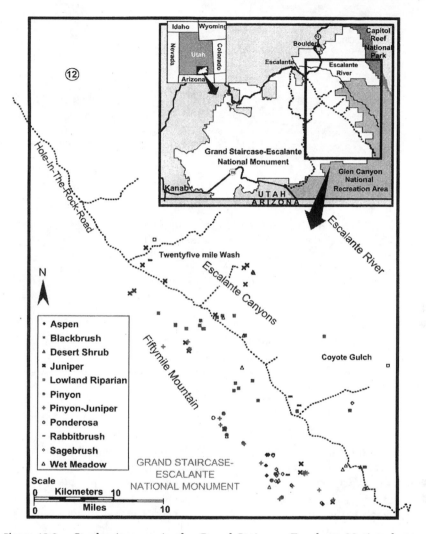

Figure 12.2.   Study site map in the Grand Staircase-Escalante National Monument, Utah.

Statistical Analysis

All statistical analyses were conducted with Systat (version 9.0; SPSS, Inc., Chicago, Illinois), as in chapter 11. In addition, t-tests were used to compare vegetation and soil characteristics between washes (and lowland depressions) that collect intermittent flooding and more xeric upland plots, and between shrub-dominated plots and tree-dominated plots (Stohlgren et al. 2001).

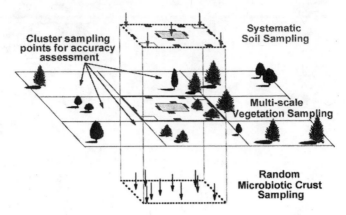

Figure 12.3. Schematic drawing of the methods used at plot sampling sites. Lesson 20. The major cost in extensive field studies is in getting crews to and from remote sites, so while they are there, they should collect more information at multiple scales and for multiple purposes.

We also used canonical correspondence analysis (CCA; PC-ORD version 3.0; see chapter 13) (ter Braak 1986, 1987a) to characterize the relationships between species composition (cover and richness) and the environmental measurements (e.g., soil characteristics, elevation). CCA is a direct gradient analysis technique that constrains the extracted pattern to linear combinations of the measured environmental variables (ter Braak 1986, 1987a). We ran CCA on two separate datasets. The first contained species richness for dominant (more than 1% cover) and subdominant (less than 1% cover) native and exotic species and cryptobiotic class. The second dataset consisted of cover data for native and exotic species, *Bromus tectorum,* and cryptobiotic crust. The species richness and cover datasets were evaluated against the same 11 environmental variables: soil carbon, inorganic carbon, nitrogen, phosphorus, potassium, sodium, magnesium, clay, elevation, and bare ground. We assessed all environmental variables for multicollinearity problems and none were found. Monte Carlo permutation tests were performed to test the significance of the first axis (ter Braak 1991). Environmental variables used in CCA were used as independent variables in stepwise forward multiple regressions (SPSS Inc. 2001) to predict native and exotic species cover and richness, *B. tectorum* cover, and cryptobiotic cover and class.

Finally, we used kriging spatial interpolation models to create contour maps of native and exotic species richness, and soil percent nitrogen and percent carbon in relation to plot location. We mapped exotic species richness and crust cover in the same way (Stohlgren et al. 2001).

Results, Discussion, and Lessons Learned

General Patterns of Native and Exotic Plant Diversity and Soil Crusts

For the ninety-seven 0.1 ha plots combined, we encountered 350 native species, 38 exotic species, and 63 specimens identified to genus. Plots, on average, contained 25.0 (±0.7; 1 SE) native plant species and 2.3 (±0.2) exotic plant species. The low standard errors demonstrate the remarkably consistent native plant richness among plots. Four plant species could not be identified because of phenological stage, desiccation, or missing parts from herbivory. Only 6 of 97 plots contained no exotic plant species, while 28 plots had three or more exotic plant species per plot (Figure 12.4).

Native plant species dominated live foliage, averaging 29.1% (±1.6%) per plot, while exotic plant species cover averaged 5.0% (±0.8%) per plot. The dominant exotic plant species was *B. tectorum,* which averaged 3.4% (±0.5%) cover, or about 68% of the exotic species cover. In this arid landscape, total bare ground (i.e., rock, bare soil, dung, etc.) averaged 68.4 % (±1.8%), while the cover of cryptobiotic crusts averaged 19.7% (±1.9%) cover. Only three plots had no cover of cryptobiotic crusts, 25 plots had less than 5% cover, and 31 plots had greater than 25% cover of cryptobiotic crusts.

As reported in the previous case studies, native and exotic plant species richness and cover varied considerably by vegetation type (Table 12.1). Mean native species richness ranged from 32.3 (±3.1) species per 0.1 ha plot in the relatively mesic, high-elevation, aspen type to 19.8 (±1.2) species per plot in the xeric, lowland blackbrush type. The maximum number of native species was also found in an aspen plot (45 species), while the most species-poor plot was in the blackbrush type (11 species). The relatively rare

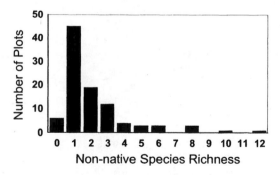

Figure 12.4.   The frequency of non-native species in 1000 m² plots in the Grand Staircase-Escalante National Monument, Utah. Adapted from Stohlgren et al. 2001.

ponderosa pine and wet meadow types were also high in native species richness (Table 12.1).

Some patterns were obvious and predictable, with greater invasions in species-rich areas. Nonnative species richness generally increased with native species richness in many vegetation types (Table 12.1). One aspen plot had 12 exotic species, while the wet meadow type averaged 8 nonnative species per plot. The lowland riparian type was also rich in exotic species. Drier vegetation types, such as the blackbrush, desert shrub, sagebrush, and pinyon-juniper types, averaged fewer than two exotic species per plot (Table 12.1). Exotic species cover was highest in the wet meadow type (27.9

Table 12.1. Mean native species richness and cover, and the number of species with less than 1% foliar cover for various vegetation types in Grand Staircase-Escalante National Monument.

Vegetation type	No. of plots	Mean species richness		Species with less than 1%	Mean percent cover	
		Native	Nonnative		Native	Exotic
Aspen	6	32.3	5.2	15.2	42.2	11.2
		(3.1)	(1.4)	(1.5)	(11.5)	(3.7)
Blackbrush	22	19.8	1.2	9.8	29.2	2.8
		(1.2)	(0.1)	(0.6)	(2.3)	(1.0)
Desert shrub	11	25.4	1.8	11.8	24.0	1.6
		(2.0)	(0.3)	(1.9)	(3.5)	(0.8)
Juniper	15	26.9	1.7	12.9	23.8	4.2
		(1.6)	(0.5)	(1.3)	(3.3)	(1.9)
Lowland riparian	2	23.5	5.0	7.5	41.4	15.5
		(0.5)	(1.0)	(3.5)	(23.9)	(5.6)
Pinyon	8	29.5	4.0	17.1	31.2	3.8
		(1.6)	(0.8)	(2.3)	(7.6)	(1.3)
Pinyon-juniper	16	24.2	1.4	12.4	24.1	1.8
		(1.0)	(0.2)	(1.0)	(3.0)	(0.8)
Ponderosa	2	31.5	1.0	21.5	47.4	2.9
		(3.5)	(0.0)	(0.5)	(16.5)	(1.5)
Rabbitbrush	4	25.8	4.2	14.0	28.7	11.6
		(1.4)	(1.4)	(1.9)	(11.5)	(6.7)
Sagebrush	8	22.8	1.6	12.6	35.7	6.3
		(1.9)	(0.4)	(1.4)	(4.9)	(2.3)
Wet meadow	3	30.3	8.0	16.0	31.5	27.9
		(4.9)	(1.2)	(6.0)	(12.2)	(5.4)
All types	97	25.0	2.3	12.7	29.1	5.0
		(0.7)	(0.2)	(0.5)	(1.6)	(0.8)

Standard error in parentheses.

Adapted from Stohlgren et al. (2001).

± 5.4% cover) and lowest in the desert shrub type (1.6 ± 0.8% cover), following moisture availability.

Conversely, soil crusts were generally more developed in xeric vegetation types. Mean crust cover and mean maximum crust development also varied considerably by vegetation type (Table 12.2). The xeric blackbrush type had the highest mean crust cover (39.9 ± 3.8% cover), while the wet meadow type averaged only 0.1 ± 0.1% crust cover. The most xeric vegetation types (e.g., rabbitbrush, desert shrub, blackbrush type) had the highest mean maximum crust development scores (greater than 10; well-developed crusts).

Soil texture and chemistry also varied by vegetation type (Table 12.2). Plant diversity tended to track mean percent carbon and nitrogen in the surface soil, with relatively high values in the mesic aspen type and wet meadow type. However, the xeric, pinyon-juniper type was also high in percent carbon and nitrogen, while most other xeric vegetation types (e.g., desert shrub, rabbitbrush, and sagebrush) were low in percent carbon and nitrogen (Table 12.2).

Mean exotic species richness was significantly greater in washes than in upland sites. Likewise, the cover of exotic species was significantly greater in washes compared to upland sites. Meanwhile, the cover of cryptobiotic crusts was significantly greater on the upland sites compared to washes. There no significant differences in native plant species richness, percent bare ground, and soil percent sand, silt, and clay, and percent nitrogen and carbon between upland and wash sites (Table 12.3).

Native and nonnative plant species richness, crust cover, and soil characteristics tested varied considerably by vegetation structure (shrublands versus forest; Table 12.4) Shrublands, on average, contained significantly fewer native and exotic species and nearly three times the cover of cryptobiotic crusts as compared to forests. Shrublands, on average, had about half the percent nitrogen and carbon, and half the silt + clay fraction in the surface soil compared to plots with trees (Table 12.4). There were no significant differences in total bare ground, native species cover, or exotic species cover between shrublands and forests, although nonnative species cover was consistently higher on plots with trees compared to plots dominated by shrubs (Table 12.4).

Relationships of Plant Species Richness
and Cover to Crust Cover and Soil Nutrients

In this case study, as in many others, several significant linear relationships emerged between plant diversity characteristics and soil characteristics (Figure 12.4). For the 97 plots combined, native species richness was significantly positively correlated to elevation ($r = 0.28$, $P < 0.005$), perhaps responding to increased precipitation and more moderate temperatures relative to the desert plateau. Native plant diversity was significantly

Table 12.2. Mean cryptobiotic crust cover, maximum crust class, percent bare ground, sand, silt, clay, carbon, and nitrogen various vegetation types in Table 12.1.

Vegetation type	No. of plots	Crust cover (%)	Maximum crust class	Bare ground (%)	Sand (%)	Silt (%)	Clay (%)	Carbon (%)	Nitrogen (%)
Aspen	6	1.3	5.8	61.6	80.1	6.6	13.4	1.89	0.13
		(0.5)	(2.9)	(4.0)	(2.2)	(1.9)	(1.1)	(0.21)	(0.02)
Blackbrush	22	39.9	11.5	75.0	85.1	4.3	10.6	0.80	0.02
		(3.8)	(1.2)	(2.1)	(2.0)	(0.9)	(1.2)	(0.13)	(0.00)
Desert shrub	11	25.1	12.6	66.2	89.5	2.8	7.7	0.53	0.02
		(5.0)	(1.6)	(4.2)	(2.0)	(1.0)	(1.1)	(0.10)	(0.00)
Juniper	15	17.0	9.7	68.2	80.6	6.3	13.1	1.09	0.04
		(3.8)	(1.6)	(7.2)	(3.6)	(1.6)	(2.0)	(0.22)	(0.01)
Lowland riparian	2	2.8	8.5	77.9	83.3	6.8	9.9	0.97	0.02
		(0.9)	(6.5)	(3.2)	(8.9)	(6.3)	(2.6)	(0.30)	(0.01)
Pinyon	8	9.5	9.8	63.8	82.1	6.4	11.5	1.08	0.05
		(2.9)	(2.6)	(5.4)	(3.3)	(2.1)	(1.6)	(0.25)	(0.02)
Pinyon-juniper	16	14.0	10.1	77.4	70.7	9.8	19.4	1.77	0.08
		(3.1)	(1.7)	(1.4)	(3.4)	(1.2)	(2.6)	(0.24)	(0.01)
Ponderosa	2	5.3	1.5	64.0	93.0	0.0	7.5	0.46	0.03
		(2.3)	(0.5)	(13.4)	(1.9)		(0.8)	(0.14)	(0.02)
Rabbitbrush	4	23.1	14.0	64.6	89.2	2.5	8.3	0.55	0.03
		(10.1)	(4.5)	(11.4)	(1.6)	(1.7)	(1.9)	(0.09)	(0.02)
Sagebrush	8	10.1	5.9	60.9	85.4	6.1	8.5	0.58	0.05
		(3.4)	(2.6)	(6.2)	(2.9)	(2.2)	(1.2)	(0.12)	(0.01)
Wet meadow	3	0.1	0.3	29.8	83.5	6.8	9.7	1.40	0.09
		(0.1)	(0.3)	(15.9)	(0.1)	(0.9)	(1.0)	(0.44)	(0.03)
All types	97	19.7	9.7	66.4	82.2	5.8	12.0	1.06	0.05
		(1.9)	(0.7)	(1.8)	(1.1)	(0.5)	(0.7)	(0.08)	(0.00)

Standard error in parentheses.

Table 12.3.   Comparison of upland (dry) plots ($n = 51$) and plots in washes ($n = 45$).

Characteristic	Upland plots	Wash plots	P
No. of native species	24.3	25.9	NS
	(0.9)	(1.0)	
No. exotic species	1.5	3.2	< 0.001
	(0.1)	(0.4)	
Native species cover	27.0	31.7	< 0.16
	(1.7)	(2.9)	
Exotic species cover	3.1	7.3	< 0.01
	(0.7)	(1.4)	
Bare ground cover	68.9	67.6	NS
	(2.7)	(2.5)	
Crypto cover	23.8	15.2	< 0.02
	(2.7)	(2.5)	

Mean values shown with standard errors in parentheses.

negatively correlated to crust cover ($r = -0.22$, $P < 0.029$), soil percent clay ($r = -0.25$, $P < 0.015$), and $\log_{10}$ carbon:nitrogen ratio ($r = -0.19$, $P < 0.07$), perhaps responding to aridity and decreased soil fertility. Exotic species richness was significantly, positively correlated to both native species richness ($r = 0.21$, $P < 0.039$) and native species cover ($r = 0.23$, $P < 0.025$). Exotic species richness was more strongly negatively correlated to the cover of cryptobiotic crusts ($r = -0.47$, $P < 0.001$). The cover of exotic plant species was also strongly negatively correlated to the cover of cryptobiotic crusts ($r = -0.39$, $P < 0.001$). Total percent nitrogen was strongly positively correlated with exotic species richness ($r = 0.40$, $P < 0.001$) and the cover of exotic plant species ($r = 0.37$, $P < 0.001$).

Scatterplots of selected variables are an effective exploratory method to show relationships among variables (Figure 12.5). The Grand Staircase-Escalante National Monument is a complex landscape with considerable variation in vegetation and soil characteristics. Still, the scatterplots show several consistent trends. Moving from right to left on the scatterplots, we can see that several soil characteristics associated with soil fertility (percent carbon, nitrogen, phosphorus, magnesium) are cross-correlated and positively associated with elevation and the moisture index. Native species richness shows positive slopes with percent nitrogen, carbon, and phosphorus, while bare ground, associated with stressful conditions for plant growth, is negatively correlated with soil nutrients. Cryptobiotic crust cover and maximum development class are also negatively correlated with soil nutrients. Meanwhile, nonnative species richness and cover appear strongly associated with increases in elevation, moisture index, and soil fertility (Figure 12.5).

Table 12.4.   Comparison of plots with a vegetation structure of shrubs ($n = 45$) or trees ($n = 49$).

Characteristic	Shrub plots	Tree plots	P
No. of native species	22.2	27.2	< 0.001
	(0.9)	(0.8)	
No. of exotic species	1.7	2.5	< 0.042
	(0.2)	(0.3)	
Crust cover	29.5	11.8	< 0.001
	(2.9)	(1.8)	
Carbon (%)	0.68	1.17	< 0.001
	(0.07)	(0.12)	
Nitrogen (%)	0.03	0.06	< 0.001
	(0.00)	(0.01)	
Sand (%)	86.6	78.3	< 0.001
	(1.2)	(1.8)	
Silt (%)	4.1	7.2	< 0.005
	(0.7)	(0.8)	
Clay (%)	9.3	14.5	< 0.001
	(0.7)	(1.2)	

Mean values shown with standard errors in parentheses.

When nonnative species cover is displayed against soil carbon and phosphorus, the prevailing pattern in more evident (Figure 12.6). Not only did nonnative species succeed better in resource-rich areas, but the mesic, locally rare, and botanically more distinctive vegetation types such as aspen and wet meadow are the most successfully invaded. The wet meadow and aspen types have the most fertile soils (highest in carbon, nitrogen, and phosphorus) and receive more precipitation because they are located at high elevations or located near natural springs (Figure 12.6). This distribution of nonnative species on this landscape is troubling. Nonnative species do not grow well in all habitats, thus we are gaining a moderate ability to predict distributions based on soils habitat data. Soil phosphorus may prove to be a powerful indicator of exotic species establishment and success (Figure 12.6).

Quantifying Patterns of Native and Exotic Species Richness and Cover

Multiple regression and path coefficient diagrams showed that native species richness was difficult to predict from the suite of environmental variables analyzed (Figure 12.7). Elevation and percent soil magnesium had "direct" (Dewey and Lu 1959), positive effects on native species richness (i.e., positive standardized partial regression coefficients). There was also an "indirect effect" of elevation on native species richness by the positive

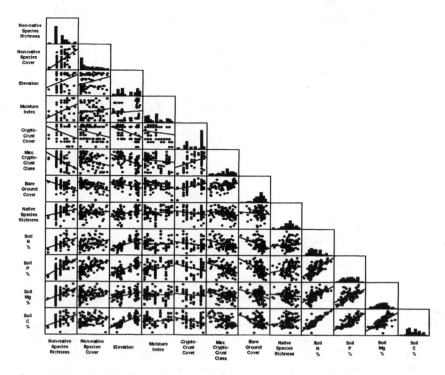

Figure 12.5.   Scatterplot of non-native species and cover, elevation, moisture index, crust characteristics, and soil characteristics for 93 plots (4 had missing data) in Grand Staircase-Escalante National Park, Utah.

correlation of elevation to soil magnesium. And the coefficient for magnesium is negative in a multiple regression, but positive in a simple regression (Figure 12.7), showing the complicated nature of multivariate models. Still, only 24% of the variation in native species richness was explained by these two variables, perhaps surrogates for moisture availability and soil fertility, and this was partly due to the extremely low variation in native species richness for all the plots (Table 12.1).

There were weaker predictions for the cover of native plant species, with less than 16% of the variance explained regardless of the variables selected. Native species richness was positively correlated to the cover of native species, but cover was highly variable. It appears that native plant species are well dispersed in a broad range of stressful to fertile habitats throughout the monument.

Nonnative species richness was far more predictable than native species richness and cover, with 52% of the variance explained by the model (Figure 12.8). Soil phosphorus, elevation, and native species richness were strong positive contributors to nonnative species richness, although the coefficient

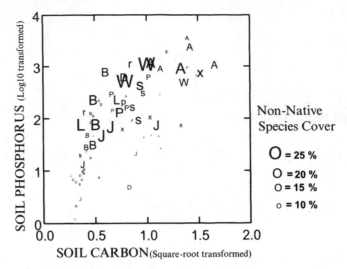

Figure 12.6. Relationship between soil $P_{(log)}$ and $C_{(square\ root)}$ and exotic species cover$_{(log)}$ by vegetation types within the study area. The size of the symbol represents the percent non-native cover class in the modified-Whittaker plot. Vegetation type, Scientific name, and codes are as follows: Aspen, "A"; Blackbrush, "B"; Desert Shrub, "D"; Juniper, "J"; Lowland riparian, "L"; Pinyon, "P"; Pinyon-Juniper, "x"; Ponderosa pine, "p"; Rabbitbrush, "r"; Sagebrush, "S"; Wet meadow, "W" from Bashkin et al. 2001.

for elevation was negative in the multiple regression. Crust cover and percent bare ground had negative direct effects on nonnative species richness. Both total plant cover and crust cover had negative effects on percent bare ground, as would be expected. High total plant cover may indicate favorable habitats for plant productivity, and high crust cover may indicate arid habitats that are less disturbed by grazing, trampling, or flooding.

Nonnative species cover was far more predictable than native species richness and cover, with 46% of the variance explained by the model (Figure 12.9) once the richness of invasive species is known. In this case study, nonnative species richness had the strongest positive effect on nonnative species cover, while percent bare ground had a strong negative correlation. There was also a positive correlation between soil phosphorus and nonnative species cover (Figure 12.9). This model is driven by the very strong positive relationship between nonnative species richness and cover ($r = 0.68$, $P < 0.001$) (Figure 12.10). Without information on the number of nonnative species in a plot (i.e., the success of nonnative species establishment), about 29% of the variation in the cover of exotic plant species can be explained (Stohlgren et al. 2001).

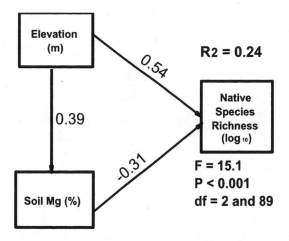

Figure 12.7.   Path analysis of native species richness and cover. From Stohlgren et al. 2001.

Limitations of Regression Analysis in Evaluating Patterns of Native Plant Diversity

Cause-and-effect relationships cannot be determined by statistical correlations. Plant ecologists hope to gain insights into various parameters associated with plant diversity, soils, climate, and other factors, and such studies may generate several testable hypotheses, but there are many sampling and design issues that limit the power of correlation studies. First, the independent variables may not be biologically related to the dependent variable. A statistically positive correlation between soil magnesium and native plant diversity may be biologically unrelated (Figure 12.6) since magnesium is not usually a limiting resource in many ecosystems. Finding a statistically significant relationship between available soil nitrogen and plant cover diversity "suggests" a possible biological relationship because nitrogen is often a limiting factor in terrestrial systems, but other possible explanations abound. Various combinations and interactions among soil phosphorus, pH, moisture, and microbes may be needed to release the available nitrogen from the total pool of nitrogen, and it might be a misleading oversimplification to suggest that the correlation statistics provided are complete explanations of complex interactions.

Second, more important variables might not have been measured (e.g., soil phosphorus, pH, moisture, and microbes) that may have provided stronger statistical and biological relationships. Tree seedlings that require certain inoculums of fungi for growth may not respond to nitrogen additions without the inoculums. Correlation studies must consider alternate expla-

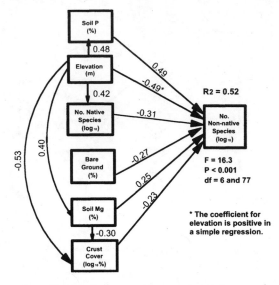

Figure 12.8.   Path analysis of non-native species richness. From Stohlgren et al. 2001).

nations for observed patterns, while recognizing some variables may have been missing or poorly measured. The amount of unexplained variation $(1 - r^2)$ may be due to the high variance of measured parameters, missing factors, or both (usually both).

Third, there are inherent problems with many statistical methods. Stepwise multiple regressions can be influenced by the order in which the variables are added and by the cross-correlations and tolerances of the various independent variables. The "signs" on the coefficients can change between

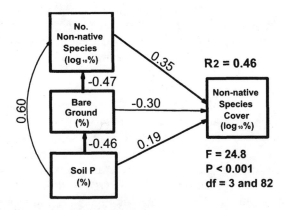

Figure 12.9.   Path analysis of non-native species cover. From Stohlgren et al. 2001.

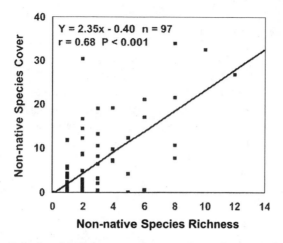

Figure 12.10.   Relationship of non-native species richness and cover. From Stohlgren et al. 2001.

simple regressions and multiple regressions. Transformed data can behave differently than raw data. Statistics are rarely infallible.

Despite these limitations, we can begin to quantitatively assess important patterns over large landscapes and for larger areas than we can test experimentally. In this case study there were several patterns describing the vulnerability of certain habitats to invasion in the study area, but there were some local exceptions. Most of the heavily invaded sites were mesic, relatively rare habitat types that were rich in native species and cover, high in soil percent nitrogen and carbon, and low in crust cover, crust development, and bare ground (Stohlgren et al. 2001). For example, the wet meadow vegetation type was the most heavily invaded habitat in the study area, with an average of 8.0 exotic species per plot and 27.9% cover of exotic species (Table 12.1). Plots in the wet meadow type ranked third in native species richness, second in soil percent nitrogen, third in soil percent carbon, and lowest in crust cover, crust development, and bare ground (Table 12.2). The aspen type had the second greatest mean exotic species richness and third highest exotic species cover. The aspen type had the highest native species richness, ranked second in native species cover, ranked first in mean soil percent nitrogen and carbon, and had the second lowest crust cover and third lowest cover of bare ground. Meanwhile, the drier blackbrush, desert shrub, and sagebrush types were relatively poorly invaded by exotic species, contained fewer native species, and generally had higher crust cover, more bare ground, and lower soil percent carbon and nitrogen (Tables 12.1 and 12.2). There are now many studies suggesting that species-rich areas may be particularly vulnerable to invasion (DeFerrari and Naiman 1994; Fox and Fox 1986; Malanson 1993; Planty-Tabacchi et al. 1996; Robinson et al. 1995; Stohlgren et al. 2002).

It is equally important in case studies to acknowledge exceptions to the observed general patterns. For example, in this study the lowland riparian type had the second highest exotic species cover and third highest exotic species richness. This vegetation type had the third highest cover of native species and had the third lowest crust cover, but unlike the other heavily invaded vegetation types, the lowland riparian plots had the highest bare ground and were lowest in soil percent nitrogen. This may be due to intermittent flooding that may reduce native species richness and total percent nitrogen in the soil or to the increasing spread of the exotic salt cedar (*Tamarix* ssp.) throughout the riparian zone. Plots in the ponderosa pine type were high in native species richness and cover, hence we would have anticipated high exotic species richness. However, few exotic species were found in these plots. We suspect that native species richness was high due to a mix of habitat-specific plant species in ponderosa pine stands (including shade-tolerant species) with habitat generalists from the surrounding vegetation types in more xeric microhabitats. High percent sand in the upper soil layers (low soil moisture-holding capacity) combined with low soil nitrogen (Table 12.1) and low light under the canopy may provide only marginal habitat for many invasive species in the ponderosa pine habitat.

Comparing the results of other observational studies, small-scale experiments, and mathematical models is one way of corroborating the generality of the observed patterns in a case study. Keeping in mind that there could be many non-mutually exclusive mechanisms involved in the invasion process, these study results corroborate mathematical models which suggest that turnover (plant mortality and replacement) may be one important mechanism in plant invasions in predominantly undisturbed areas. In most of the habitats in the central United States that we have surveyed, about half of the plant species averaged less than 1% foliar cover (Table 12.1) (Stohlgren et al. 1999b). Low frequency and low cover may indicate high turnover in space and time, creating many opportunities for both native and exotic plants. There was a significant positive relationship between exotic species richness and the number of species with less than 1% cover ($r = 0.20$, $P < 0.045$). There was an even stronger significant positive relationship between total native species richness and the number of species with less than 1% cover ($r = 0.72$, $P < 0.001$). This could indicate that high species richness and high turnover may go hand in hand, which generally follows the theory of May (1973), who postulated that diverse systems would be particularly susceptible to high turnover. Likewise, Huston and DeAngelis (1994) showed that many species can coexist in spatially heterogeneous areas as long as nutrients and light are not limiting. Seed sources for exotic plant species are often readily available in the soil seed bank or via the matrix of riparian zones and roads that are corridors for invasion (Stohlgren et al. 1998d), and from other ubiquitous dispersers of invasive species (e.g., wildlife, wind, and livestock). Still, plant turnover in undisturbed areas may result in only moderate frequency and relatively low cover of invasive plant species

(Stohlgren et al. 1999b), so seed availability (Tilman 1997) and plant turnover in harsh environments may be only part of the story.

The second general pattern evident in this case study is that disturbance of soil crusts may greatly accelerate the invasion process, a claim well studied by others (DeFalco 1995; Howell 1998; Larsen 1995). Intact soil crusts often present a physical barrier to invasive species establishment and growth by preempting space. We found that exotic species richness was strongly negatively correlated to the cover of cryptobiotic soil crusts ($r = -0.47$, $P < 0.001$). Howell (1998) and Larsen (1995) showed that germination of native *Stipa* was not affected by soil crust cover, while germination of *B. tectorum* was inhibited by intact crusts. *Bromus* germination was stimulated when crusts were broken, but left in place. Once the crusts are disturbed by grazing, trampling, or vehicle use, resources, particularly nitrogen, become temporarily available for the establishment of native or exotic plant species (Belnap 1995, 1996). The moderate frequency and relatively low cover of invasive species resulting from plant turnover may predispose the greater landscape to disturbance-enhanced invasion. Plant turnover throughout a landscape may provide widely dispersed seed sources and seed reserves in soils. Subsequent trampling by livestock or recreationists can produce moderately high exotic species richness and cover (Table 12.2 and 12.3) such that previously rare disturbances such as fire could become more commonplace, further enhancing the invasion. Increased invasion of exotic species could lead to extirpation of native plant species, but this has not been measured in this system.

Multiple well-designed case studies may begin to shift paradigms in field ecology. In contrast to previous theories about invasion, this study and our other studies (Stohlgren et al. 1998d, 1999a) suggest that it might be difficult for species-rich areas to completely monopolize resources in fertile sites (Case 1990) to maintain stability (Tilman et al. 1996) and to resist invasion (Elton 1958). The invasion is taking place in disturbed and undisturbed habitats throughout many continental landscapes (Stohlgren et al. 1999a,b). This case study showed that nonnative plant species have successfully invaded Grand Staircase-Escalante National Monument. Over this broad landscape, 94% of the plots in the study area have been invaded by at least one exotic plant species. There are several reasons for concern. First, exotic plant species are invading hot spots of native plant diversity and rare/unique habitats in this arid landscape. These habitats often contain our most treasured botanical resources (i.e., rare species). Second, habitats that were vulnerable to invasion by several exotic plant species had higher cover of exotic species. This sets the stage for the local replacement of native species by exotic species. Third, continued disturbance of fragile cryptobiotic crusts by livestock and recreationists may facilitate the further invasion of exotic plant species. The long recovery times of damaged crust may provide ample opportunity for invasive exotic plants to gain a foothold on the landscape.

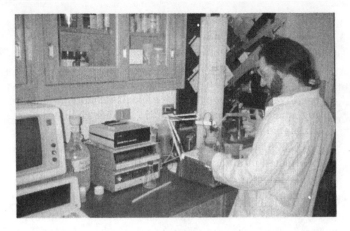

Figure 12.11.   Lesson 21. Understanding patterns of plant diversity requires an understanding of soils and geology. Befriend a soils ecologist.

One at a time, the results of case studies often do not make a compelling general story. However, several recent landscape-scale studies are reporting an alarming and consistent pattern in the Central Grasslands (Stohlgren et al. 1998d), Rocky Mountains (Stohlgren 1999b; Stohlgren et al. 1997a), and in portions of the Grand Staircase-Escalante National Monument (this case study). Where resources are plentiful for native plant species (high light, soil percent nitrogen, and water), exotic plant species are also plentiful (Figure 12.11). We found strong positive relationships between total percent nitrogen and exotic species richness and cover, and mesic sites were generally more invaded than xeric sites (Tables 12.2–12.4 and Figure 12.5). The easiest explanation is that native and exotic plant species take advantage of the same types of resources (Stohlgren et al. 1998d, 1999b) at the same time or at different times. Resources for invading species must be available, even in species-rich areas. In short, broad-scale integrated surveys help to quantify patterns of plant diversity and generate hypotheses.

PART IV

MODELING PATTERNS OF PLANT DIVERSITY

13

Nonspatial Statistical Modeling
of Plant Diversity

In this chapter I provide selected examples of nonspatial statistical modeling of plant diversity. I focus on commonly used techniques including correlation and simple regression, multiple regression, path coefficient analysis, canonical correspondence analysis (CCA), regression tree analysis, and logistic regression. Most of these techniques have been introduced briefly in case study examples earlier in the book. Here I provide more details and considerations in their applications. Each of these procedures is covered in far greater depth in textbooks on biostatistics (Krebs 1999; Neter et al. 1990; ter Braak 1991; Zar 1974; and others). I use data from published studies to illustrate typical applications and interpretations of results to address several commonly asked questions from students. There are many other statistical models that I mention in passing that may deserve serious consideration for different study objectives or to complement the few examples I provide. At the end of the chapter I provide suggestions and recommendations, including having clear analysis objectives in mind, testing multiple techniques, and considering spatial analysis following nonspatial modeling of plant diversity.

Correlations and Simple Regression

Correlations test for linear mathematic relationships between two variables. Simple regression also tests for relationships between two variables, but with one variable being an independent variable (x, or predictor variable) and one a dependent variable (y). Typical questions are as follows:

- Is there a significant positive relationship between native species richness and nonnative species richness in a set of independent, randomly selected plots?
- Can I estimate exotic species richness by measuring native species richness in a series of plots?
- Is there a statistically significant relationship between total soil nitrogen or phosphorus and either native or nonnative species richness?

One approach is to use the scatter diagram and cross-correlation procedures available in most spreadsheet and statistical software programs to plot and calculate the relationships among variables.

Simple correlations and regressions test whether the slope of the calculated regression line between two variables is significantly different from zero (Figure 13.1). In this example, a significant positive relationship was found between native and nonnative species richness ($r = 0.26$, $P < 0.02$). However, less than 7% of the variance (i.e., $r^2 = 0.07$) is explained by this relationship, suggesting that other factors are likely involved and that natural variation in the system is substantial. Many points fell outside the 80% confidence intervals. Many other factors could have also influenced the statistical relationship, such as the extent of environmental gradients and soils covered by the suite of plots; various levels of disturbance, herbivory, and competition experienced by the species in the plots; and any other peculiar historical effects and land use legacies of the plots.

The other regressions depicted may add insights to the dataset. There were significant positive relationships between nitrogen and phosphorus, and between nonnative species richness and both nitrogen and phosphorus (Figure 13.1). The relationship between native species richness and nitrogen was not significant at $\alpha = 0.05$, but the slope was positive, and there is an 88% chance ($1 - r$) of making a type II error (i.e., rejecting a hypothesis when it should have been accepted as true). It was interesting to note the very strong

Reminder: The data for some variables may not be normally distributed. They will need to be transformed prior to analysis. In the cases below, nonnative species richness (LEXOSP) and native species richness (LNATSP) were log-transformed prior to analysis as $\log_{10}(x + 1)$ (Figure 13.1). It is also presumed that the assumptions of regression analysis are met (random, independent sample points, homogeneity of variance, and no obvious pattern in the residuals) (Neter et al. 1991). Other types of transformations may be necessary for some variables, such as square-root transformations and arc-sine transformations.

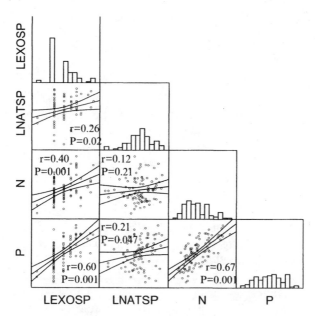

Figure 13.1.   Scatter diagram for 92 1000 m² vegetation plots in Grand Staircase-Escalante National Monument, Utah. Non-native species richness (LEXOSP) and native species richness (LNATSP) were log-transformed prior to analysis as $\log_{10}(X+1)$. The 80% confidence intervals are shown around the regression equations.

positive relationship between soil nitrogen and phosphorus in this arid eco-system, and that there were stronger relationships of soil nitrogen and phos-phorus to nonnative species richness than to native species richness.

These data should be cautiously interpreted. There are strong positive relationships between the selected soil characteristics and native and non-native plant species richness. That is, the measured attributes of increased soil fertility are positively correlated to increased richness of native and nonnative species in the sample plots. The low statistical P-values suggest that the probabilities of the positive relationships resulting from chance are very remote. However, from the last chapter, remember that statistical corre-lations do not determine cause and effect, and other unmeasured variables may be more biologically relevant in producing the observed patterns. In addition, it has been well established throughout this book that spatial scale greatly affects species richness. So the results may only apply to the specific scales tested (i.e., 1000 m² plots in these vegetation types, under similar envi-ronmental conditions and histories). Still, the observed patterns make bio-logical sense, the observed patterns have been demonstrated in many other studies (Brown and Peet 2003; Bruno et al. 2004; Keeley et al. 2003; Stohlgren et al. 1998d, 1999a, 2000b, 2001), and the strong statistical relationships should probably not be discounted without convincing contradictory data and experiments in these vegetation types at these spatial scales.

Multiple Regression Analysis

In this next case study we measured important predictors of native species richness and total species richness in forty-two 1000 m² plots on 14 transects that were randomly selected across the landscape in Rocky Mountain National Park, Colorado (Table 13.1). Stepwise forward regression analysis showed that most of the environmental factors such as aspect (degrees from due south), slope, elevation, photosynthetically active radiation (PAR) in the understory, and soil depth were important predictors of species richness.

For example, 59% and 66% of the variation in native and exotic plant species richness, respectively, could be explained by aspect, slope, elevation, PAR, and soil depth. Elevation was the only negatively correlated variable in the models for native species and total species richness. The 66% of the variance in exotic species richness ($\log_{10}$-transformed data) could be explained by positive relationships with PAR and the number of native species (Table 13.1). The T- and P-values showed that all individual variables used in the regressions were significant (i.e., the slopes were signifi-

Table 13.1.   Multiple regression results for the forty-two 1000 m² vegetation plots in ecotones and homogeneous forests in Rocky Mountain National Park, Colorado.

Dependent variable/ predictors	Coefficient	T	P	Model F, R^2, P
No. of native species				
Constant	91.2	4.42	0.001	$F = 10.5$
Aspect	0.08	2.61	0.013	$R^2 = 0.59$
Slope	0.48	2.68	0.011	$P < 0.001$
Elevation	−0.03	−4.76	0.001	df = 5 and 37
PAR	23.2	3.26	0.002	
Soil depth	0.6	2.81	0.008	
No. of exotic species ($\log_{10}$)				$F = 37.5$
Constant	−0.4	−5.16	0.001	$R^2 = 0.66$
PAR	0.29	2.16	0.032	$P < 0.001$
No. of native species	0.01	6.59	0.001	df = 2 and 40
Total no. of species				
Constant	95.8	4.34	0.001	$F = 10.9$
Aspect	0.09	2.72	0.010	$R^2 = 0.60$
Slope	0.50	2.60	0.013	$P < 0.001$
Elevation	−0.037	−4.76	0.001	df = 5 and 37
PAR	25.8	3.37	0.002	
Soil depth	0.63	2.84	0.007	

Adapted from Stohlgren et al. (2001).

cant from zero). These data can be cautiously interpreted to suggest that species richness is strongly positively correlated with more light (PAR under the canopy), warmer temperatures (i.e., indicated by a negative correlation to elevation), and deeper soils. The strongest predictor of nonnative (exotic) species richness was native species richness ($T = 6.59$, $P < 0.001$). However, there are several caveats to the analysis. Just as with simple correlation analysis, cause and effect cannot be determined, and the results can be greatly affected by many aspects of the sample design. For example, this suite of 42 plots may behave differently than another group of plots spread over a larger area. The spatial distribution of plots may cover less-extensive environmental gradients, thus reducing the amount of variation explained in the models. There is no guarantee that the same relationships would hold in severely dry years, following wildfires, or if local environmental conditions change with respect to herbivory, pathogens, contaminants, or other factors.

Path Analysis

Path coefficient analysis (Dewey and Lu 1959) provides a graphical representation of multiple regression data and evaluates the direct and indirect relationships of the abiotic and biotic factors to nonnative species richness and cover. Path analysis is a special case of structural equation modeling (Pugesek et al. 2003). In brief, structural equation modeling is a general technique for multivariate analysis. One such application is path analysis (or "causal analysis"), which tests potential causal relationships among variables with linear equations. In many cases, path analysis and structured equation modeling cannot be used to definitively establish causation between independent variables and the dependent variable, but these approaches often are helpful in describing relationships among multiple factors (Pugesek et al. 2003).

The procedure begins with forward stepwise regression, the most widely used multiple regression model (Neter et al. 1990), to include only those independent variables that are significant predictors of the dependent variable. Usually only those variables significant at $P < 0.15$ are included in the models. In this example we regressed the richness of nonnative plant species (dependent variable) with data on the richness and cover of native species, cover of cryptobiotic crusts, elevation, and soil percent phosphorus (independent variables).

In this example, the independent variables explain 46% of the variation in the richness of nonnative plant species (at the 0.1 ha plot scale). Soil percent phosphorus and elevation had the strongest direct effects on nonnative species richness, as denoted by the magnitude of the standardized partial regression coefficients (Figure 13.2). The signs of the standardized partial regression coefficients may be reversed compared to simple regressions—and

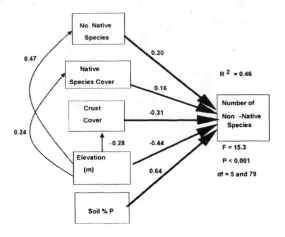

Figure 13.2.   Example of a path coefficient diagram derived from 85 plots in Grand Staircase-Escalante National Monument, Utah. Direct arrows to non-native species richness include standardized partial regression coefficient values, while arrows between environmental variables are simple correlation coefficients. R^2 is the adjusted coefficient of multiple determination. Data from Stohlgren et al. 2001.

these should be checked prior to interpretation of results. In this case, a simple regression between elevation and nonnative species richness reveals a positive correlation ($r = 0.33$, $P = 0.002$), so the direct effect can be considered positive. There were also positive direct effects with the number and cover of native species, and positive correlations between elevation and the richness and cover of native species. We can consider the "indirect effects" of elevation on nonnative species richness by the indirect effects on native species richness and cover. Meanwhile, the cover of cryptobiotic crusts had a negative direct effect on nonnative species richness (it was negative in simple regression too), and elevation was negatively correlated to crust cover, which we consider an indirect effect.

This stepwise multiple regression model may not always result in the "best" regression model for all comparisons (see Neter et al. 1990, pp. 452–453), and the reported relationships should always be examined to see if they agree with field observations. Path coefficient analysis simply displays the multiple regression analyses to assess direct and indirect effects of environmental factors on the dependent variable, with significant ($P < 0.05$) simple correlation coefficients (indirect influences) shown among environmental variables. The residual factors from the stepwise linear regressions can be easily calculated as $R_{(x)} = \sqrt{(1 - R^2)}$ (Dewey and Lu 1959; Stohlgren and Bachand 1997). The same caveats of multiple regressions listed above apply here. Some researchers prefer tabular presentations of results (Table 13.1), while others prefer to diagram results as in Figure 13.2.

Canonical Correspondence Analysis

Canonical correspondence analysis is used to characterize complex relationships between biotic and abiotic variables (CANOCO, version 3.12) (ter Braak 1987b, 1991). Commonly species composition data from many vegetation plots (e.g., 30 to 100 or more plots) constitute one input file, while environmental data for each plot constitute a second input file. The species data can be individual species cover/abundance data or they can be lumped by functional group, genetic makeup, structure class, or some other class. CCA is a direct gradient analysis technique that constrains the extracted pattern of vegetation to a linear combination of the measured environmental variables (ter Braak 1986, 1987a), and it is a proven, robust method for describing species-environment relationships [see Palmer (1993) for a complete review; Reed et al. 1993]. In this example from Utah, we evaluated the relationships among native and exotic species richness, cryptobiotic crust cover, and environmental measurements including elevation, percent bare ground, and soil nutrients (nitrogen, carbon, inorganic nitrogen, magnesium, and phosphorus) (Figure 13.3). All the default options were selected in the CCA (i.e., no special weighting of species or environmental factors and no samples were excluded from the analysis). It is important to assess all environmental variables for multicollinearity problems. High "tolerance" values in independent variables in a multiple regression are one indication of potential multicollinearity between variables. Generally the software packages will also allow the option of conducting Monte Carlo permutation tests (99 random permutations) to test the

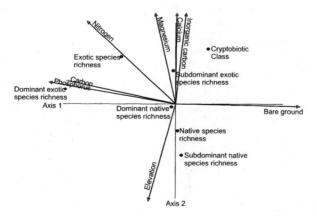

Figure 13.3.   Canonical Correspondence Analysis ordination for native and exotic species richness and cryptobiotic class. The "dominant" designation refers to the number of species with >1% cover. The "subdominant" designation refers to the number of species with <1% cover. Total richness is the total number of species regardless of cover. Data from Bashkin et al. 2002.

significance of the first canonical axis (ter Braak 1991). This provides additional evidence of the likelihood of a statistically robust model.

Canonical correspondence analysis found that the centroids of native and exotic species richness were separated by many soil characteristics (Figure 13.3). The centroid for native species richness was located closer to the center of the diagram, corresponding to low elevation and less fertile soils, but perhaps broadly distributed throughout the diagram (and common in the landscape). The centroid for nonnative species richness was located in higher soil nutrient areas lying directly on the soil nitrogen gradient. The centroid for dominant exotic species richness (the number of exotic species with cover greater than 1%) was also linked to a soil nutrient gradient most closely associated with increasing soil carbon and phosphorus.

In this example, the output that accompanied the diagram showed that the first two canonical axes explained 36.9% of the cumulative variation in the species groups. The top three environmental factors that correlated significantly ($P < 0.05$) with canonical axis 1 included soil phosphorus ($r = -0.84$), bare ground ($r = 0.71$), and elevation ($r = -0.67$). The top three environmental factors that correlated significantly ($P < 0.05$) with canonical axis 2 included soil magnesium ($r = 0.54$), soil inorganic carbon ($r = 0.54$), and elevation ($r = -0.44$). Monte Carlo permutation tests showed that the first canonical axis was significant (eigenvalue = 0.029, $P < 0.01$). This type of analysis can be carefully interpreted to suggest that the centroid for the number of exotic species is closely associated with increasing soil nitrogen and the centroid for dominant exotic species (number of nonnative species where cover is greater than 1%) is closely associated with increasing soil carbon and phosphorus (high resource environments) (Figure 13.2) in low elevation areas. In contrast, native species richness was greatest in higher elevation sites with more moderate soil nutrients (or perhaps a more widespread pattern from fertile to nonfertile soils). Together with other data on the success of plant invasions in rare, species-rich vegetation types on fertile soils, these patterns generally showed that invading nonnative species were more successful establishing in more fertile habitats in the landscape.

In another example of CCA, we examined the patterns of dominant forest understory species associated with various overstory tree dominants along 14 vegetation transects in Rocky Mountain National Park, Colorado (Stohlgren et al. 2001). We found that dominant understory species distributions at landscape scales were described generally by gradients of elevation and available light (undercanopy PAR). The first two ordination axes explained 64% of the variance in vegetation patterns. Environmental factors that correlated significantly ($P < 0.001$) to the first ordination axis included elevation ($r = -0.94$) and PAR ($r = 0.64$). Environmental factors that significantly correlated to the second ordination axis included aspect (degrees from due south; $r = -0.60$, $P < 0.001$) and slope ($r = 0.46$, $P < 0.002$). Monte Carlo permutation tests showed that the first canonical axis was highly significant (eigenvalue = 0.84, F-ratio = 3.4, $P < 0.01$). In this study we mapped the domain of understory species associated with the particular overstory tree species by connecting perimeter plots from the same vegetation type on the CCA plot. Connecting the perimeter plots from only the homogeneous plots (Figure 13.4, top) resulted in small domains for the five dominant understory species associated with the spruce-fir and limber pine vegetation types.

In this example (Figure 13.4, top), only plots in homogenous units were used in the analysis and domains of the five dominant understory species associated with specific overstory species were confined in environmental

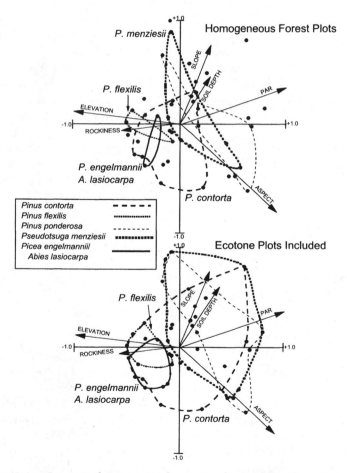

Figure 13.4.   Canonical Correspondence Analysis ordination. Top diagram shows the domain of the dominant five understory species per plot excluding the ecotone plots, while the bottom diagram includes the ecotone plots. Lesson 22. Excluding heterogeneous areas from landscape-scale studies of plant diversity could create very misleading representations of species distributions and species-environment relationships.

space, though there was much overlap among the Douglas-fir, ponderosa pine, and lodgepole pine types. The domain of understory species associated with lodgepole pine was fairly large, as was the domain for the understory species of Douglas-fir and ponderosa pine types.

However, when the ecotone plots are included (i.e., heterogeneous habitats between homogeneous stands), the potential domain of dominant understory species was greatly expanded for all forest types (Figure 13.4, bottom). The domain of the dominant understory plants associated with Douglas-fir more than doubled on the CCA plot. In this study we learned that confining sampling to more homogeneous forest stands could be very misleading in terms of the distributions and environmental conditions supporting associated understory species (Stohlgren et al. 2001). Heterogeneous areas may make up a large portion of any landscape, and species coexistence and persistence might be greatest under heterogeneous forest conditions.

Canonical correspondence analysis has the advantage of assessing abiotic and biotic features of the environment simultaneously. Like other ordination techniques, it has weaknesses too. The outcome is heavily dependent on many factors, including the size of the vegetation plots used, the intensity and pattern of plot placement, how well and which factors are measured, and the magnitude and heterogeneity of the environmental gradients underlying the observed plots. Linear combinations of variables may not adequately represent nonlinearities and thresholds in the data, the centroids of depicted dependent variables may not represent the true variability around the displayed mean, and domains built by connecting the dots only provide a superficial picture of complex species (or species-group) distributions. As with other correlation-based statistical approaches, no cause-and-effect relationships can be assumed or implied. Still, the approach and statistics used here help to generate testable hypotheses while adding to our understanding of patterns of plant diversity in a nonspatial way.

Regression Tree Analysis

Regression tree analysis is a nonparametric approach used to describe multiple variables in a dichotomous, two-dimensional framework. It is similar to multiple regression, in that many independent variables can be related to a single dependent variable. It is also similar to cluster analysis in separating groups of plots with similar characteristics. An example exploratory null hypothesis might be: There is no difference in the nonnative species richness per plot as related to site characteristics such as native species richness, soil nitrogen, and elevation.

Regression tree analysis begins with all cases (plots) in one cluster and splits the data attribute by attribute into a hierarchical binary tree. The terminus of each branch represents a cluster whose members are more similar to each other than to members of the twin cluster (SPSS Inc. 2001). All default options were

> *Reminder:* Most complex procedures in commercially available statistical packages have many options and default settings. Plant ecologists need a sound understanding of the algorithms used, the assumptions of the models, and the effects of various options, including the default options used in the statistical procedures.

used to determine stopping rules and loss functions. This creates a parsimonious model with a simple branching structure in most cases.

Regression and classification trees have been used to create theoretical, predictive decision trees for introducing nonnative species into an area (Lee 2001; Reichard and Hamilton 1997). Regression tree analysis is used to identify key independent variables, but not as a predictive tool.

In this example, regression tree analysis was used on raw data and on $\log_{10}$-transformed data to characterize general trends in the role of life cycles in the invasion of nonnative plant species. Raw data performed the same as $\log_{10}$-transformed data, so raw data were used for ease in interpretation. This sample of 293 plots in Colorado averaged 1.9 nonnative annuals per 0.1 ha plot (Figure 13.5). The plots averaged 1.9 nonnative annuals per 0.1 ha plot. Almost 72% of the plots averaged less than one nonnative species. However, if there were more than four native annual species per plot, then the plots averaged 4.4 nonnative annual species. If there were less than 16 native perennial species on the plot, the average nonnative annual species richness increased to 8.5 species per plot. Thus very high richness of native perennials was negatively associated with the nonnative species richness.

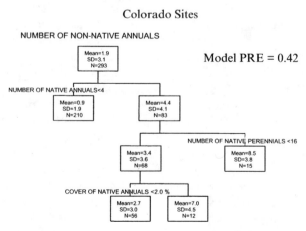

Figure 13.5.   Regression tree analysis of 293 plots in Colorado. The independent variables included in the model were the richness and cover of native annual and native perennial species, to predict the number of non-native annual species in 0.1 ha plots.

In another branch of the model, high cover of native annuals was positively associated with the number of nonnative annual species per plot. About 42% of the variation is explained by the independent variables [proportion reduced error (PRE) values, which are similar to R^2 values].

In this example, the hypothesis was that the cover and richness of native annual and perennial species might help assess the vulnerability of a site to invasion by nonnative species, especially nonnative annual species. High native annual species richness on a site might indicate seasonally variable climate, frequent disturbance, and less competition from larger perennial species compared to sites with low richness of native annual species. Such sites might be more vulnerable to invasion by a greater variety of nonnative annual species. That was the case for these 293 plots in Colorado (Figure 13.5), but cause-and-effect relationships associated with this pattern could only be confirmed by carefully crafted experiments.

Regression tree diagrams are easily understood by most practitioners and students. They have a great advantage over complex multiple regression and path coefficient analysis techniques in that regard. The disadvantages of the approach have to do with the dichotomous nature of the defining algorithms. The branching pattern may mask complex thresholds in the data or they may conveniently split a gradual cline in the data. Again, as with other regression techniques, it doesn't "predict" or determine cause-and-effect relationships, it merely quantifies patterns.

Logistic Regression

Logistic regression is a multivariate approach commonly used to identify significant variables useful in describing the "presence" of a given species in a given suite of plots based on environmental characteristics. In this example we looked at the patterns of a three successful invasive species in the Grand Staircase-Escalante National Monument, Utah: *Bromus tectorum*, a dominant generalist (annual grass) found in 114 of 142 plots used in the study; *Erodium cicutarium*, a codominant generalist (annual herb) found in 14 plots; and *Poa pratensis*, a specialist (perennial sod-forming grass) found in 14 plots (Table 13.2) (Chong 2002). Each study plot was assigned with a binary value for each of the three nonnative species based on its presence within the plot (species present = 1, absent = 0). Stepwise forward logistic regression was used to identify significant predicting variables for the particular species. Independent variables included elevation, latitude, longitude, native species richness, percent soil sand, percent bare ground, percent woody vegetation cover, percent perennial vegetation, soil phosphorus, soil nitrogen, soil carbon, maximum cryptobiotic crust development, average crust cover, and moisture index.

In this example, the logistic regression models for the three species were significant ($P < 0.001$), but the variables in the model explained only 36–59%

Table 13.2.   Summary of logistic regression analyses for three selected exotic species in Grand Staircase-Escalante National Monument, Utah.

Variable	Estimate	Standard error	t-ratio	P
a. *Bromus tectorum*, dominant generalist ($n = 35$)				
1. Soil phosphorus	3.38	1.26	2.68	0.01
2. Elevation	−0.002	0.001	−1.64	0.10
3. UTM north	−0.00	0.00	−1.90	0.06
4. UTM east	0.00	0.00	1.27	0.20
5. Native species richness	0.06	0.05	1.72	0.24
6. Percent sand	−0.02	0.02	−0.77	0.44
7. Bare ground	−0.33	0.18	1.78	0.08
8. Percent woody cover	−2.81	1.63	−1.67	0.10
Whole model McFadden's $p^2 = 0.36$	$P < 0.001$		Total correct 80.3%.	
b. *Erodium cicutarium*, codominant generalist ($n = 135$)				
1. Percent perennial species	−10.5	4.65	−2.25	0.024
2. UTM north	−0.00	0.00	−3.08	0.002
3. Elevation	−0.01	0.01	−4.10	0.000
4. Soil phosphorus	6.10	2.07	2.95	0.003
5. UTM east	−0.00	0.00	−2.38	0.017
Whole model McFadden's $p^2 = 0.53$	$P< 0.001$		Total correct 90.2%	
c. *Poa pratensis*, habitat specialist ($n = 142$)				
1. Moisture index	0.37	0.13	2.88	0.004
2. Maximum crust development class	−0.43	0.19	−2.23	0.026
3. Topographic position	−2.94	1.37	−2.15	0.032
4. Bare ground	−0.56	0.21	−2.61	0.009
Whole model McFadden's $p^2 = 0.59$	$P < 0.001$		Total correet 88.3%.	

of the variation (McFadden's ρ^2). Despite this variation, the models had a very high "prediction success" of between 80% and 90% (total correct; Table 13.2).

We conducted two-sample t-tests to compare the differences in means of each variable between plots invaded by selected nonnative species and plots not invaded (Table 13.3). For example, plots invaded with *E. cicutarium* were 284 m lower in elevation, had 3.6 fewer native species per 0.1 ha plot, and had 5.4% more sand (fraction) than uninvaded plots. Plots invaded by *B. tectorum* had significantly higher native species cover and higher soil percent nitrogen and phosphorus compared to uninvaded plots.

Logistic regression analyses have the benefit of quantifying the current range distribution of target species. Such analyses assume that the pattern of sampling captured the entire range of the species in question and the environmental gradients associated with the species distribution. These assumptions are rarely met in ecological studies because small sample sizes prohibit extensive sampling of the full environmental gradients covered by the vast majority of species. Like other regression techniques, cause-and-effect relationships cannot be presumed or implied.

Table 13.3.   Mean differences in variables (mean value in "present" plots minus mean value for "absent" plots) for the three invasive plant species in tested with logistic regressions.

Species/characteristic	Bromus tectorum	Erodium cicutarium	Poa pratensis
No. of present plots	114	14	14
Moisture index			5.8
Topographic position			−0.5
UTM north		−13450	
UTM east	15692		
Elevation (m)		−284	221
Native species		−3.6	
Native cover	0.8		
Bare ground percent	−2.5	−1.3	−3.4
Maximum crust class			−2.7
Percent crust cover			−2.8
Percent woody cover	−0.1		−0.2
Percent perennia1	−0.1	−0.1	
Percent graminoid	0.06		0.07
Percent soil sand		5.4	5.2
Percent soil carbon			
Percent soil nitrogen	0.13	0.07	
Percent soil phosphosus	0.33		0.39

Suggestions and Recommendations

It is important to clearly articulate analysis objectives prior to statistical tests and modeling because there are many possible approaches to the nonspatial modeling of plant diversity. Begin by designing the study in such a way that the right field measurements are made and the appropriate ancillary data are collected. Appropriate statistical techniques can only be selected after study objectives, hypotheses, and specific questions have been clearly stated relative to the collected data. A "shotgun" approach, where various techniques are tested until desired or statistically significant results are achieved, is cost inefficient and inappropriate in many cases. Regardless of the specific study objectives, it is always wise to consult a statistics text (Neter et al. 1990; Zar 1974; and others) and a statistician at the design phase of the study. It is often equally important to test analysis techniques much like you would test field techniques. However, never assume that all the statistics in software packages contain the proper algorithms for your specific objectives. One student calculated standard deviations in a popular software package only to discover later that the software package calculated the variance and standard deviation for a "population" rather than a "sample." Population data are rare in ecological field studies, while simple computer programs are common on field ecologists' computers.

There are many other aspects of nonspatial modeling and model selection that are beyond the scope of this text (see Burnham and Anderson 2002; Stohlgren et al. 2005a). Often it is important to try multiple approaches to modeling the same data, not as a shotgun approach, but to evaluate what the models do and don't tell you and how various models may complement each other. Various assumptions and the advantages and disadvantages of various models have to be stated and explored. It is also important to consider using spatial analysis in conjunction with nonspatial models. Spatial models are detailed in subsequent chapters.

14

Spatial Analysis and Modeling

The Issue

Data on plant diversity are inherently spatial. Spatial analysis helps us determine where plant diversity is high or low. As we shall see, understanding spatial autocorrelation is important for theoretical reasons and for conservation. There are various spatial techniques available to map and model plant diversity or the distributions of selected species at landscape scales. The most rudimentary of maps are point distributions or mapped patches of known occurrences (Figure 14.1). Generalized, broad-scale maps of species distributions are often more of an art than a science. Usually only large, contiguous areas are mapped by connecting the outermost dots of point distribution maps. Alternatively, point location maps may show more exact data about species presence without attempting to display contiguous populations (Figure 14.1). Assuming the locations are reasonably current and accurate, and the map is at an appropriate scale to guide new observations from field crews, such a map might be immediately useful to assess changes in rare plant populations or to control specific populations of noxious weeds. However, there may be no way of knowing how well the landscape was surveyed or if many other populations exist but are not yet mapped. Likewise, time may have passed since the initial data were gathered or since the map was created and the population in question may have disappeared into the seed bank, or by dying. Occurrence

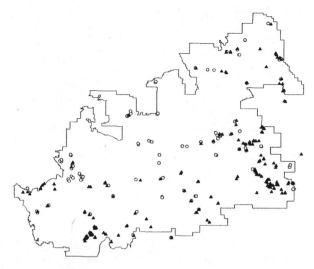

Figure 14.1. A generalized point distribution map of *Bromus tectorum*, cheatgrass, based on the presence (solid triangles) and absence (open circles) of the species in 0.1 ha plots in Grand Staircase-Escalante National Monument, Utah (unpublished data).

maps rarely show cover or abundance data for the points or other data to assess the probability of occurrence, cover, or abundance across the landscape in question.

Range and point distribution maps are more useful when combined with ancillary data on foliar cover, topographic variables (e.g., slope, aspect, and elevation), and soil characteristics (e.g., texture, chemistry). Field crews can then analyze the correlations between occurrences and environmental factors to aid in finding other populations of concern. If most of the species of concern are found in low-elevation riparian zones with high soil fertility rather than in high-elevation, upland sites of low soil fertility, this may help guide future surveys (Stohlgren et al. 1998d). However, as Daubenmire (1968) points out, finding a species present in a particular habitat tells you something about where the species can become established; not finding a species in a particular habitat tells you nothing (i.e., the species may be hidden in the seed bank or may invade there soon). As we know, two habitats may have similar environmental attributes, but different species composition (see chapter 2). Likewise, in our invasive species example, sites with different environmental constraints may be invaded by the same species. Or if two sites have similar environmental attributes, but one site is directly adjacent to a seed source, it may be invaded before the more remote site. Understanding the spatial relationships among many factors is an important prerequisite to understanding patterns of invasion, rarity, endemism, and persistence.

The crux of the issue is that only a small portion of any landscape can be affordably sampled (usually less than 1%), so we had better be pretty good at estimating what is in the much larger unsampled area. Having mapped points or patches, or after collecting plant diversity data and ancillary information on many plots, several questions remain (Chong et al. 2001). From a small sample of points, how can we accurately model the patterns of plant diversity or the distribution of individual species? How much of the variation in the spatial patterns of plant diversity or the distribution of individual species can be explained by the model? How much uncertainty is there in the spatial representation of model outputs?

A Simple Example of Kriging

Most spatially explicit data can be analyzed with simple kriging algorithms available in many statistical software packages. The basic inputs are x and y coordinates and several data points of a single biotic or abiotic resource of interest. The objective is to produce a simple contour-like map of the resource of interest, for example, producing isoclines of native species richness. In this example of simple kriging, data on native and nonnative (exotic) species richness and soil percent nitrogen and carbon were gathered from twenty-four 0.1 ha plots in a 754 ha area in Rocky Mountain National Park, Colorado. The UTM coordinated for each plot were obtained in the field with a GPS. We relied on a standard statistical package (SPSS Inc. 1997), using the PLOT routine with kriging selected as the "smoother."

The resulting graphical output shows isoclines of the biotic or abiotic variables in coordinate (x, y) space (Figure 14.2). It was apparent by inspection that there were coinciding patterns of native plant diversity and soil quality (percent nitrogen and percent carbon) in the study area (Figure 14.2). It was also apparent that nonnative species richness was tracking native species richness and soil quality.

There are many caveats with this simple kriging example. The models are affected by the accuracy and completeness of the field data (e.g., plot size and shape, the number and placement of plots, the accuracy of the GPS unit and plant taxonomist in the field, etc.). The models do not quantify the relationships among native and nonnative species richness and soil variables, they simply display them. As with other statistical models, no cause-and-effect relationship can be determined or assumed. There are also underlying statistical considerations that must be examined. The algorithms used by the software package have defaults preset for the number of neighboring points considered in the kriging function, and the boundary conditions and distribution of sample points in the kriged area can influence results (Figure 14.2). More importantly, there is no estimate of the variance explained or the level of uncertainty with these simple models, and there would be a missed opportunity if the statistical relationships among the variables were not ex-

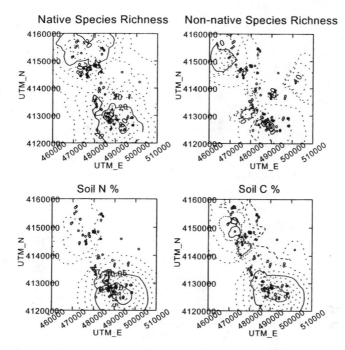

Figure 14.2.   Kriging diagrams of native and exotic species richness, soil % C, and soil % N over the study area. UTM coordinates are on the X and Y axes.

plored and quantified. That is the realm of the more complex kriging example that follows.

A More Complex Approach to Kriging

Quantifying the patterns of native and nonnative plant species over large landscapes could be very challenging. In this example we wanted to see if habitats with high native plant species diversity areas were more successfully invaded by nonnative plant species than species-poor areas (Stohlgren et al. 1998d, 1999a, 2001), and whether rare, wetter habitats high in soil fertility were particularly vulnerable to invasion (Chong et al. 2001). Spatial models developed from multiscale field data can provide important information on patterns of plant invasions so that resource managers can conduct strategic control efforts (Kalkhan and Stohlgren 2000). Full-coverage maps of the environment are needed to extrapolate information from points to landscapes (Reich and Bravo 1998). Spatial statistics (or geostatistics) provide a means to develop spatial models that can be used to correlate coarse-scale geographical data with multiscale field measurements of biotic and abiotic variables (Kalkhan and Stohlgren 2000). In this section I provide a brief introduction to the general concepts of simple spatial interpolation models and provide some preliminary results from our spatial modeling approach as an example of data analysis and synthesis.

Example Methods

In this example from Chong et al. (2001), we used modified Whittaker vegetation data (ninety-four 1000 m² plots) from a 54,000 ha portion of Rocky Mountain National Park. Sample points were located based on stratified random sampling in vegetation cover types ranging from wet meadow to alpine tundra [the procedure is described in Stohlgren et al. (1997a)]. This example is used to demonstrate preliminary spatial models to interpolate patterns of native and nonnative species richness and the probability of encountering a nonnative plant species in 30 m × 30 m cells throughout the study area. The 94 plots represent a sampling intensity of 1 plot per 575 ha (i.e., less than 0.2% of the landscape was sampled).

The geographic information system database used to develop the models contained several coverages of independent variables thought to influence the variability in species richness and the presence of exotic species (Chong et al. 2001). These included a 30 m resolution digital elevation model (DEM; Department of Interior, U.S. Geological Survey), which was used to create a 30 m grid overlay of percent slope and aspect (GRID; ArcInfo, ESRI 1997). The database also included 30 m resolution overlays of Landsat Thematic Mapper ™bands 1 through 7. The point coverage of the sample data was used to extract point estimates of elevation, slope, aspect, and the digital numbers associated with the seven Landsat bands (Table 14.1).

Geostatistical Analysis

The process of developing spatial models for the distribution and abundance of plant species contains three repeating steps:

1. The acquisition of new or previously collected, spatially explicit data, collected carefully to link to remotely sensed data.
2. The generation of spatial coverages of specific resources such as the distribution and abundance of the taxa or soils using geostatistical algorithms. The coverages should include assessments of the relative accuracy of the spatial interpolations.
3. Subsequent data are collected to refine the model and improve the accuracy of the model projections. This iterative sampling and modeling greatly improves the models.

Acquiring field data appropriate for spatial modeling of plant diversity in complex terrain can be difficult and expensive. Adequate data must be collected in common and rare habitats, and along extensive environmental gradients throughout the landscape to provide accurate projections across

Table 14.1.   Summary statistics of data used in modeling species richness (native and nonnative species per 1 m² plot) and the presence of exotic species in a 54,000 ha area of Rocky Mountain National Park, Colorado.

Variable	Mean	Std. Dev.
No. of native species	7.2	4.7
No, of nonnative species	0.6	1.0
Elevation (m)	2778	271.8
Slope (%)	13.6	8.5
Aspect	97.5	54.7
Band 1	60.5	9.1
Band 2	26.2	5.7
Band 3	26.9	9.0
Band 4	60.7	15.6
Band 5	71.5	27.8
Band 6	154.6	17.3
Band 7	33.6	17.0

Adapted from Chong et al. (2001).

unsampled regions. Unbiased field data are preferred in most cases, but subjective sampling may be needed to capture the extreme tails of environmental gradients that would be missed by random sampling with the small sample sizes typical in many studies. Data from previous studies, even if the data have bias associated with the sample design, may be useful in model development.

Spatial models begin by incorporating various geographic information system data such as remote sensing, vegetation type, soil characteristics (e.g., texture, nutrients), and topographic variables. In this example, multiple regression analysis was first used to explore the variability in species richness and the presence of nonnative species as a function of the geographical location, elevation, slope, aspect, and Landsat TM bands 1–7 (Figure 14.3).

Stepwise regression is used to identify a linear combination of independent variables to include in a trend surface model of geographical variables and measures of species richness to describe large-scale spatial variability in the study area (Chong et al. 2001; Kallas 1997; Metzger 1997). The regression coefficients associated with the trend surface component of the model are calculated as in a typical multiple regression (Table 14.2).

The algorithms used in this example include ordinary least squares, Gaussian least squares, and autoregressive models that are described in Kallas (1997), Metzger (1997), and Chong et al. (2001). Producing the "error term" for the model is very important because the data may or may not be spatially correlated with neighboring data (Kallas 1997; Metzger 1997). Residuals of the regression models are computed and used for modeling their semivariograms, with model parameters estimated using weighted least

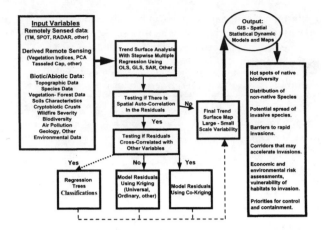

Figure 14.3   Flow diagram of statistical procedures. Adapted from Kalkhan 2002.

squares (Cressie 1985). The residuals are also analyzed for spatial autocorrelation and cross-correlation (Bonham et al. 1995; Czaplewski and Reich 1993; Reich et al. 1994) with the geographic variables (Figure 14.3). Inverse distance sampling is used to define the spatial weights matrix, and estimates of the residuals are obtained using ordinary kriging [as described in Chong et al. (2001)].

If the included variables are spatially correlated with the variable of interest, this information can be used to improve estimates (Isaaks and Srivastiva 1980) by interpolating values for points not measured (Robertson 1987). The use of auxiliary information in spatial prediction is referred to

Table 14.2.   Regression model used to explain the large-scale spatial variability of the number of native species (in a 30 m × 30 m cell) in a 54,000 ha area of Rocky Mountain National Park, Colorado.

Variable	Coefficient	P
Intercept	−454.3	0.030
x coordinate	−0.0002	0.001
y coordinate	0.0001	0.015
Band 1	0.348	0.001
Band 3	−0.394	0.001
Band 5	0.107	0.001
Band 6	0.145	0.001
Band 7	−0.168	0.002
Elevation (m)	0.006	0.001

$R^2 = 0.21$, standard error $= 4.2$, $n = 940$ plots (1 m^2)
    The x, y coordinates are in meters (UTM coordinates).

as cokriging (Chong et al. 2001). I will provide only limited details of kriging and cokriging here, leaving the bulk of the topic, including equations, variance estimates, and validation techniques, to other textbooks (e.g., Cressie 1991) and Web pages (http://www.warnercnr.colostate.edu/frws/people/faculty/reich.html) dedicated to spatial analyses.

An important aspect of the process is to assess the spatial autocorrelation or cross-correlation of the residuals from the regressions (Figure 14.3). To account for spatial autocorrelation in the residuals of the model developed to describe large-scale spatial variability, we might model the small-scale spatial variability (i.e., spatial noise) using the cokriging model [see Chong et al. (2001) for model details]. One of the appealing features of cokriging is that the auxiliary information does not have to be collected at the same data points as the variable of interest. This allows us to combine remote sensing and field data to provide a full coverage map with a higher resolution than would have been possible by using remote sensing or field data alone. In essence, predictive models provide information on large-scale spatial variability, while field data provide information on small-scale spatial variability (Chong et al. 2001).

Cokriging involves additional computational steps that are beyond the scope of this chapter. In short, the residuals of the model describing the large-scale spatial variability are analyzed for anisotropy (spatial autocorrelation changes with direction) prior to fitting the cokriging model. The residuals are also evaluated for the presence of spatial cross-correlation (Bonham et al. 1995; Czaplewski and Reich 1993; Reich et al. 1994) with the independent variables included in the trend surface model, and variables for which only data associated with field plot locations are available (Chong et al. 2001). Complete coverage of the variables associated with the field data is usually not available in the geographic information system database because one can only afford to sample points. If no spatial cross-correlation is detected, the residuals can be modeled using ordinary kriging, otherwise the residuals are modeled using cokriging. Where large datasets from the field are possible, regression tree analysis, a much more expedient computational method, can be substituted for cokriging (Figure 14.3).

Spatial Integration

Spatial integration is used to model field data over any specified geographical region (e.g., point- and plot-level field data, management unit, watershed, region) to obtain a point estimate and associated standard error of prediction (Chong et al. 2001). This is accomplished by integrating the three-dimensional response surface representing the variable of interest over the area of interest and dividing by the area. The spatially modeled response surfaces can be represented as a grid in ARC/INFO (ESRI 1997) as a finite number of grid cells of uniform size (i.e., 30 m × 30 m). Specific mathematical models (the details of which are beyond the scope of this book) can be used to develop point estimates of a resource in some bounded region, with the point estimate containing a calculated variance (Chong et al. 2001). The spatial correlation is estimated using the appropriate variogram function. Resource managers can use the error maps to determine what areas warrant further field data collection to increase model accuracy.

In this example, estimates of native and nonnative species richness and the presence of nonnative species are obtained by adding the regression estimates based on elevation, slope, aspect, etc., and the estimated residuals computed using ordinary kriging. Kriging was carried out using the four nearest neighbors in this example.

The modified residuals kriging models are then cross-validated to assess the variability in the prediction errors (Chong et al. 2001). Cross-validation includes deleting one or more observations from the dataset and predicting the deleted observation using the remaining observations in the dataset. This process is repeated for all observations in the dataset. Summary statistics of the estimated values can then be computed and mapped as the average differences between observed and expected values (Havesi et al. 1992).

Example Results

Regression Models

In this example from 94 sets of ten 0.1 m^2 plots in a 54,000 ha area of Rocky Mountain National Park, Colorado, the first multiple regression model explained only 21% of the variation in the native species richness based on location, elevation, and Landsat TM bands 1, 3, 5, 6, and 7 (Table 14.2). The small amount of variation explained in predicting native species richness might be due to many reasons. Native species have adapted to a wide range of conditions and environmental gradients. The model might have been improved if forest canopy cover, soil moisture, and soil fertility were included in the model or if more plots had been established covering greater environmental gradients. Still, the models were effective at quantifying the large number of native species in the northern and western portions of the study area (Chong et al. 2001). The significant Landsat TM bands provide information about differences in the vegetation and soils throughout the study area and their influence on the richness of native and nonnative species (Jensen 1996).

In this example, the regression model for nonnative species richness explained 31% of the variation in nonnative species richness based on geographical location, elevation, slope, aspect, and Landsat TM bands 2, 3, 5, 6, and 7 (Table 14.3). Thus there are other factors (e.g., soil moisture, soil fertility) or other models (e.g., cokriging, regression trees) that might help explain more of the variation in nonnative species richness. Still, nonnative species richness (Table 14.3) may be more easily predicted than native species richness in the study area (Table 14.2).

Table 14.3. Regression model used to describe the
large-scale spatial variability in the nonnative species
richness (in 1 m² areas) in a 54,000 ha area of Rocky
Mountain National Park, Colorado.

Variable	Coefficient	P
Intercept	83.42	0.065
x coordinate	0.00003	0.001
y coordinate	−0.00002	0.039
Band 2	−0.0966	0.001
Band 3	0.0560	0.007
Band 5	0.0386	0.001
Band 6	−0.0082	0.074
Band 7	−0.0528	0.001
Elevation (m)	−0.0011	0.001
Slope (%)	0.0069	0.081
Aspect	−0.0011	0.066
No. of native species	0.0803	0.001

$R^2 = 0.31$, standard error = 0.84, $n = 940$ plots (1 m²).
   The x, y coordinates are in meters (UTM coordinates).

An important feature of the two models is that general patterns could be
detected that matched field observations. Nonnative species were more
prevalent in the southern and eastern portions of the study area and at lower
elevations. The positive correlation with slope and the negative correlation
with aspect indicate that nonnative species are more prevalent on steeper
and more southerly exposures. The positive correlation with the number of
native species may indicate that the nonnative species are invading areas
with high native plant species richness. This result agrees with the findings
in similar habitats (Kalkhan and Stohlgren 2000; Stohlgren et al. 1998d,
1999a, 2001).

In the next aspect of this example, the regression model to predict the
probability of occurrence of nonnative species (in a 30 m × 30 m cell) was
similar to the one developed for nonnative species richness (Table 14.4),
with a preliminary model explaining 38% of the variation.

As with the previous models, the residuals were analyzed for spatial
autocorrelation and cross-correlation. In this example, the residuals of the
regression models were positively spatially autocorrelated at the $a = 0.05$
level of significance. This suggests that kriging was an appropriate model
to run (Figure 14.3). No significant cross-correlation was observed between
the residuals and the independent variables used in developing the models
(Chong et al. 2001), so cokriging would have been inappropriate for this
dataset.

Table 14.4.   Regression model used to describe the
large-scale spatial variability in the probability of the
presence of nonnative species (in 1 m^2 areas) in a 54,000
ha area of Rocky Mountain National Park, Colorado.

Variable	Coefficient	P
Intercept	38.89	0.05
x coordinate	0.000	0.001
y coordinate	−0.000	0.04
Band2	−0.045	0.001
Band 3	0.029	0.001
Band 5	0.015	0.001
Band 7	−0.019	0.001
Elevation (m)	−0.001	0.001
Slope (%)	0.007	0.001
Aspect	−0.001	0.05
No. of native species	0.037	0.001

$R^2 = 0.38$, standard error $=0.38$, $n = 940$ plots (1 m^2).
The x, y coordinates are in meters (UTM coordinates).

Kriging Results

In this example, kriging was used to create a generalized map of plant species
richness in 30 m × 30 m cells in a 54,000 ha area of Rocky Mountain National
Park, Colorado (Figure 14.4). The greatest species richness occurred in the
low-elevation areas in the eastern and southwestern portions of the study area.

Similar kriging models, with slight variations, were used to describe
native species richness (exponential semivariogram model), nonnative species richness (Gaussian model), and the probability of nonnative species
occurrence in a 30 m × 30 m cell. Model parameter estimates of the semivariograms for the three models provide an instructive comparison (Table 14.5).

The large range associated with the residuals for native species richness
suggests the presence of large-scale spatial continuity in the number of native
species across the study area (Chong et al. 2001). This indicated broad-scale
adaptations to many habitats and general patterns correlated to elevation
and forest zones (Peet 1981), with fewer plant species occupying thickly
forested and high-elevation areas. In contrast, the small range associated with
the nonnative species models indicates that the nonnative species occur in
small patches throughout the study area (e.g., open areas, aspen stands,
meadows). A large nugget effect relative to the sill for a model reflected high
variation within patches.

There are clear benefits to spatial interpolation models over multiple
regression models where spatial autocorrelation obviously exists in the
dataset (Table 14.6). For example, the kriging model for the number of native species explained 62% of the variation, while the regression model
explained only 21% of the variation (Table 14.6).

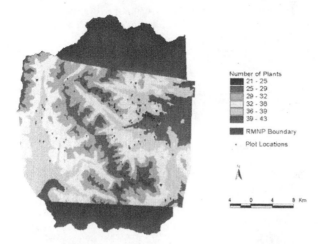

Figure 14.4.   Generalized map of plant species richness in 30 m x 30 m cells (smoothed with a minimum mapping unit of about 150 ha) in a 54,000 ha area of Rocky Mountain National Park, Colorado (Chong 2002).

Kriging the residuals improved the model by reducing the relative mean squared error by 53%. The modified residual kriging model for nonnative species richness had a relative mean squared error of 0.50 ($R^2 = 0.51$). This reduced the relative mean squared error to 29%. Similar mean squared errors were observed for the probability of occurrence model for nonnative species. The larger errors associated with the nonnative species models may have been due to the small-scale spatial heterogeneity associated with the patchy distributions of nonnative species. This was commonly observed in the field survey. This small-scale spatial heterogeneity makes it difficult to predict the spatial variability in the probability of occurrence or number of nonnative species at the 1000 m^2 plot scale, but it may help guide land managers to specific habitats that are easily invaded (Chong et al. 2001).

Table 14.5.   Parameter estimates of the semivariograms used to describe the spatial continuity in the residuals.

Regression model	Nugget	Sill	Range	Semivariogram model
Native species richness	15.94	1450	32,705,000	Exponential
Nonnative richness	0.76	1.12	151.4	Gaussian
Probability of nonnative species	0.08	0.19	215.1	Gaussian

Adapted from Chong et al. (2001).

Table 14.6.   Relative mean square errors associated with estimating the number of native and exotic species and the probability of observing an exotic species in a 30 m × 30 m cell.

Regression model	Estimation technique	Relative mean square error	R^2
Native species richness	Regression	17.8	0.21
	Kriging	*8.4*	*0.62*
Nonnative species richness	Regression	0.70	0.31
	Kriging	*0.50*	*0.51*
Probability of a nonnative species in a plot	Regression	0.14	0.38
	Kriging	*0.12*	*0.47*

Validating Spatial Models and Assessing Model "Uncertainty"

All spatial models must be validated and assessed for uncertainty. Just as plant ecologists would never (or rarely) think of presenting a histogram of means without standard error bars, landscape ecologists and modelers should not display modeled maps without information on validation and uncertainty.

Validating a model can occur in several ways, such as collecting independent samples and comparing those "observed" values to modeled or "expected" values. This is particularly important in poorly sampled areas of the landscape or where natural variability is great in space or time. The new data can then be used to improve the model in an iterative process. Alternatively, subsets of the data can be removed, the model rerun, and the observed and expected values of the data subset can be examined. This process can be automated and repeated in a Monte Carlo process to create a map of standard errors (Figure 14.5).

In this example, standard errors were lowest in the vicinity of the sample plots and greatest near the borders of the study area where sampling intensity was low (Figure 14.5). Where sampling intensity was zero (i.e., no field plots in the vicinity), the standard errors are not modeled, creating a blank area—perhaps a target of future sampling efforts. Even in the worst-case scenario, the standard errors only represent a small portion of the mean (approximately 5–10% of the mean, on average), so there is fairly high confidence in the overall map, with less certainty around the borders (Figure 14.5).

Discussion and Lessons Learned

Spatial analyses provide important tools for investigating patterns of native and nonnative plant diversity in large areas (Figure 14.4). Kriging and cokriging

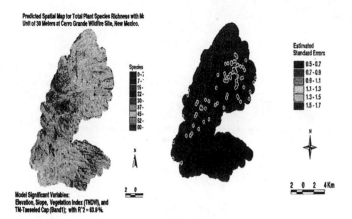

Figure 14.5.   Map of native species richness per 0.1 ha plot (left) and standard errors of the estimate (right) in the Cerra Grande burn area near Los Alamos, New Mexico. From M. Kalkhan, used with permission.

models are not limited to questions of species richness and probability of occurrence. The spatial patterns of abiotic resources such as soil fertility, soil moisture, understory solar radiation, and canopy cover can be modeled in similar ways to begin to develop a mechanistic understanding of the patterns revealed by plot-based vegetation surveys. Alternative models that might also be explored by plant and landscape ecologists include individual-based reaction-diffusion and spatially explicit simulation models, but these too have many untested assumptions and limitations (Higgins et al. 1996). Kriging and cokriging are based on current species locations, so no direct assumptions are made about past or future distributions, dispersal, or autecology (see Kot et al. 1996).

Many improvements can be made to the example shown here. Improved remote sensing data are needed to accurately map small, rare, important habitats. For example, the use of full-coverage, fine-scale variables [e.g., SPOT imagery, multispectral imagery, light detection and ranging (LIDAR)] would be a valuable addition to landscape-scale spatial models. Adding fine-resolution plot data from multiscale vegetation plots is also important (Kalkhan and Stohlgren 1999). For example, the spatial correlations from twenty-five 0.1 ha plots (Table 14.7) can be very different from the spatial autocorrelations from 10 dispersed 1 m^2 subplots nested within the 0.1 ha plots (Table 14.8). The spatial correlations of the large plots captured the large-scale variation in patterns of vegetation types in the study area.

The spatial correlations of the small subplots capture small-scale variation in the study area, with an emphasis on patterns within vegetation types in the landscape. Clearly far more spatial correlations can be found with subplots that capture local homogeneity and heterogeneity. In any case, investigators are obliged to evaluate the effects of scale on study results, and nested-scale plots are ideally suited to address such questions.

Table 14.7. Spatial correlations from twenty-five 0.1 ha plots in Rocky
Mountain National Park, Colorado.

Variable	No. of native species	No. of nonnative species	Native species cover	Nonnative species cover
Elevation (m)	−0.02			
Slope	0.16		−0.63	
Aspect				
No. of native species				
No. of nonnative species				
Native species cover				
Nonnative species cover				
Soil percent carbon	−0.12			
Soil percent nitrogen	−0.12			0.09
Soil percent sand	0.10			−0.09
Soil percent silt	−0.10			
Soil percent clay	−0.10			−0.06

Adapted from Kalkhan and Stohlgren (1999).

Table 14.8. Spatial correlations from two hundred fifty 1 m² subplots
(10 subplots per 0.1 ha plot) in 25 plots in Rocky Mountain National Park,
Colorado.

Variable	No. of native species	No. of nonnative species	Native species cover	Nonnative species cover
Elevation (m)	−0.05	−0.13	−0.19	−0.14
Slope	−0.12	−0.16	−0.23	−0.11
Aspect	−0.15	−0.08	−0.18	−0.11
No. of native species	0.38	0.17	0.19	0.12
No. of nonnative species	0.17	0.20	0.19	0.31
Native species cover	0.20	0.19	0.40	0.24
Nonnative species cover	0.12	0.31	0.24	0.37
Soil percent carbon	−0.06	0.09	0.06	0.10
Soil percent nitrogen	0.04	0.11	0.20	0.13
Soil percent sand	−0.07	−0.16	−0.24	−0.19
Soil percent silt	0.07	0.22	0.27	−0.27
Soil percent clay	0.05	0.07	0.17	0.09

Adapted from Kalkhan and Stohlgren (1999).

The multiphase, multiscale sampling and modeling methods employed in these examples are easily scaled to smaller or larger areas. For example, if more detailed information is needed on a smaller portion of the 54,000 ha area used in this example, the models can be rerun for a subset of the data (Figure 14.6).

In this example, finer detailed maps can be created displaying the estimated number of nonnative species in each 30 m × 30 m cell in the study area. These types of models can assess the patterns across management units, biomes, and regions, or for taxonomic groups (e.g., birds and butterflies) or groups of plants (e.g., functional groups, rare genotypes).

The modeling techniques can become more automated for developing maps of uncertainty (Figure 14.5). Likewise, new field data collected on aspects of plant diversity patterns can be quickly inserted into the models (Figure 14.2) (Chong et al. 2001; Kalkhan and Stohlgren 2000).

As a practical matter, maps of resources and associated levels of uncertainty are routinely requested by resource managers to set priorities for conservation and to target nonnative plant species for control efforts. In addition, such maps could be used to identify areas of potential migration or corridors of invasion, such as roads and riparian zones, that might be targeted for control efforts (Greenberg et al. 1997; Stohlgren et al. 1998d).

Additional variables, such as additional soil characteristics and high-resolution mapping of vegetation structure, will enhance the models' pre-

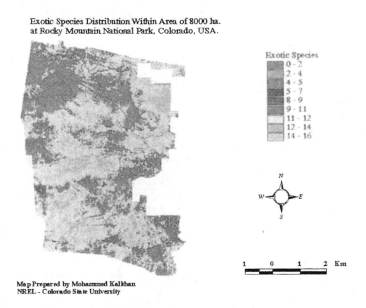

Figure 14.6. The potential distribution of non-native plant species for a 9000-ha section of Rocky Mountain National Park From Chong et al. 2001.

Figure 14.7.   Lesson 23. Spatial models can be used to better understand
the effects of grazing, natural and prescribed fire, and rapid climate
change on patterns of plant diversity. Since most studies of plant diversity rely on collecting sparse, spatially explicit data, modeling tools such
as these will become increasingly important to plant ecologists. This is
another reason to befriend a spatial statistician and modeler. (Photograph
by author)

dictive capabilities (Kalkhan and Stohlgren 2000). The combination of spatial statistics and stepwise multiple regressions greatly increased the predictive capabilities of our models for estimating the numbers of native and
nonnative species and the probability of nonnative species distributions. One
of the strengths of these modeling approaches is the ability to develop maps
of "uncertainty" based on subsampling the data with Monte Carlo simulations (Kalkhan and Stohlgren 2000). This provides land managers with a
spatial representation of the confidence of the model and the completeness
of the plot data. The multiphase sampling approach (i.e., data from ground-
truth plots, aerial photos, and Landsat TM images) provides additional ways
to assess vegetation classification accuracy and where more ground-truth
plots are needed (Kalkhan et al. 1998). Similar models can be developed for
individual species in more restricted areas (with a greater density of sample
points) to better understand their ecology and patterns of establishment,
growth, and spread (Figure 14.7).

PART V

MONITORING PLANT DIVERSITY

15

Concepts for Assessing Temporal Changes in Plant Diversity

The Issue

There are several considerations in developing appropriate strategies for monitoring plant diversity at landscape scales. Many plant ecologists and land managers are rightly concerned with current and rapidly emerging issues as well as long-term, chronic issues involving trends in plant diversity and the threats to that diversity. Thus various strategies for monitoring plant diversity might be considered.

One strategy, the "current and emerging issues strategy," might focus entirely on currently recognized or potential threats primarily to guide immediate, generally localized management decisions. Examples of local, current threats include the immediate effects of fire, grazing, land use changes, restoration activities, and efforts to assess the threat of a recently arrived, very invasive plant species or disease, or habitat loss related to threatened and endangered plant species. Such a strategy usually involves a few plots or transects in relatively small areas over a relatively short time period, where significant changes in plant diversity in the study plots are virtually guaranteed. The advantages of concentrating effort on current and emerging issues include (1) responding to urgent issues, (2) designing such a monitoring effort is relatively less expensive compared to the "chronic issues strategy," and (3) such studies are easy to administer. Disadvantages of a current and emerging issues strategy include (1) the data are gathered with relatively few plots on relatively few species in localized areas, so large areas go unmonitored;

and (2) subtle changes in plant diversity from chronic stresses and other changes over larger areas of the landscape scale are less likely to be detected when sampling is spatially limited.

The second strategy, the "chronic issues strategy," might focus long-term monitoring efforts on chronic stresses to plant diversity at landscape and regional scales. Climate change, nitrogen additions from air pollution, succession, slow migrations of many nonnative species, and changes in disturbance regimes are examples of broad-based, long-term stresses to natural landscapes. Here a network of long-term vegetation plots or transects could be established in areas most vulnerable to rapid vegetation change. For climate change, long-term vegetation transects can be monitored at the upper and lower tree lines and across various vegetation ecotones (i.e., the gradients between two vegetation types) (Stohlgren et al. 1998b,e). For air pollution, long-term vegetation plots can be established in nitrogen-limited vegetation types (e.g., tundra, dry meadows;) (Stohlgren et al. 1997a,c). Likewise, a network of plots can be established throughout a natural area and in adjacent areas to monitor the spread of invasive species (Chong et al. 2001; Stohlgren et al. 1997a, 1999a). The advantages of a chronic issues strategy are that (1) a targeted approach is more likely to detect subtle changes, and sometimes unanticipated changes, from long-term, landscape- to regional-scale chronic stresses; (2) the design can investigate probable causes (e.g., transects from tree line to the tundra may detect climate changes); and (3) data can be directly incorporated into landscape-scale predictive models (Chong et al. 2001; Kalkhan and Stohlgren 2000). The disadvantages of the chronic issues strategy are that (1) it is expensive—sets of long-term plots and transects are difficult to maintain—and there is no guarantee of detecting significant short-term change; (2) only a small portion of any landscape can be affordably sampled, and change may not necessarily take place first at study plot locations; and (3) fully evaluating temporal and spatial variation with small sample sizes is problematic at landscape scales.

Obviously a combination of strategies is needed, as landscape ecologists often are faced with current/emerging and chronic issues. The overwhelming advantage of a two-pronged approach is that multiple needs can be addressed for multiple stresses at multiple scales. The disadvantage is that costs are admittedly high and the optimum allocation of effort among current/emerging and chronic issues must be empirically determined (see below).

Past and Present Sampling Design Considerations

Three families of studies adapted from Likens (1989) may (with many assumptions and caveats) lend themselves to time series analysis that addresses long-term changes in plant diversity: retrospective studies, chronosequence studies that substitute space for time, and long-term plant diversity monitoring. Regardless of the approach, well-designed plant diversity monitoring programs will accurately quantify the level of completeness of plant diversity sampling, describe the distribution and abundance of species of concern (e.g., locally rare native species and invasive nonnative species), and develop an unbiased series of plots suitable for monitoring urgent and chronic threats.

Retrospective Studies

Retrospective studies try to infer past vegetation composition based on pollen stratigraphy and macrofossils (Davis 1989). For example, Green (1982, 1983) used pollen stratigraphy and time series analysis to study succession following fire in the forests of southwest Nova Scotia. However, there may not be a direct relationship between the past pollen and macrofossil record and past plant diversity. Not all species produce long-lasting pollen, rare species in protected microhabitats may not contribute to the pollen record, the exact size of a pollen catchment is generally unknown, and the pollen record may have been diluted or influenced greatly by upwind areas (Brubaker 1986, 1988; Davis 1983; Jacobson et al. 1987).

Retrospective studies generally provide better information on the change of dominant species, but they cannot easily quantify the level of completeness of plant diversity sampling at any one time. For example, pack rat middens (i.e., urine-solidified macrofossils in caves) provide concrete evidence of plant species occurrence within 50–100 m of a cave entrance at a given decade or century in the past, depending on the dating method used. Since caves are located in only a subset of topographic landforms and geologic types, inferences are limited to those areas. Tree and shrub species, which are major components of biomass, but minor components of the local flora, are selected for nesting materials over herbaceous species, so the historical macrofossil record is biased toward woody species. Locally and regionally rare plant species are frequently missed, and it is difficult to accurately quantify the abundance of species. The main problem with retrospective studies is that it is often impossible to assess what species, or how many species, were missed at each sampling point. Thus the validation of current and future trends in plant species diversity still requires long-term vegetation plots (Stohlgren 1994).

Chronosequence Studies

Chronosequence studies substitute space for time and attempt to infer a temporal trend by studying different-age sites, allowing relatively convenient and inexpensive sampling (Stohlgren 1994). They may show general trends in plant diversity with succession, generate hypotheses about patterns and mechanisms, and complement long-term plant diversity studies (Pickett 1989). For example, Christensen and Peet (1984) compared five successional age classes of upland forests in North Carolina. In their study they provide important insights to secondary succession and changes in plant diversity that would otherwise had required more than 80 years of study. Duffy and Meier (1992) studied the diversity of herbaceous understory plants in a chronosequence of forest stands in the southern Appalachians to show that plant diversity was far greater in old-growth forests compared to approximately 90-year-old clear-cut stands. In both studies, the general trends were reasonably detected and supported, even if some (or many) species were missed by the size of plots and number of plots used.

However, chronosequence studies rely heavily on enormous, untested assumptions that include the sites having the same initial conditions; comparable boundary conditions such as drought cycles, insect outbreaks, and vegetation dynamics; and similar priority effects such as past ecosystem processes on the sites (Pickett 1989). For this reason, chronosequence studies are receiving increased scrutiny by the scientific community (Elliott and Loftis 1993; Pickett 1989). High spatial and temporal variability in plant species distributions and past disturbances make it difficult to assume initial conditions are similar with respect to many aspects of plant diversity. Thus an unbiased series of plots suitable for future long-term monitoring are needed to validate chronosequence studies.

Monitoring with Permanent Plots

Establishing and maintaining long-term vegetation plots for plant diversity monitoring is time consuming and, by definition, expensive. Yet measurements can be made uniformly over time with testable precision and accuracy. Such measurements (e.g., changes in species richness, composition, cover, frequency) provide valuable data for spatiotemporal analyses (see chapter 18). The power of long-term, landscape-scale plant diversity studies increases when combined with short-term experiments that can determine cause-and-effect relationships, retrospective studies that can help interpret the rate and magnitude of past change, and chronosequence studies that can describe general temporal trends. Long-term studies are essential to detect, assess, and validate predicted plant diversity changes at landscape scales, but they too are not without constraints and limitations (Stohlgren 1994).

Generic plant diversity monitoring studies might include the following:

- A stratified random sampling design with adequate replication in common and rare, but important, habitat types.
- A multiphase design that is used to link plot-level survey information to remotely sensed data on the broader unsampled landscape.
- Nested intensity sampling (see chapter 17) and cluster sampling designs to reduce cost and improve information gain.
- A strong geographic modeling component to help guide future survey sampling sites by identifying hot spots of plant diversity, unique plant assemblages, and areas requiring more studies.

- Species additions documented in standardized databases (e.g., the International Taxonomic Information System; http://www.itis.usda .gov/) and compared to current lists.
- Voucher specimens collected and accessioned by experienced plant taxonomists. Species recorded from the vegetation plots should be associated with the systematic collection of important habitat and geographic characteristics. In some cases (e.g., *Phragmites*), genotypes may require identification and monitoring.
- Ancillary data collected to document location and geographic characteristics at each sample site using systematic techniques and modern technology.
- Assessments of completeness including species-area curves, species accumulation curves, and other field methods to assess completeness. For example, Chiarucci et al. (2001) used Michaelis-Menton models and nonparametric estimation methods to estimate the completeness of flora in a 431 ha nature reserve.
- Synthesis and analysis of results to make valid extrapolations of species diversity, abundance, and distribution with an emphasis on local native and nonnative plant diversity in relation to adjacent lands.
- Modeling the distribution and abundance of species of concern, such as rare species, nonnative species, and those species associated with habitats of special concern.
- Quantifying observer bias if multiple field crews are used in space or time and providing consistent training to and overlap of field crews.
- Overlapping and calibrating new methods to evaluate bias and sampling error if there are changes in methods over time.

Examples of Temporal Plant Diversity Studies

There are many examples of long-term vegetation studies (Stohlgren 1994), but few are designed to assess long-term changes in rare and common species at landscape or regional scales. Most monitoring efforts have been designed to study temporal changes in dominant plant species in plots or along transects.

Smithsonian Institution/Man and the Biosphere
Biological Diversity Program

The Smithsonian Institution/Man and the Biosphere Biological Diversity Program has a 20-year history of establishing permanent vegetation plots in representative tropical forests and in other Biosphere Reserve areas worldwide to inventory and monitor changes in biological diversity (Dallmeier 1992). Often a representative 1 ha plot, or small set of 1 ha plots, is established and each tree larger than 10 cm DBH is mapped, tagged, and identified. The plots are recensused annually.

The permanent plots were particularly useful in surveys before and after Hurricane Hugo in the Luquillo Biosphere Reserve, Puerto Rico (Dallmeier 1992; Dallmeier et al. 1992). The plots clearly measured the cataclysmic changes in tree diversity and replacement by simplified early successional species assemblages after the hurricane. However, this study relied on a few large plots that the hurricane conveniently hit.

The major underlying assumption of sampling a single or a few large plots is that the plots are typical, or representative, of the broader, unsampled area. Krebs (1989) refers to this sampling design as "judgmental sampling" because the investigator selects study sites on the basis of experience. He points out that this may give the correct results, but statisticians reject this type of sampling because it cannot be evaluated by theorems of probability theory. Because plots are expensive to establish, survey, and monitor, establishment of replicate plots is usually restricted (Kareiva and Anderson 1988). The take-home message by Berkowitz et al. (1989) bears repeating—that results from long-term study plots may be difficult to extrapolate because of (1) fundamental differences between the study system and surrounding areas; (2) unknowable boundary conditions or the occurrence of unique events; (3) unknown uncertainty or bias in the results due to poor replication; (4) the nature of the underlying process/phenomenon (i.e., different perturbations may act on the system in different ways through time); and (5) inappropriate methods or poor data.

Recently the Smithsonian program has been augmenting the larger plots with smaller (0.1 ha), nested-scale plots similar to the modified Whittaker plot. Ten 0.1 ha plots were placed in tropical forests in the Urambaba River area of Peru. Almost 500 tree species (or morphological types when specimens could not be identified to species) were documented in all the plots, with about half the species represented by a single individual (Mistry et al. 1999). The plots have not yet been remeasured, but they have the potential to shed light on the spatial and temporal patterns of rare tree species demography in tropical forests.

Selected Federal Plant Monitoring Programs

There are many inventory and monitoring methodologies for plant diversity in use today on public lands. Two such programs of national methodologies include the U.S. Forest Service's Forest Health Monitoring Program (Burkman and Hertel 1992; Conkling and Byers 1992) and the U.S. Army's Land Condition and Trend Analysis Program (Diersing et al. 1992).

The Forest Health Monitoring/Forest Inventory and Analysis Program was described in chapter 9 (Messer et al. 1991; Riitters et al. 1992). The monitoring program consists of a nationwide, uniform distribution of sample plots providing a large, unbiased sample of the nation's forests (1 plot/63,942 ha). The unbiased, systematic design and well-tested methods have the potential to provide long-term, multiscale data on temporal trends in plant diversity at local, regional, and national scales.

Subsets of the plots are remeasured each year in a roving panel design. The multiscale methodologies of the monitoring program can be expanded or simplified to local, regional, and national patterns in plant diversity. It can be expanded to cover nonforested types, rare vegetation types, pre- and post-disturbance sites, or to develop a series of long-term plots to evaluate the probable arrival of invasive plants, animals, and diseases. The plots can be simplified by reducing the number of 168 m^2 subplots measured in small habitats (e.g., small patches of weeds or wetlands, thin riparian zones, krummholtz patches).

The U.S. Army's monitoring program is designed to standardize inventory and monitoring procedures for plants and wildlife on 4.9 million ha of public land. Their permanent plots (100 m × 6 m transects) generally are allocated in a stratified random pattern based on soils, color reflectance categories, and vegetative cover using a geographic information system (Diersing et al. 1992). Plant cover is often measured with the line intercept technique or with Daubenmire-type transect and plot methods (see chapter 2), neither of which may provide accurate estimates of plant cover, except for the most common species (see chapter 7) (Stohlgren et al. 1998c). Plant species richness often is collected on the entire 600 m^2 plot, which provides excellent data on plant diversity on many Department of Defense installations throughout the United States. Again, the data have not yet been synthesized to assess temporal variation, but the potential for synthesis is huge.

Scott (1998) provided a sampling design (without data) for monitoring vegetation over time with partial replacement of sample clusters, and cluster sampling of four 0.001 ha plots for tree regeneration nested in 0.1 ha plots where only overstory trees are measured. These 0.1 ha plots were clustered and placed 50 m apart on the vortices of a square. Comparisons with other designs and real-world data may be the only way to effectively test this and other designs. In another example, Matlack (1994) studied vegetation dynamics of the forest edge using fifty 1 m^2 plots at 0, 5, 10, 20, and 40 m from the forest edge into the forest, and collected soil depth and pH, light, temperature, and relative humidity measurements along transects on 14 sites in southeastern Pennsylvania and northeastern Delaware. The changes in the 1 m^2 plots may or may not be indicative of the cumulative changes at larger spatial scales, and many species may be missed (see chapter 7).

Site-Specific Studies

There are many examples of site-specific studies designed to evaluate changes in plant diversity over time. Several agencies, nongovernment organizations, and researchers/students have established long-term vegetation plots. I give examples with a variety of designs and techniques to show the breadth of such studies.

Long-Term Observations

Abandoned agricultural or grazing fields have offered ecologists many opportunities for chronosequence studies as a somewhat weak substitute for long-term observations. As mentioned above, the initial conditions of the fields are rarely measured and site differences are assumed negligible. However, Bazzaz (1975), with a chronosequence of fields in Illinois, and Tilman (1988), with a chronosequence of fields in Minnesota, found increasing plant species richness 40 and 22 years after abandonment, respectively. It is intuitive that plant diversity might increase initially in many abandoned fields, but may decrease substantially if or when forest tree canopies begin to dominate the site. It is embarrassing that long-term studies of plot diversity with 0.1 ha or larger plots are so noticeably absent in the literature (Rosenzweig and Ziv 1999).

Fire Effects Monitoring

Several agencies, nongovernment organizations, and researchers/students have established long-term vegetation plots. Many national parks and monuments have established 0.1 ha plots for pre- and postfire monitoring. The field methods often include line intercept methods or small quadrats in the plots. Complete plant species lists are not always obtained since the primary objectives of the plots are to measure pre- and postfire fuel conditions and tree species demography. Plot locations generally are subjectively selected, but gaining important pre- and postfire effects data is still valuable in assessing whether specific prescribed fire objectives are met in monitoring plots.

Eric Petterson (1999) followed plant diversity for 2 years after fire in 0.1 ha modified Whittaker plots in several vegetation types in the Colorado Front Range. He found that plant species richness and cover changed locally (i.e., at the plot scale) after fire, but changed very little at the landscape scale (i.e., all plots combined). Control (unburned) plots also showed that species richness commonly changed 10–20% between years, and up to 40% in one plot—despite the same field crews recording data on carefully marked and relocated plots.

Monitoring Changes in the Distribution of Forest Tree Species: A More Detailed Example

Monitoring changes in the distribution of dominant forest tree species sounds easy, and learning how to monitor changes in dominant species may provide valuable insights into monitoring locally rare and patchily distributed species. In this next example I demonstrate one of many possible methodologies to begin to estimate the scope of the problem of monitoring vegetation changes of long-lived species. The first step in such an inquiry is to quantify the current patterns of tree species distributions in various growth phases (e.g., seedlings, saplings, and mature trees) and the associated environmental conditions under which they survive (Stohlgren et al. 1998b).

Because ecotones often represent the physiological or competitive limit of distribution of species, they serve to define a species' local distribution. Ecotones are ideal places to monitor changes in tree species distributions and changes in competitive advantage due to rapid climate changes, nitrogen deposition, fire, and other perturbations and disturbances. For example, changes in the upper tree line give one indication that forest distributions can change rapidly. In the Rocky Mountains of Colorado, for example, the upper tree line averaged about 500 m lower than its present position during the Pinedale glaciation (22,400 to 12,200 years ago). There is evidence that forest boundaries in the Front Range underwent periods of elevation advancement and retreat over the past 8000 or more years (Nichols 1982). The upper tree line advanced up to 300 m above its present position during a warm period (3000–5000 years B.P.) (Madole 1976). A cooling in the climate during the past 2500 years has forced the tree line down to its present location, perhaps with a more rapid decline during the Little Ice Age between 1550 and 1800 A.D. (Nichols 1982). Exactly how lower- and middle-elevation tree species migrated or adapted in response to climate change is problematic. The paleoecological record is clear that forest tree species distributions are not stable in space or time (Madole 1976; Nichols 1982).

To measure changes in forest distributions in the coming decades, we established fourteen 80–480 m long, 20 m wide vegetation transects, randomly selected throughout Rocky Mountain National Park, Colorado. We quantified vegetation dominance (basal area of dominant tree species) and environmental factors, including elevation, slope, aspect, PAR, summer soil moisture, and soil depth and texture across each ecotone (Stohlgren and Bachand 1997; Stohlgren et al. 1998a, 2000b). The general objective was to monitor species-environment relationships (i.e., the abiotic and biotic controls on forest distribution) focusing on ecotones as a prerequisite for measuring and understanding trends in herbaceous plant species richness over time (see chapter 16). Specific objectives were to (1) monitor the relationships of vegetation dominance and environmental factors across current forest ecotones, and (2) develop empirical models relating specific species and neighboring forest tree species limits to environmental variables at various growth stages. A secondary objective was to maintain a series of unbiased, long-term study transects to allow future ecologists to monitor vegetation change and validate empirical models in the decades and centuries to come.

The study area ranged from 1800 m to nearly 3000 m in elevation, with a wide array of vegetation, including shortgrass steppe, shrub steppe, montane pine and fir forests, subalpine forests, and alpine tundra (Allen et al.

1991; Peet 1981). Dominant tree species generally from low to high elevations include ponderosa pine, Douglas-fir, lodgepole pine, aspen, blue spruce (*Picea pungens*), spruce/fir mix (*Picea engelmannii/Abies lasiocarpa*), and limber pine (*Pinus flexilis*).

All transects were located in an unbiased manner using a stratified random sampling design based on the vegetation type map in the park and a geographic information system. At each study site the transect was established across the major environmental gradient (i.e., elevation or aspect) (Figure 15.1). In each 20 m × 20 m plot, slope (clinometer; to the nearest degree), aspect (compass; to the nearest degree from due south), and elevation (altimeter calibrated with a Trimball Pathfinder GPS and differential corrections for the three survey stakes) were measured. All trees were identified to species, numbered, measured (diameter and height), and mapped. Each 20 m × 20 m plot included systematic assessments of particle size analysis, soil depth, surface rockiness, PAR in the understory, and summer soil moisture [see Stohlgren and Bachand (1997) for details].

Canonical correspondence analysis (CANOCO, version 3.12) (ter Braak 1987b, 1991) was used to characterize the relationship between forest composition and dominance (i.e., basal area by species) and environmental measures (e.g., elevation, slope and aspect, PAR index, soil depth, surface rockiness, summer soil moisture, and soil sand, silt, and clay fractions) (Figure 15.2).

In this example, mature trees of each species have been constrained to environmentally different sites than seedlings of the same species (Figure 15.2). Two variables, soil moisture and elevation, were the key factors in describing the distributions of the various species and life stages, but seedlings generally occurred at lower elevations and drier sites or microsites than mature trees of the same species. Saplings and tall seedlings were found in intermediate environmental spaces.

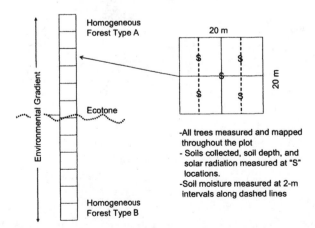

Figure 15.1. Ecotone transect sampling scheme. Adapted from Stohlgren and Bachand 1997.

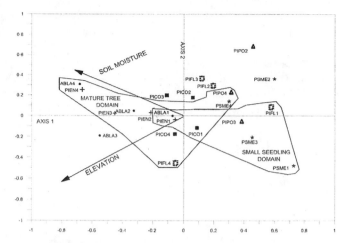

Figure 15.2.   Examples of different species-environment relationships for mature trees and seedlings of the same species in Rocky Mountain National Park, Colorado (from Stohlgren et al. 1999a). Ponderosa pine = PIPO, Douglas fir = PSME, lodgepole pine = PICO, aspen = POTR, Engelmann spruce = PIEN, subalpine fir = ABLA, and limber pine = PIFL. Small seedlings end in "1," large seedlings end in "2," saplings end in "3," and mature trees end in "4." Lesson 24. Assessing species-environment relationships and species distributions requires the investigation of multiple life stages—seedlings may have very different requirements for growth and survivorship than mature individuals.

The value of long-term transects or plots as "vegetation time capsules" cannot be understated. Because the transects or plots are permanently marked and accurately georeferenced, ecologists will be able to return to these same sites 50 or 100 years from now. The transects in this case study also provide a means to assess changes in herbaceous species (see chapter 16).

Constraints and Limitations of Long-Term Plant Diversity Studies

Assessing temporal changes in plant diversity at landscape scales is expensive. Long-term vegetation plots have to be monitored in many habitats over many years. Landscape-scale inventorying and monitoring must provide consistent measurements of the immigration and extirpation of many locally rare species. This requires well-trained taxonomists, maintenance of a data management system, a strong modeling capability (see chapter 18), and equally strong analysis and synthesis capabilities.

Inventorying and monitoring of plant diversity in large areas over long time periods is still in its infancy, and it is hampered most by inadequate sampling procedures (Stohlgren 1994; Stohlgren et al. 1997a, 1998c). To date, there is little standardization of techniques in landscape-scale, long-term studies on plant diversity. Field methods used are often site and study specific. Critical analyses of sampling techniques and designs are just now surfacing (Stohlgren et al. 1995a, 1997b, 1998c). Sampling design theory for landscape diversity is not well advanced (Fortin et al. 1989; Legendre and Fortin 1989). Progress in sampling design theory has been hampered by the historically poor understanding of complex, heterogeneous landscapes (Stohlgren 1994), an overemphasis on common species, and the inertia of landscape ecologists to change sampling approaches.

Understanding and Accepting Complex, Heterogeneous Landscapes

Plant ecologists have been slow to accept the two major characteristics of plant diversity patterns in natural landscapes. First, natural landscapes have various ages of substrates at multiple spatial scales. Mixes of various types of small-scale soil actions from invertebrate galleries, ground mammals, tree falls, and erosion pathways are continuously overlaid on larger-scale soil disturbance actions such as fire, flooding, landslides, and organic layer development through succession. Second, plant ecologists have been slow to accept a strict Gleasonian view of species distributions (Gleason 1926), with species responding individualistically to local environmental conditions (e.g., microclimate, topography, soils, disturbance, etc.) (Figure 15.3).

Figure 15.3.   Patchy distributions of species in subalpine meadows near Grand Teton National Park, Wyoming.

Plant ecologists have been preoccupied with classifying plant communities, developing vegetation type maps, and delineating higher-order groupings of dominant species distributions [e.g., giant sequoia (*Sequoiadendron giganteum*) groves; aspen patches], functional groups (e.g., annual grasslands, shrublands, coniferous or deciduous forests), or biomes (e.g., deserts, grasslands). We have accepted this "vegetation type" view of the world as though the world consists of mostly (or nearly all) homogeneous units.

Once plant ecologists fully accept and embrace the inherent complexities of heterogeneous landscapes, some theoretical and practical sampling constraints can be addressed. Hinds (1984) stated that "field sampling variability is both large and poorly understood in relation to the rest of any long-term-effects detection effort." Poorly replicated permanent plots may be put in atypical locations (Stohlgren 1994). Detecting changes in plant diversity following a catastrophic wildfire may be easier than detecting subtle changes in plant diversity caused by changes in climate or chance.

Overcoming the Overemphasis on Common Species

Characterizing changes in the abundance or cover of common, dominant species is more difficult than it sounds. Annual variations in climate, reproduction, mortality, and microscale disturbances, along with measurement errors, provide serious challenges to monitoring common species. However, these challenges pale in comparison to monitoring the presence, abundance, and cover of locally rare (i.e., most) plant species at landscape scales. The primary constraint of plant diversity monitoring is quantifying the completeness of sampling in space and time (see chapter 18).

To obtain reliable estimates of species richness based on survey data, we assume that all species are detected and that the detectability of species does not vary among samples (Boulinier et al. 1998). We define the detectability of a species as the probability of detecting at least one individual of a given species in a particular sample unit, given that individuals of the species are present during the sampling session (Boulinier et al. 1998). Note that a species' detection probability depends on both its population size and the behavioral attributes of individuals relative to the sampling methods employed (Nichols 1992).

Unfortunately the assumption of equal detectability among species, or the assumption that the likelihood of detection, given presence, equals one is almost never true. As a consequence, we need methods that account for species that are present but not detected. A number of different methods have been proposed [reviewed in Bunge and Fitzpatrick (1993)] that are based on estimating the "missing" species based on the observed patterns of species occurrences across samples. Yet such approaches are also limited by the sampling design, size and number of plots, and the pattern of sampling in complex landscapes. And how does one account for the seed bank at all sites, rapid species turnover at some sites, and the species undetected in the larger unsampled portions of the landscape or region? Measuring changes in plant diversity is a young science.

Inertia of Plant Ecologists to Change Sampling Approaches

The greatest constraint to monitoring changes in plant diversity may be the inertia of plant ecologists to change from sampling approaches designed to delineate a few dominant species to those that monitor locally and regionally rare species. As long as plant ecologists stick to using a few small quadrats (usually along a line) in presumed homogeneous units, and over only a few years, we will gain few insights into plant diversity changes over time. Ecologists have to consider larger and more plots in homogeneous and heterogeneous areas, and for several years or decades, to understand landscape-scale and regional changes in plant diversity.

16

Case Study

*Monitoring Shifts in Plant Diversity
in Response to Climate Change*

The Issue

Improved sampling designs are needed to detect, monitor, and predict plant migrations and plant diversity changes caused by climate change and human activities (Adler and Lauenroth 2003; Diaz et al. 2004; Stohlgren et al. 2005a). A methodology based on multiscale vegetation plots established across forest ecotones might provide baseline data on patterns of plant diversity, invasion of nonnative plant species, and plant migrations at landscape scales (Stohlgren et al. 2001).

Paleoecological evidence clearly shows that plant species migrate long distances in response to climate change (Kullman 1996; Sykes and Prentice 1996; Woods and Davis 1989). What surprises many plant ecologists are the modeled rapid rates of migration, with the pollen record of some tree species presumably migrating up to a kilometer per year, coinciding with climate warming in the Holocene (Pitelka et al. 1997). In recent decades, emphasis on plant migrations has shifted to the rate of spread of nonnative invasive plant species. Eurasian cheatgrass (*Bromus tectorum*), for example, spread to over 200,000 km² in about 40 years (Mack 1996; Pitelka et al. 1997). Even more rapid plant migrations may be likely in the near future in response to accelerated climate change (Foley 1994) and land use changes (Stohlgren et al. 1998e).

Managers of national parks and other natural areas are justly concerned about rapid plant migrations. First, plant species could migrate from designated "protected" areas in preserves to less protected areas outside preserves

(Peters and Darling 1985). Second, some plant species may be squeezed into smaller or fewer habitats, such as higher elevation sites or wetland habitats. Third, more sedentary species could be outcompeted by rapidly invading species, such as nonnative plants. Given the ubiquitous threats of habitat destruction, altered disturbance regimes, climate change, and nitrogen deposition from air pollution, vegetation monitoring in nature preserves must be capable of detecting, monitoring, and predicting changes in species composition and plant migrations (Stohlgren et al. 2001).

Background and Sampling Considerations

As discussed in chapter 15, many vegetation sampling programs may be poorly designed to detect subtle changes. Vegetation plots are often too small (commonly less than 3 m²) (Kareiva and Anderson 1988) or are placed in more stable homogeneous plant communities (Barbour et al. 1987; Daubenmire 1968; Mueller-Dombois and Ellenberg 1974) rather than developing an understanding of species-specific migrations and the disassociation of plant communities in space and time (Gleason 1926). Oddly, many ecologists interested in "gradient analysis" subjectively placed vegetation plots in more homogeneous areas (Allen et al. 1991; Peet 1981; Whittaker 1967), avoiding local environmental gradients between homogeneous and heterogeneous or periphery areas of a vegetation type (Stohlgren and Bachand 1997). However, peripheral (i.e., marginal) areas of a population and mixed species stands may have different patterns of establishment, growth, and survival than homogeneous (or core) areas of a population (Neilson 1991; Stohlgren 1992, 1993). Placing vegetation plots only in homogeneous areas may exaggerate the differences between plant communities (Stohlgren and Bachand 1997; Stohlgren et al. 1998a) and reduce our ability to monitor and predict plant species migrations, which are more likely to occur first at the physiological boundaries of the species.

To understand plant species distributions and migration, and changes in plant diversity at landscape scales, ecologists are now focusing more attention on ecotones (Figure 16.1) (Cornelius and Reynolds 1991; Gillison 1970; Gosz 1993; Hansen and di Castri 1992; Holland et al. 1991; Risser 1993; Rusek 1993; Weisberg and Baker 1995; Wiens et al. 1985). In 1995 Risser echoed a plea by Wiens et al. (1985) 10 years earlier that argued that we must understand the steepness of environmental gradients between homogeneous vegetation associations before we can understand the dynamic nature of landscapes (Stohlgren et al. 2001).

Most ecotone studies were not designed to directly answer questions about plant migrations or plant diversity patterns at landscape scales. Usually only one transect or ecotone was measured (e.g., Giesler et al. 1998; Kieft et al. 1998; Martinez and Fuentes 1993; Montaña et al. 1990; Wesser and Armbruster 1991). In other studies with replicate transects, emphasis was placed

Figure 16.1.   An example ecotone in Rocky Mountain National Park, Colorado. The transect includes Ponderosa pine in the foreground, yielding upslope to Douglas-fir forests—most ecotones are patchy gradients of change rather than obvious boundaries.

on the boundaries of a few dominant plant species (Puyravaud et al. 1994; Rusek 1993; Stohlgren and Bachand 1997; Stohlgren et al. 1998a). One study assessed α and β diversity relative to water drawdown on the shores of one reservoir (i.e., one ecotone) in central Tanzania (Backéus 1993). My colleagues and I found no studies that were designed to monitor species richness patterns, nonnative species invasions, and the direction and magnitude of species migrations with replicate ecotones in a landscape (Stohlgren et al. 2001).

Our goal in this case study was to design a long-term vegetation moni-toring program to (1) assess changes in plant diversity at landscape scales, (2) detect species migrations in response to rapid environmental change, and (3) monitor the invasion of nonnative plant species in the forests of Rocky Mountain National Park, Colorado (Stohlgren et al. 2001). Such a design would be broadly applied to gradients of pollution, land use change, and plant invasions.

Case Study Area

The Colorado Front Range in the southern Rocky Mountains spans an ele-vation range from 1600 m to more than 4300 m. This mountain region em-braces a wide array of vegetation, including shortgrass steppe, shrub steppe, montane pine and fir forests, subalpine coniferous forests, and alpine tun-dra. Latitudinal and elevation arrangements of species distributions have been attributed to temperature and precipitation, as typically influenced by elevation and topography (Allen et al. 1991; Peet 1981, 1988). Common tree species include ponderosa pine (2320–3170 m), Douglas-fir (2370–3213 m), lodgepole pine (2380–3480 m), aspen (2350–3500 m), Engelmann spruce (2530–3710 m), subalpine fir (2530–3710 m), and limber pine (2620–3,560 m). Paleoecological evidence suggests that forest ecotones in the Rocky Moun-tains of Colorado may be sensitive indicators of climate change, since the distributions of tree species probably changed faster than soil development in many areas (Madole 1976; Markgraf and Scott 1981; Stohlgren and Bachand 1997).

Case Study Methods

Vegetation Sampling and Environmental Measurements

Fourteen transects (120–480 m long, 20 m wide) consisting of 146 forest plots (20 m × 20 m) were established between 1992 and 1995 in Rocky Mountain National Park, Colorado, to assess tree species distributions (Table 16.1) (Stohlgren and Bachand 1997; Stohlgren et al. 1998a, 2000b).

Since the origination of the present stands, these areas have remained undisturbed [see Stohlgren and Bachand (1997) and Stohlgren et al. (2001) for details]. At each study site, the transects were established across major environmental gradients (i.e., elevation or aspect) (Table 16.1) such that both terminal sections of each transect were in homogeneous stands of their re-spective forest types, with 70% of the basal area in the dominant tree spe-cies (Figure 16.2). Thus we define the ecotone as the transition zone between the two relatively homogeneous forest types.

Table 16.1.   Ecotone transects in Rocky Mountain National Park, Colorado.

Transect name	Vegetation type	Vegetation type 2	Elevation (m)	Slope (%)	Aspect (degrees)	Estimated age (years)
Aspenglen	Pin pon	Pin con	2530–2610	4.0–9.0	135	180
Deer Ridge	Pin pon	Pin con	2740–2800	5.0–8.0	90	130
Upr. Beaver	Pin pon	Pin con	2530–2550	6.0–7.0	360	130
Bear Lake	Pin con	Picea-Abies	2870	9.0–14.0	0–22	90
Hitchen's	Pin con	Picea-Abies	2960–2990	7.0–14.0	90	110
Lawn Lake	Pin con	Pin fle	3020–3080	5.0–7.0	270	130
Meeker Dr.	Pin con	Pin fle	3000–3050	5.0–7.0	90	100
Wild Basin	Pin con	Pin fie	2980–3080	4.0–7.0	180	120
South Lt.	Pin pon	Pse men	2614–2628	2.0–20.0	360	90
Eagle Cliff	Pin pon	Pse men	2471–2540	4.0–27.0	20	70
High Drive	Pin pon	Pse men	2926–2942	26.0–35.0	90	240
Aspen Brook	Pse men	Pin con	2727–2764	6.0–29.0	270	110
Emld Mtn.	Pse men	Pin con	2746–2762	5.0–16.0	360	110
Thunder Mtn.	Pse men	Pin con	2620–2630	6.0–24.0	90	95

Pin pon = *Pintis ponderosa*; Pin con = *Pinus contorta*; Pse men = *Pseudotsuga menzesii*; Picea-Abies = spruce/fir, Pin fle = *Ptnus flerilis*

Adapted from Stohlgren et al. (2001).

Environmental factors that were measured along the ecotone transects in that previous study (Stohlgren and Bachand 1997) included (1) average soil depth from 144 sites in each 20 m × 20 m forest plot; (2) average surface rockiness (i.e., the percent of soil depth measurements that were impenetrable due to surface rock); and (3) average PAR (400–700 nm; in $\mu mol/m^2/sec$; Decagon Sunfleck Ceptometer Model SF-80, Decagon Devices, Pullman, Washington) beneath the forest canopy between 10 A.M. and 2 P.M. on cloud-free days in late June or early July. The ceptometer was held horizontally 1 m above ground level and five readings (80 sensors per reading) were taken at four cardinal directions in each quadrant and at the center of each 10 m × 10 m subplot, then averaged for each plot (Figure 16.2). Measurements were also taken in canopy openings to calculate a PAR index (i.e., average plot reading divided by the maximum open-canopy reading for the day), interpreted as an index of light available to understory vegetation in the various plots. At the terminal ends (comparatively homogeneous forest stands) and center (ecotone) of the 14 transects, a 20 m × 50 m modified Whittaker nested vegetation sampling plot was established (Figures 7.5 and 16.2).

Statistical Analysis in the Case Study

We used Jaccard's coefficient to quantify plant species overlap between vegetation types (Krebs 1989, chapter 4) as shown earlier. We used CCA (CANOCO, version 3.12) (ter Braak 1987b, 1991) to characterize the relation-

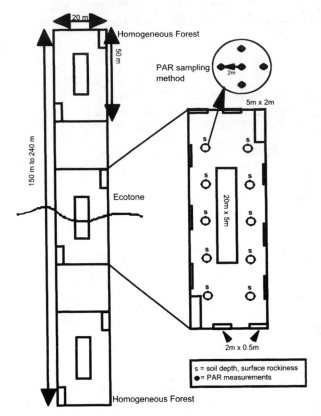

Figure 16.2.   Ecotone plot design with modified-Whittaker plots at the ends and middle of the transect. From Stohlgren et al. 2001.

ship between understory composition (cover of the dominant five species) and environmental measurements (e.g., slope, aspect, elevation, PAR index, soil depth, surface rockiness). Aspect was converted to a linear scalar as the absolute value of 180 degrees minus the azimuth reading such that values ranged from 0 degrees (due south) to 180 degrees (due north). We assessed all environmental variables for multicollinearity problems and none were found. Monte Carlo permutation tests (99 random permutations) were performed to test the significance of the first canonical axis (ter Braak 1991).

As in other case studies, we used stepwise forward multiple regressions (SPSS Inc. 1997) to assess the ability of environmental factors to predict native, exotic, and total species richness. Environmental factors included slope, aspect, elevation, PAR index, surface rockiness, and soil depth. The forward linear regression models included only variables meeting the $P <$ 0.15 criterion (Neter et al. 1990). Data were assessed for normality prior to analysis (Zar 1974), and only exotic species richness required $\log_{10}$ transformation. Finally, we used distance-weighted least squares models (SPSS

Inc. 1997) to create three-dimensional displays of native and exotic species richness in relation to elevation and PAR.

Results, Discussion, and Lessons Learned

Species Richness Patterns in Homogenous Forest Plots and Ecotone Plots

Large plots quickly differentiated patterns in native and exotic species richness by ecotone type and vegetation type (Table 16.2). Mean native species richness in 1000 m² plots ranged from 15.3 (±2.6) species per plot in homogeneous, high-elevation lodgepole pine plots to 47.7 (±2.3) species per plot in homogeneous, low-elevation ponderosa pine plots. Likewise, higher-elevation plots in ecotones generally had fewer species than lower-elevation ecotone plots (Table 16.2).

The large 1000 m² plots also showed that mean exotic species richness was highest in low-elevation vegetation types, ranging from zero species per plot in Douglas-fir to lodgepole pine transects, to 2.3 (±0.7) species per plot in homogeneous and ecotone plots in ponderosa pine to Douglas-fir transects (Table 16.2). For all ecotone plots combined, mean exotic species richness was 10% greater than in homogeneous forest types combined, but this difference was not significant. Total species richness patterns mimicked native species richness patterns due to the small numbers of exotic species (Table 16.2). Higher mean richness of native, exotic, and total species coincided with higher ($P <$ 0.1) mean PAR values in ecotone plots (Table 16.2).

Species Overlap Between Vegetation Types

Plant species overlap between pairs of 1000 m² plots along all ecotone types was high. Ponderosa pine plots shared 65% of the plant species found in the ecotone plots on the ponderosa pine to Douglas-fir transects. The mean species overlap between ecotone plots and the homogeneous forest types was almost as high on the other transects, ranging from 44% to 65%. Species overlap between pairs of homogeneous forest plots on the transects ranged from 35% to 56% (Table 16.3).

Combining the species lists for the homogeneous plots by forest type revealed that 59% of the species found in the ponderosa pine type were also found in the Douglas-fir type. The lodgepole pine type shared almost half the plant species found in the ponderosa pine and Douglas-fir types, but shared less than one-third of its species with the high-elevation limber pine type.

Species Affinities to Individual Vegetation Types

Plant species showed weak affinities to overstory vegetation types (Figure 16.3). Only 49 (27.8%) of the 176 plant species encountered had strong

Table 16.2.   Mean number of native, exotic, total. and unique plant species in 1000 m²
plots in various ecotone and forest types.

Ecotone type/vegetation	No. of plots	No. of native species		No. of exotic species		PAR	
		Mean	S. E.	Mean	S. E.	Mean	S. E.
Ponderosa to *Douglas-fir*							
Ponderosa	3	47.7	(2.3)	2.3	(0.7)	0.55	(0.11)
Douglas-fir	3	36.7	(7.2)	1.7	(0.7)	0.61	(0.12)
Ecotone	3	41.7	(3.8)	2.3	(0.7)	0.76	(0.13)
Ponderosa to *Lodgepole*							
Ponderosa	3	40.3	(10.5)	2.0	(1.0)	0.56	(0.12)
Lodgepole	3	40.4	(8.1)	0.7	(0.7)	0.31	(0.06)
Ecotone	3	43.7	(9.9)	2.0	(1.5)	0.56	(0.13)
Douglas-fir to Lodgepole							
Douglas-fir	3	35.0	(4.4)	0.0	(0.0)	0.28	(0.12)
Lodgepole	3	21.0	(1.2)	0.0	(0.0)	0.43	(0.08)
Ecotone	3	30.0	(3.6)	0.0	(0.0)	0.54	(0.14)
Lodgepole to *Spruce-fir*							
Lodgepole	2	27.5	(3.5)	1.0	(1.0)	0.44	(0.16)
Spruce-fir	2	26.0	(0.0)	0.5	(0.5)	0.31	(0.04)
Ecotone	2	19.5	(3.5)	0.0	(0.0)	0.34	(0.01)
Lodgepole to *Limber*							
Lodgepole	3	15.3	(2.6)	0.0	(0.0)	0.35	(0.07)
Limber	3	25.7	(7.2)	0.3	(0.3)	0.60	(0.06)
Ecotone	3	25.0	(6.1)	0.3	(0.3)	0.54	(0.08)
All types to ecotones							
All types	28	30.9	(2.4)	0.9	(0.2)	0.45	(0.04)
Ecotones	14	32.9	(3.4)	1.0	(0.4)	0.56	(0.04)

Mean undercanopy photosynthetically active radiation (PAR) index by ecotone and forest type.

affinities to individual vegetation types (based on 0.1 ha plot data). However, this included 48 locally rare species that were found in only one plot
while sampling, while 23 plant species were unique to the low-elevation
ponderosa pine plots. Fifty-two species were found in two vegetation types,
with 24 species shared between low-elevation ponderosa pine and Douglas-
fir plots. More than 72% of the plant species were found in two or more vegetation types, with 43% of the plant species found in three or more vegetation
types (Figure 16.3).

Canonical Correspondence Analysis of Ecotones

Based on CCA of the 14 vegetation transects, we found that dominant understory species distributions at landscape scales were described generally
by gradients of elevation and available light (undercanopy PAR). The first

Table 16.3.   Mean species overlap between "homogeneous" vegetation types based on pairs of 0.1 ha plots.

Forest type	Douglas-fir	Lodgepole	Spruce-fir	Limber
Ponderosa	47% (10%)	35% (7%)	—	—
Douglas-fir		44% (6%)	—	—
Lodgepole			56% (16%)	48% (12%)

Standard errors in parentheses.

Adapted from Stohlgren et al. (2000).

two ordination axes explained 64% of the variance in vegetation patterns. Environmental factors that correlated significantly ($P < 0.001$) to the first ordination axis included elevation ($r = -0.94$) and PAR ($r = 0.64$). Environmental factors that significantly correlated to the second ordination axis included aspect (degrees from due south; $r = -0.60$, $P<0.001$) and slope ($r = 0.46$, $P<0.002$). The correlation matrix showed that elevation was significantly correlated to aspect ($r = -0.51$, $P < 0.001$). Monte Carlo permutation tests showed that the first canonical axis was highly significant (eigenvalue = 0.84, F-ratio = 3.4, $P < 0.01$).

We mapped the domain of understory species associated with the particular overstory tree species by connecting perimeter plots from the same vegetation type on the CCA plot. Connecting the perimeter plots from only the homogeneous plots (Figure 16.4, top) resulted in small domains for the five dominant understory species associated with the spruce-fir and limber pine vegetation types.

The domain of understory species associated with lodgepole pine is fairly large, as is the domain for the understory species of Douglas-fir (*Pseudotsuga menziesii*) and ponderosa pine (*Pinus ponderosa*). However, when the ecotone plots are included, the potential domain of dominant understory spe-

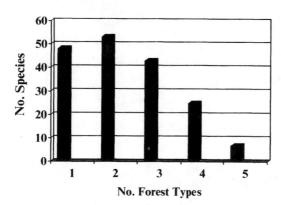

Figure 16.3   Species affinities to individual vegetation types. Adapted from Stohlgren et al. 2000.

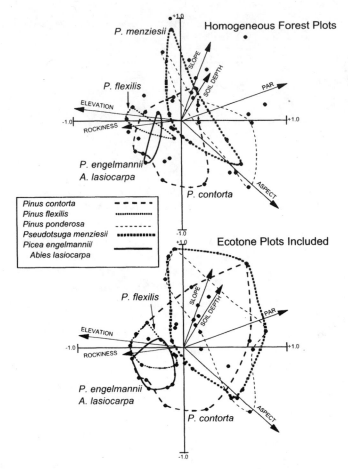

Figure 16.4. Canonical Correspondence Analysis ordination. Top diagram shows the domain of the dominant five understory species per plot excluding the ecotone plots, while the bottom diagram includes the ecotone plots.

cies is greatly expanded for all forest types (Figure 16.4, bottom). The domain of the dominant understory plants associated with Douglas-fir more than doubled on the CCA plot. Adding subdominant plant species would have increased the domain of each forestry type, as few understory species are unique to one overstory type (Figure 16.3).

Predictors of Species Richness at Landscape Scales

Multiple regression analysis showed that most of the environmental factors we measured were important predictors of native species richness and total species richness in 1000 m² plots spread across the landscape (Table 16.4). For example, 59% and 60% of the variation in native and exotic plant species richness, respectively, could be explained by aspect, slope, elevation, PAR, and soil depth. Elevation was the only negatively correlated variable in both cases. Almost 66% of the variation in exotic species richness ($\log_{10}$-transformed data) could be explained by positive relationships with PAR and the number of native species (Table 16.4). The t- and P-values showed that all individual variables used in the regressions were significant.

Frequencies of Plant Species in 1000 m² Plots

We found a steeply declining curve of plant species frequencies in 1000 m² plots. Forty-eight of the 176 species encountered were not found in multiple plots (Figure 16.5).

Common juniper (*Juniperus communis* L.) was found in 36 of 42 plots. Douglas-fir was found in 30 of the 42 plots, despite the fact that only 6 of 14 transects were anchored in Douglas-fir communities. Seedlings of limber pine, which is usually considered a high-elevation species, were found in 57% of the plots, while aspen was found in 43% of the plots. The nonnative species cheatgrass (*B. tectorum*) was found in 11 plots, and 18 plots contained at least one nonnative species (Figure 16.5).

Table 16.4.   Multiple regression results for the 42 plots combined.

Dependent variable/ predictors	Coefficient	T	P	Model F, R^2 P
No. of native species				
Constant	91.21	4.42	0.001	$F = 10.5$
Aspect	0.08	2.61	0.013	$R^2 = 0.59$
Slope	0.48	2.68	0.011	$P < 0.001$
Elevation	−0.03	−4.76	0.001	
PAR	23.25	3.26	0.002	
Soil depth	0.58	2.81	0.008	
No. of exotic species ($\log_{10}$)				
Constant	−0.38	−5.16	0.001	$F = 37.5$
PAR	0.29	2.16	0.032	$R^2 = 0.66$
No. of native species	0.014	6.59	0.001	$P < 0.001$
Total no. of species				
Constant	95.82	4.34	0.001	$F = 10.9$
Aspect	0.09	2.72	0.010	$R^2 = 0.60$
Slope	0.50	2.60	0.013	$P < 0.001$
Elevation	−0.04	−4.76	0.001	
PAR	25.79	3.37	0.002	
Soil depth	0.63	2.84	0.007	

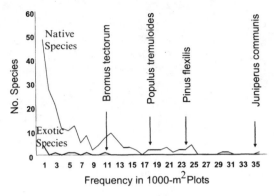

Figure 16.5.   Frequencies of plant species in 1000-m² plots with selected species shown. Adapted from Stohlgren et al. 2000.

Patterns of Native and Exotic Species Richness at Landscape Scales

The distance-weighted least squares (three-dimensional) models showed important differences in the distributions of native and exotic species at landscape scales (Figure 16.6). Elevation and available light (PAR) were significant ($P < 0.001$) predictors of both native and exotic species richness ($R^2 = 0.40$ and $R^2 = 0.33$, respectively). Native species richness remained high at high elevations as long as there was adequate light (Figure 16.6, top), whereas exotic species richness sharply declined with increasing elevation, regardless of the available light (Figure 16.6, bottom).

Baseline Data for Monitoring Changes in Plant Diversity at Landscape Scales

Monitoring changes in plant diversity at landscape scales will not be easy because (1) patterns of species richness and cover are determined by multiple environmental factors; (2) plant species have low affinities to overstory vegetation types; and (3) understanding and predicting species-specific responses to environmental change and multiple stresses may seem hopelessly complex. Still, this series of ecotone transects, initially established for the analysis of tree distribution changes (Stohlgren et al. 1998a), may be ideally suited to monitor long-term changes in plant diversity, as illustrated below.

There is little doubt that patterns of plant species richness are determined by many environmental factors. Climate is likely a driving force for overstory species at large spatial scales (Table 16.3 and Figure 16.6), as evidenced by broad responses of species to elevation, which is correlated to temperature, slope, and aspect, all of which are correlated to soil moisture (Peet 1981; Stohlgren and Bachand 1997). However, superimposed on broad-scale cli-

NATIVE SPECIES

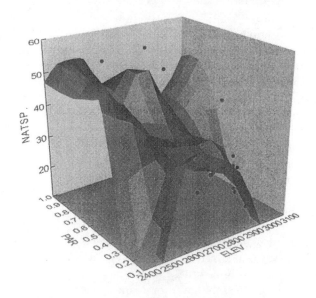

EXOTIC SPECIES

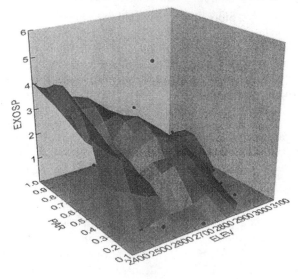

Figure 16.6. Patterns of native (NATSP) and exotic (EXOSP) species richness in 1000-m² plots with respect to elevation and PAR, using a distance-weighted least squares model. A plus sign (+) indicates an ecotone plot, and asterisks (*) indicate homogeneous forest plots.

matic gradients are small-scale gradients characterized by patches of light (Table 16.2), pockets of fertile soil (Stohlgren et al. 1997b), and zones of high soil moisture (Stohlgren et al. 1997c). Disturbances that occur at multiple scales, such as fire and insect outbreaks, influence local and landscape scale patterns of diversity (Stohlgren et al. 1997b,c). Monitoring plant species richness along major environmental gradients, along with a series of monitoring plots in disturbed areas, can provide the baseline data necessary to isolate the effects of climate change from successional change.

No vegetation monitoring system of plots and transects will be able to capture all aspects of plant diversity change. However, some subdominant species, such as aspen, are captured surprisingly well in the forty-two 1000 m² plots in randomly placed transects. Aspen was thought to be relatively rare on the landscape, representing 2% of the forest cover on broad-scale vegetation maps (Stohlgren et al. 1997c). Aspen stands provide critical habitat for many species of plants, butterflies, and birds. Of major concern to resource managers was one recent study that showed virtually no aspen regeneration in large stands in the Estes Valley due to excessive browsing by large populations of elk in Rocky Mountain National Park (Baker et al. 1997). Broader studies showed that many other large aspen stands in the park and surrounding forests had more aspen regeneration (Kaye 2002; Suzuki 1997; Suzuki et al. 1999), but little is known about the contribution of small stands and scattered aspen trees to the persistence of aspen. We found that 43% of the plots surveyed contained aspen trees or saplings (Figure 16.5). These broadly distributed trees and saplings may serve as a vast reservoir of potentially large clones after a fire or insect outbreak (Hadley and Veblen 1993). Long-term monitoring of aspen in these randomly located plots will provide a landscape perspective of the persistence of aspen and associated plants and animals.

Detecting Species Migrations in Response to Rapid Environmental Change

This case study also aids the selection of "indicator species" to assess environmental change. It is rather obvious that common juniper, found in 83% of the plots and in five vegetation types, may be less sensitive to environmental change than a narrowly distributed plant species like western yellow paintbrush (*Castilleja occidentalis* Torr.), which occurred primarily in the limber pine type. Monitoring the distribution changes of the latter species along elevation and moisture gradients in the limber pine transects may be a sensitive indicator of climate change. Changes in the frequency distributions of several species (Figure 16.5) linked to environmental data (e.g., soil characteristics, PAR, topography) may help ecologists select appropriate indicators for specific, anticipated environmental changes.

Another unexpected finding was that limber pine was found in 57% of the plots and in all five vegetation types. Limber pine, generally considered a high-elevation species in the Colorado Rockies, was presumably more common in the low elevations of the park in the early Holocene (Markgraf and Scott 1981). Low-elevation populations still persist on rocky cliffs in the nearby plains, and our finding of limber pine throughout the elevation range attests to its broad environmental and physiological range. Monitoring the establishment and growth of limber pine into higher elevation tundra sites may indicate a response to long-term annual warmer and wetter conditions (Baker and Weisberg 1995), while expanding limber pine populations in lower elevation habitats may indicate local cooler and wetter summer conditions in response to land use change (Stohlgren et al. 1998e).

Detecting, monitoring, and predicting the migrations of most species will also not be easy. It is becoming increasingly clear that modeling vegetation change must consider distribution shifts of individual species (Bartlein et al. 1997; Woodward 1993). Simple vegetation change models that are based on homogeneous plant communities (e.g., Peters and Darling 1985; Romme and Turner 1991) and biomes (Crumley 1993; Neilson 1995) may poorly reflect potential changes in plant diversity. The broad distributions of overstory species in the Front Range (Peet 1981; Stohlgren and Bachand 1997; Stohlgren et al. 1998a) and understory species (Peet 1981; this study, see Figures 16.2–16.4) suggest that plant species respond to the range and variation of microsite conditions more than "average" precipitation, temperature, and potential evapotranspiration (assumed in most vegetation change models). The low affinity of understory to overstory vegetation (Figure 16.5) further suggests that (1) any perceived "plant communities" are loosely organized around a few dominant species or small segments of multiple environmental gradients (Figure 16.4, top); (2) the majority of the landscape is a mix of transient, locally rare understory species and mixed species overstory stands (Stohlgren et al. 1998a,c); and (3) individualistic responses to environmental change likely have dominated the landscape in the past and will likely dominate in the future (Bazzaz 1996; Davis 1991). Identifying broad suites of indicator species will require additional research and experimentation, but a mix of species with broad and narrow ranges may be useful in quantifying local and landscape-scale changes.

Monitoring the Invasion of Nonnative Plant Species

Monitoring nonnative plant invasions provides a means to assess changes in native plant diversity and plant migrations. For example, 43% of the plots contained at least one exotic species (Figure 16.5). They are currently confined to low-elevation sites with fairly high light levels in the understory, but that could change. Elevation and PAR were strong predictors of exotic species richness at landscape scales (Figure 16.6, bottom).

However, nonnative species declined rapidly with increasing elevation, even when light was available, suggesting that temperature may control the distributions of many nonnative species in these mountains. Increasing temperatures in the high elevations could facilitate the spread of mid- and lower-elevation plant species from Mediterranean areas throughout the world. Exotic annuals, in particular, are more confined to lower-elevation areas relative to exotic annual species (Figure 16.7). Native annuals were continuous along the elevation gradient.

The use of long transects that cross major environmental gradients is ideal for identifying long-term trends. Several invasive plant species are particularly worthy of future monitoring. Cheatgrass was found in 26% of the plots and is known to spread excessively after fire (Stohlgren et al. 2001). After decades of fire suppression and fuel buildup in forests, fires and the continued spread of cheatgrass are inevitable. Common dandelion (*Taraxacum officinale*) has already spread from low-elevation ponderosa pine stands to high-elevation limber pine stands. Changes in the frequency and cover of other exotic plant species will be easy to detect with this system of long-term multiscale monitoring plots.

Implications to Theorists and Practitioners

Theorists are making important inroads in models of ecotone structure (Milne et al. 1996) and edge detection (Fortin 1994). Our work provides a significant challenge to theorists to model changes in both overstory structure and understory species diversity at the same sites. Phase transition theory must be expanded to include more than simple "forest to grassland ecotones" and two-dimensional transition zones. Multiple environmental gradients (Table 16.3) and nonlinear responses to them (Figures 16.4 and 16.6) may have to be addressed with a combination of statistical models (e.g., Fortin 1994; Milne et al. 1996) and ecosystem process models (e.g., Coughenour 1993). New structural equation modeling techniques (e.g., Grace and Pugesek 1997)

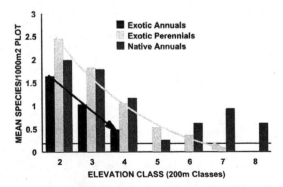

Figure 16.7.   Number of exotic (non-native) annual and perennial species and native annual species in 0.1 ha plots at 200 m elevation classes in Rocky Mountain National Park, Colorado (the first class begins at 1600 m).

appear well suited for these complex systems. The spatial processes of seed dispersal and herbivory are equally complex for modelers (Stohlgren et al. 1997b). Current multivariate techniques such as CCA and multiple regression are hard-pressed to describe patterns of plant diversity at multiple spatial scales. Developing accurate spatial and temporal models of changes in plant species diversity will be a significant endeavor.

Because ecotones and heterogeneous forests can occupy large portions of many landscapes, we propose that monitoring ecotones may significantly contribute to our understanding of landscape-scale patterns of plant diversity in three ways. First, ecotones may play crucial roles in expanding the physiological ranges of many plant species (Figure 16.4, bottom) and fostering species composition overlap among communities (Figure 16.2 and Table 16.4). Bazzaz (1996, p. 228) fully recognized that different genotypes could be found in "patches differing in levels of environmental resources due to their different tolerances of those resources" and that the spatial distributions of genotypes profoundly affects the way "communities as a whole" respond to environmental change. By exploiting ecotones, plant species can greatly expand their domain (Figure 16.4, bottom) and extend the physiological tolerance of certain genotypes. This process may be especially important if the landscape contains large heterogeneous areas relative to the amount of core, homogeneous forest stands (Stohlgren and Bachand 1997; Stohlgren et al. 1998a, 2000b).

Second, high species overlap among ecotones and donor communities, and low affinities of understory species to overstory forest types (Figure 16.3) may predispose vegetation types and biomes to rapidly respond to environmental change. Questions arise whether species will be able to readily adapt to climate change (Bartlein et al. 1997). The ability of a plant species to migrate following climate change is determined by dispersal modes, recruitment and growth patterns, competitive relationships, and general trends in environmental factors (Bazzaz 1996). In our study area, ponderosa pine, a superior invading species over the past 6000 years, was found in 75% of the ecotone plots and 50% of the homogeneous forest plots. Invasive exotic grasses and forbs also established more frequently in ecotone plots (Table 16.2). Ecotones and heterogeneous sites may be the breeding grounds for successful migrators.

Third, the same ecotone characteristics that expand the physiological ranges of many plant species may also foster species composition overlap among communities, improve the ability of plant species to migrate in response to rapid environmental change, and help buffer plant species from extirpation and extinction (Gosz 1991). For example, limber pine, which was more common in the early Holocene (Markgraf and Scott 1981) and is now a predominantly high-elevation species with low-elevation refugia, was found in 67% of the ecotone plots and 57% of the homogeneous forest plots. Thus ecotones and heterogeneous sites may serve as refugia for genotypes that were common under previous climate scenarios. Ecotones may play important roles in maintaining plant diversity at landscape scales and prolonging species persistence in rapidly changing environments.

Monitoring changes in plant diversity requires monitoring many individual species. Gleason (1926) was right! The fact that understory species have very low affinity to overstory types (Figure 16.2 and 16.3), with more than 72% of the plant species found in two or more vegetation types, soundly supports Gleason's individualistic concept. Previous research in the study area showed that seedlings, saplings, and trees of all the dominant conifer species were found throughout the forests, and that large single-species stands were rare at landscape scales (Stohlgren et al. 1998a, 2000b). Replicate transects along several environmental gradients may provide the means to monitor plant diversity and species migrations at landscape scales.

PART VI

RESEARCH NEEDS

17

Case Study

Testing a Nested-Intensity Sampling Design

The Issue

Throughout this text I have stressed that plant ecologists need cost-efficient, accurate, and precise systems to inventory plant diversity at landscape scales. Still, the variability inherent in all natural landscapes poses a significant challenge in designing accurate and complete inventories of plant diversity given typical costs constraints (Barnett and Stohlgren 2003). One option is to establish many large plots across the landscape, or maybe twice as many smaller plots across the landscape for the same cost. However, the fewer larger plots may miss key habitats or rare features on the landscape, while the smaller plots may miss many locally rare species at each site but capture more of the potential environmental gradient and some rare habitats. All designs have cost limitations.

Methods such as "double sampling," that increase sample size and efficiency, are not new to ecology. Range ecologists have long used ocular estimates of biomass and subsamples of clipping and weighing actual biomass to improve biomass estimates of large grazing allotments (Ahmed et al. 1983; Bonham 1989). Similarly Kalkhan et al. (1998) used double sampling to efficiently estimate the accuracy of vegetation type characterizations from a Landsat TM classification map. A few ground plots were used to correct for misclassification errors between Landsat TM data and aerial photographs. The variety of sampling intensity applied in double sampling provides managers with an effective means for quantifying landscape-scale variation, as sample size can be increased at a reduced cost.

Background and Sampling Considerations

Limited budgets and large landscapes make accurate and useful inventories difficult to design and present managers interested in landscape conditions and trends with several trade-offs (Barnett and Stohlgren 2003; Nusser and Goebel 1997; Stohlgren 1994):

- Large plots may provide a good picture of local vegetation conditions, but the use of these thorough plots may limit the total number of plots that can be placed on the landscape. Quality information at localized points may sacrifice understanding of landscape patterns.
- More small plots may increase the quality of broad-scale pattern description, but travel time increases and information collected at each site decreases relative to a few large plots. Spatial extent is increased at the expense of detailed understanding at specific sites.
- Simple, extensive sampling designs with single-scale plots may provide precise data, but the designs may not be cost effective for ecologists and land managers. Intricate methods for inventory are often not used by managers because of their complexity, but they may provide more accurate data at reduced costs.

Can a "nested-intensity design" address the challenges of the trade-offs above? In this case study we used three plot designs—a large multiscale plot, a smaller multiscale plot, and a single-scale plot—to assess the advantages and disadvantages associated with the integrated use of three vegetation sampling plot designs (Barnett and Stohlgren 2003).

An Experimental Approach

Case Study Site Description

The sampling took place in aspen vegetation on a ranch on the western slope of the spine-like Sangre de Cristo mountain range in south-central Colorado. The Culebra group, the southern subsection of the Sangre de Cristo, stretches north for 50 km from the New Mexico border. The elevation ranged from approximately 2600 m to the top of 4267 m Culebra Peak. Forested vegetation types include ponderosa pine, Douglas-fir, Engelmann spruce, subalpine fir, and aspen. Tundra and an abundance of alpine rock dominate the higher elevations. Sampling was limited to the aspen vegetation type, which covers approximately 10,000 ha of the 30,000 ha ranch. Aspen supports floral (Stohlgren et al. 1997c), butterfly (Simonson et al. 2001), and bird (DeByle and Winokur 1985) diversity, which is otherwise rare on the landscape. The success of aspen regeneration is a concern in similar regions across the Rocky Mountains (Barnett and Stohlgren 2001; Romme et al. 1995).

Plot Designs and Vegetation Sampling

We stratified aspen vegetation using high-resolution aerial photographs (1:24,000) from 1997 (Barnett and Stohlgren 2003). Random points were located within the aspen stands and assigned to one of three vegetation sampling methods. The nested-intensity design included three vegetation sampling methods as follows.

We established eight modified Whittaker plots as previously described (Figure 5.1 and Figure 17.1a). In the ten 1 m^2 subplots, we identified all vascular plant species and estimated the average height and cover to the nearest percent for each species. We also recorded the species presence in the 10 m^2 and 100 m^2 subplots, and in the 1000 m^2 plot.

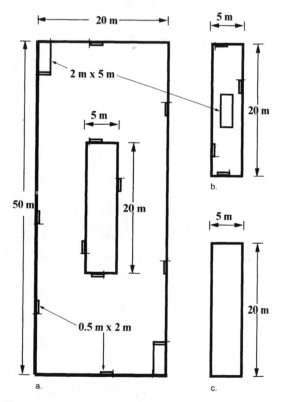

Figure 17.1.   The various plot designs tested in this case study. From Barnett and Stohlgren 2003, with kind permission of Kluwer Academic Publishers.

Based on the design of the modified Whittaker plot, we developed the 100 m², nested "Intensive" plot (Figure 17.1b). The Intensive plot contained four 1 m² subplots and one 10 m² subplot, both nested within the 100 m² plot. In each of the 1 m² subplots, we identified all vascular plant species and estimated the cover to the nearest percent for each species. In the 10 m² subplot and 100 m² plot we recorded the presence of all vascular plant species. We placed 15 Intensive plots across the landscape, intermixed with the modified Whittaker plots.

We established thirty 100 m² "Extensive" plots dispersed among other sample design locations. Species presence was recorded in the plot, but no evaluation of subplots or species cover occurred in this plot (Figure 17.1).

We evaluated aspen stand structure (in half of every 100 m² plot) to better describe local aspen regeneration. In the modified Whittaker and Intensive plots, the diameter at breast height (DBH; in centimeters) was recorded for every tree. Less information on tree diameters was collected at Extensive plot locations. For all stems more than 2 m tall, we recorded whether the tree was alive or dead, was larger or smaller than 10 cm DBH, and the average percent browse. For all stems less than 2 m tall, the number of stems and whether the tree was alive or dead was recorded.

Statistical Analyses

We compared plant species richness, total species detected, and total unique species captured between the three plot techniques (Barnett and Stohlgren 2003). A "unique species" was defined as a plant species detected or captured only by one particular sample design. Moran's I was used to check the species richness of the 100 m² plots for spatial autocorrelation using spatial applications (Reich and Bravo 1998) in the S-Plus statistical package (MathSoft Inc. 1999). All other analyses, except where noted, were carried out using Systat (SPSS Inc. 2001).

To better evaluate travel time and sampling time, we frequently sampled with more than one design in a single day. To calculate the total time required to complete the entire sampling of each design, including travel time, we subtracted the time to sample each plot in a single day from the length of the day to yield the time required for travel. Travel time was divided equally among the number of plots sampled that day. Travel time and sample time for each design were totaled within a given day and across the number of total sampling days. The average time required to sample with each design was also calculated and compared using ANOVA.

We compared the average species richness per plot and the average species richness in the 100 m² subplot (modified Whittaker plot design) or plot (Intensive and Extensive plot designs) from each plot type using analysis of covariance (ANOCOVA). We also plotted the number of species in the 100 m² plots against each other to examine the potential bias of the techniques.

We compared modified Whittaker and Intensive plot plant richness totals at the 1 m² and 10 m² subplots with *t*-tests. Then we calculated the species accumulation curves for each design. We also created species accumulation curves that described the contribution of each sampling technique to the total species list generated by all three techniques combined. For example, one iteration initiated the curve with the species detected using the modified Whittaker plots. The second section of the curve was defined by the new species accumulated with the addition of the Intensive plots on the landscape. The third portion of the curve was dictated by the addition of species detected by the Extensive plots and not previously detected by the other techniques. Using this method we built total species accumulation curves in several combinations: the Intensive plots, Extensive plots, modified Whittaker plots; and the Extensive plots, modified Whittaker plots, and Intensive plots. All curves were created using Estimate-S (Colwell 1997) as described earlier.

Species-Area Curves

Species-area curves were calculated for the modified Whittaker and Intensive sampling methods, but not for the Extensive design which contained no nested subplots. To generate species-area curves at each appropriate sampling location, we calculated the mean number of species in the 1 m² and 10 m² subplots, and the total number of species in the 100 m² subplot and 1000 m² plot for each modified Whittaker plot. Similarly we calculated the mean number of species in the 1 m² subplots and the total number of species in the 10 m² subplot and 100 m² plot of the Intensive plots (Barnett and Stohlgren 2003).

We tested the now familiar power model of the form ($S = cAz$), log-log form ($\log S = z\log A + \log c$), and semilog form ($\log S = A + z\log c$), as described in chapter 4. We used a method of confrontational modeling (Burnham and Anderson 2002) to select the model that most accurately described the plant species richness data collected by the modified Whittaker and Intensive designs. Model parameters were estimated by minimizing the sum of squared errors. The maximum log-likelihood for each observation was calculated and used to calculate small sample Akaike Information Criteria (AIC), Δr, and the Akaike weights for each model (Burnham and Anderson 2002). The model associated with the lower small-sample AIC indicates the model that most accurately describes the data. Δr is an indicator of models that have sufficient support given the data, and should be considered; models with a $\Delta r \leq 2$ should be considered. Akaike weight accounts for uncertainty in the order of model ranking, given errors in the data, and frames the model selection as a probability (Burnham and Anderson 2002).

Average species-area curves were then fitted for the modified Whittaker and Intensive sampling techniques based on the model that best described the data as noted above. The slopes of these lines were compared using the general linear model in the SAS statistical package (SAS Institute 1998). Using the model that best described the curves, we derived species-area curves and the slopes of the curves for each modified Whittaker and Intensive plot location based on the subplot and plot data.

Based on data from the modified Whittaker and Intensive designs, we created a multiple linear regression to describe the slope of species-area curves at the modified Whittaker and Intensive plots in an attempt to estimate species-area curve slopes at locations sampled with the single-scale Extensive plots. We assessed the validity of the model by iteratively removing one modified Whittaker plot from the model at a time, rebuilding the model, and then comparing the observed and predicted slopes for the removed plot (repeated five times). To estimate the slope of the species-area curve at the single-scale Extensive plots, we incorporated data collected at the Extensive plots into the model that described the slope of the modified Whittaker and Intensive plot species-area curves (Barnett and Stohlgren 2003).

To fulfill the requirements of the model that described the slope of species-area curves at the modified Whittaker and Intensive plots, it was necessary to predict the DBH of live aspen stems in the Extensive plots based on the number of live stems both larger and smaller than 2 m. Because the DBH of live aspen stems was not collected at the Extensive plots, we developed another model with the modified Whittaker and Intensive plot data that predicted live DBH (adj. $R^2 = 0.89$, $P < 0.001$) using the number of live stems larger and smaller than 2 m. We collected the number of stems smaller and larger than 2 m in the Extensive plots and were able to estimate live DBH at the Extensive plot with the DBH model created using the modified Whittaker and Intensive aspen data. Lastly, multiple linear regression was used to generate a model that described the number of species in 100 m² plots in both the modified Whittaker and Intensive plot designs (Barnett and Stohlgren 2003).

Results, Discussion, and Lessons Learned

There were obvious and subtle differences in this test of sampling designs. The modified Whittaker plots took longer to sample, but they captured more species than the smaller, simpler sampling techniques (Barnett and Stohlgren 2003). However, sampling all of the plots of each technique required similar time when travel was included, and each technique accumulated a similar number of total species (Tables 17.1 and 17.2).

Despite these equalities, the modified Whittaker technique still collected more unique species than the other two techniques. Regression models based on combinations of these techniques proved to be useful in predicting data not collected at the simpler, Extensive plot. For this case study, we presented

Table 17.1. The total number of plots sampled and the time required to sample each plot of the nested intensity design as sampled in the Sangre de Cristo Mountains, Colorado.

Characteristic	Modified Whittaker plot 1000 m²	Intensive plot 100 m²	Extensive riot 100 m²
Number of plots	8	15	28
Mean hours per plot at a site	6.7 (0.9) a	2.8 (0.9) b	0.5 (0.1) c
Total sampling time (hours)	53.6	42	13.1
Total travel time (hours)	27.5	41.9	57.7
Total sampling and travel time (hours)	81.8	83.9	70.8

The standard error of the mean appears in parentheses.
   The subscripts show significant differences between labeled values, $a = 0.05$.

Adapted from Barnett and Stohlgren (2001).

results in three sections: plot type comparisons, comparisons of species-area curves, and predictive models. A Moran's I value of -0.06 suggested that species richness in the 100 m² plot or subplot was spatially independent, so traditional statistical procedures were applied in the analyses (Barnett and Stohlgren 2003).

How the Various Plot Designs
Captured Plant Diversity

The larger modified Whittaker plots required greater sampling time than the Intensive plots, which took longer to sample than an average Extensive plot ($P < 0.001$) (Table 17.1). However, sampling all plots of each design required a similar amount of time (Table 17.1). A greater number of Extensive plots and the travel time associated with sampling each plot greatly increased the time required to sample the methods with larger sample sizes.

The larger modified Whittaker plots (1000 m²) also detected greater mean species per plot than both the Intensive (100 m²) and Extensive (100 m²) plot types ($P < 0.001$) (Table 17.2). Because the various plots were not iteratively sampled at the same location, site variation could have affected these results. However, when compared at the 100 m² plot size, no statistical difference ($P = 0.19$) separated the three sample techniques (Table 17.2). Furthermore, scatterplots of species richness in the 100 m² plots or subplots suggested there was no systematic bias associated with the designs.

Despite different sample sizes and total sample area, the three test designs captured similar numbers of cumulative species (Table 17.2). However, the modified Whittaker plots detected more unique species than the other two designs. Of the species detected by the modified Whittaker plot, 25.8% were unique species, compared to 14.7% and 14.4% captured by the Intensive and Extensive plots, respectively (Barnett and Stohlgren 2003).

Table 17.2. Comparisons of the three different vegetation sampling techniques used as components of the nested intensity inventory in the Sangre de Cristo Mountains, Colorado.

Characteristic	Modified Whittaker plot 1000 m²	Intensive plot 100 m²	Extensive riot 100 m²
Maximum plot area (m²)	1000	100	100
Total area sampled (m²)	8000	1500	2800
Mean species richness	56 (8.8)$_a$	30 (2.7)$_b$	25 (1.4)$_b$
Mean 100 m² species richness	29 (1.9)	30 (2.7)	25 (1.4)
Cumulative species richness	122	112	122
Cumulative exotic species richness	7	6	7
No. of unique species	32	18	14
No. of unique exotic species	2	1	1

The standard error of the mean appears in parentheses.
Different subscripts show significant difference between values, $\alpha = 0.05$.

The species accumulation curve of the modified Whittaker plot increased at a faster rate than the curves of the smaller designs (Figure 17.2). The total number of new species increased with the addition of new plots (Figure 17.2). Early plots contributed more new species to the species lists than later plots that detect many species previously measured by other plots.

We also determined the contribution of each design to the total species list accrued from the suite of the designs. When we fit this accumulation by alternating the order of each design's contribution to the species list, we found that (1) the shape of the curve was significantly affected by the order of the inclusion; and (2) adding large-area plots, at any time, caused significant jumps in species accumulation curves (Figure 17.3).

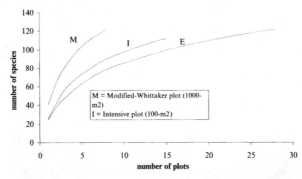

Figure 17.2. Species-accumulation curves for Modified-Whittaker (W), Intensive (I), and extensive (E) plot types. From Barnett and Stohlgren 2003, with kind permission of Kluwer Academic Publishers.

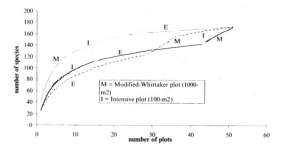

Figure 17.3.   Species-accumulation curves with various combinations of Modified-Whittaker (W), Intensive (I), and extensive (E) plot types. From Barnett and Stohlgren 2003, with kind permission of Kluwer Academic Publishers.

Species-Area Curves and Predictive Models

Model selection criteria metrics (Burnham and Anderson 2002) suggested that the power function provided the best description of the plant species richness data collected (Akaike weight = 0.99) (Table 17.3).

The general linear model used to compare the average slopes of the species-area curves generated by the modified Whittaker and Intensive sampling techniques indicated that the slopes of these two lines were not significantly different ($P = 0.98$) and therefore the multiscale richness data from the modified Whittaker and Intensive designs could be pooled in an effort to predict the slope of unmeasured species-area curves at the Extensive plots.

Ancillary data, species richness in 100 m² plots or subplots, and diameter of live aspen stems allowed us to estimate the slopes of the species-area curves of the modified Whittaker and Intensive sample designs ($R^2 = 0.64$, $P < 0.001$) and proved to be 96% accurate when validated. We found that 70% of the variation in the slopes of the species-area curves could be explained by the longitude (UTMx), aspect, species in 100 m² subplots, and aspen stocking (total DBH). We also found that 85% of the variation in species richness in 100 m² plots could be explained by the species richness in 1 m² plots, longitude (UTMx), and aspect.

Table 17.3.   Confrontation model selection metrics for selection of a species-area curve to fit multiscale data from modified Whittaker and Intensive design vegetation sampling techniques.

Model	Small sample AIC	Δr	Akaike weight
Exponential model	556	296	0.00007
Power model	537	277	0.99

Adapted from Barnett and Stohlgren (2003).

Plant ecologists may not have sufficient time or funds to adequately describe plant diversity throughout the landscape with an abundance of large, multiscale plots (Tables 17.1 and 17.2). Alternatively, simple, small-plot sampling may be quicker to measure, but may not adequately represent landscape conditions (Tables 17.1 and 17.2) (Stohlgren et al. 1998c) or be an efficient way to capture plant species richness. Combining these alternatives in a single inventory effort incorporates the advantages of plot types, provides insights into the effect of plot size on inventory efforts, and may be an efficient means for increasing inventory quality and usefulness, as described below.

Advantages and Disadvantages of the Three Designs

The plant diversity sampling techniques described in this case study possessed specific advantages and disadvantages. The methods differed in their ability to accurately quantify local or landscape-scale patterns of plant diversity, but the advantage of one design frequently corrected for the disadvantage of another design (Barnett and Stohlgren 2003).

The modified Whittaker design quantified species richness with greater accuracy and efficiency than the other designs. It captured greater species richness per plot than the other the two techniques (Table 17.2), and when comparing species lists by technique, the modified Whittaker design detected more unique species than the other two designs (Table 17.2). Furthermore, the modified Whittaker plot accumulated species at a faster rate than other designs (Figure 17.2), and the large 1000 m² plot accounted for 14 of the 30 unique species, and 2 of these were unique exotic species. These results emphasize the ability of the larger plot size to detect locally rare and patchy species not captured in small plots.

The four sample scales nested within the large modified Whittaker plot ensure accurate species-area relationships that describe the diversity of the local area. The slope of the species-area curve built from the three scales of the Intensive design suggested the number of species in the 1000 m² area of the modified Whittaker plot could be estimated with the Intensive plot data. Using the modified Whittaker plot remains important, as the Intensive plot can predict the richness but not the identity of those extra species captured by the modified Whittaker technique. Not only was the modified Whittaker plot able to add two unique exotic species to the total species list, in both cases the exotic species were only detected in the 1000 m² plot. This reemphasizes that the large plot size is necessary for the early detection of exotic plant invasion and the presence of patchily distributed and locally rare species (Stohlgren et al. 1998c, this case study).

The modified Whittaker design also has disadvantages. The design is expensive and time consuming. The costs associated with the modified Whittaker might limit the number of plots, thus increasing the probability of missing important locations across the landscape, reducing the environmental gradient sampled, and making extrapolation to unsampled areas difficult. Furthermore, expense and time limitations might limit the frequency of resampling sites, thus sacrificing the early detection of vegetation trends (Barnett and Stohlgren 2003; Stohlgren 1994).

The smaller, multiscale Intensive plot addresses the disadvantages of the modified Whittaker technique. Both the smaller size and fewer subplots as compared to the modified Whittaker design allowed many Intensive plots to be distributed across the study area for a cost (sample and travel time) (Table 17.1) similar to the modified Whittaker plots. The greater number of plots facilitated better coverage of complex environmental gradients and a greater chance of locating rare microhabitats that host unique species. Despite the effort to account for landscape-scale variation with a greater number of plots, the Intensive plot design maintained multiscale sampling features. This combination allowed for diversity characterizations with species-area curves at Extensive sample locations and made the Intensive design effective for calibrating other sampling techniques. For example, the Intensive design increased the accuracy of the species-area curves developed to describe the sampled aspen vegetation by sampling at many points. The choice of the species-area curve model and multiscale extrapolations to the single-scale Extensive plot data may have been less accurate if formulated on the data from the eight modified Whittaker plots alone (Barnett and Stohlgren 2003).

The exclusion of 1000 m² measurements caused most of the disadvantages associated with the Intensive plot. The Intensive plot detected fewer species per plot and fewer unique species than the larger modified Whittaker plot (Table 17.2). Decreasing the number of subplots may have reduced the accuracy of the curve used to describe diversity at particular locations and across the entire vegetation type. Furthermore, maintaining the multiscale component still required significant sampling time and therefore limited the number of plots and landscape-scale heterogeneity sampled.

The Extensive plot, also characterized by the 100 m² plot, but without nested subplots, increased the ability of this study to account for a large degree of spatial heterogeneity across the landscape. The inclusion of this design added one exotic species to the total species list (Table 17.2), added additional patches of aspen regeneration (Table 17.4), and established a landscape-scale baseline of data suitable for detecting early and isolated change on the landscape. A greater number of monitoring locations is more likely to detect localized disturbance, exotic plant invasion, or aspen regeneration, as processes that influence change may have patchy effects (Baker 1989; Barnett and Stohlgren 2001; Green 1989; Turner 1989). We were able to increase the value of this widely distributed Extensive plot data. Because the simplified data were similar to the data collected at the multiscale plots, statistical models developed using the data from modified Whittaker and Intensive plots were able to estimate species-area relationships at the Extensive plot sites.

Table 17.4.   The characteristics of the aspen stand structure as sampled with the nested-intensity sampling design in the Sangre de Cristo mountains of Colorado.

	Modified Whittaker 1000 m²	Intensive plot 100 m²	Extensive plot 100 m²
Number of plots	8	15	27
Mean stems/ha- >2-m tall, >10-cm dbh	1125 (320.6)	867 (180.7)	941 (118.8)
Mean stems/ha- >2-m tall, >10-cm dbh	1675 (549.0)	1027 (279.4)	1237 (243.4)
Minimum stems/ha- >2-m tall, >10-cm dbh	400	0	0
Maximum stems/ha- >2-m tall, >10-cm dbh	4800	2800	4000
Number of plots with >2-m tall, >10-cm dbh	8	11	11
Number of plots with no >2-m tall, >10-cm dbh	0	4	9

Adapted from Barnett and Stohlgren (2003).

The Extensive plot also had disadvantages. The smaller size of the plot missed species at sample points (Table 17.2), thus limiting completeness, and captured fewer unique species than the larger modified Whittaker design (Table 17.2). With no 1 m² subplots, species cover could not be precisely recorded in the Extensive plots. Simply collecting the number and identity of species does not provide cover estimates necessary for detecting species-specific increases or decreases. For example, assessing changes in cover of the nonnative species *Poa pratensis* in a particular plot and across many plots may alert managers to the site-specific spread of the nonnative species (Barnett and Stohlgren 2003).

Eliminating the subplots from the Extensive design limited the measure of species richness to a single scale and provided little insight into species-area relationships. This exception limits the usefulness of the information collected at each location, especially when compared to other plot locations that describe species richness at a variety of scales. Finally, collecting less information at a specific location increased the cost of travel time and therefore the cost for the information return at a local point (Barnett and Stohlgren 2003).

Benefits of Nested Intensity Designs

Of the three vegetation sampling designs, the modified Whittaker plot contributed the most detailed and accurate data at just a few sample locations. The advantages and the proven ability of the modified Whittaker design (Table 17.2) (Stohlgren et al. 1998a,c, 1999a, 2000a) to quantify plant species diversity at landscape scales might prompt managers to assess landscape patterns of plant diversity using only such a large and multiscale design. Other large, nested plot designs (e.g., Keeley et al. 1995) would probably have worked equally well. If we dedicated all the funding and time available for this inventory effort to sampling with large, nested plots, approximately 24 plots could have been placed on the landscape. Given the ability of large plot designs to capture greater species richness, and the steep modified Whittaker species accumulation curve (Figure 17.2), 24 modified Whittaker plots may have been the best option if simply generating the largest plot-based species list was the inventory objective. However, most managers must consider entire landscapes. The increased sample of 51 plots using all three sample designs may provide a more complete picture of landscape condition and variation, especially if the data collected at the smaller and single-scale plots can be leveraged as with traditional double sampling techniques (Ahmed et al. 1983).

Similar to classical double sampling techniques, we leveraged the study design by relating data collected with fast and simple designs to more detailed data. We demonstrated the power of the nested intensity design by detecting hot spots of species richness across the aspen landscape. Using a multiple regression model parameterized with data from the multiscale modified Whittaker and Intensive plots, we estimated species-area curves for each Extensive plot location. Generation of species-area curves for every location allowed each plot to be ranked and spatially mapped according to the steepness of the slope or the richness at each location. Species richness may vary across scales (Gaston 2000), so the slope of a curve is a powerful tool for detecting local hot spots of diversity (Connor and McCoy 1979; Rosenzweig 1995; Stohlgren et al. 1997b). Detecting hot spots of diversity on the landscape may be important for setting priorities for preservation of important habitats or landscapes (Margules and Pressey 2000; Myers et al. 2000), directing management activities (Mulder et al. 1999; Nusser and Goebel 1997), or locating regions or plots important for inventory and monitoring purposes (Stohlgren et al. 1997b, 2000a).

There are many possibilities to leverage the effectiveness of a nested intensity design. For example, we used the nested intensity approach to evaluate the spatial distribution of successful aspen regeneration. We collected more detail on stand structure at the modified Whittaker and Intensive plots than the Extensive plots, but the quantity of pole-like stems indicative of recent regeneration was recorded in all plots. Using methods similar to the species richness predictions, we were able to estimate the diameter of live and dead stems in Extensive plots. Similar manipulations could be attempted to increase the spatial understanding of aspen stand structure and regeneration for both baseline inventory and monitoring purposes. Alternatively, a manager not concerned with patchy processes such as herbivory and fire, which affect the spatial distribution of aspen regeneration (Romme et al. 1995), might simply be interested in the average number of regenerating stems on a landscape. This question may not require a large sample size facilitated by the nested intensity design, as Monte Carlo simulations suggested that only 10–15 plots would be needed to indicate that there was an average of approximately 1300 regenerating stems per hectare on this landscape. The ability of nested intensity techniques to evaluate the condition of aspen demonstrates that such a system can be used to address a variety of species, but may not be necessary for particular species in all cases.

Assessing Cumulative Species in an Area: Lessons Learned

Examining species accumulation curves allowed us to further evaluate the contribution of each sampling design to the nested intensity design. The steeper curve of the modified Whittaker design (Figure 17.2) indicated these plots accumulate species at a faster rate than the other designs and reflects the ability of the larger modified Whittaker plot to detect more species per plot than the smaller designs (Table 17.2). The Intensive plots produced a species accumulation curve steeper than the Extensive plots (Figure 17.2). This difference must be due to the greater species diversity at some of the locations where the Intensive plots were sampled, as the Intensive and Extensive plots were the same size. We expect the trajectory of these curves would become quite similar with continued sampling that would extend the curves and dampen the effect of hot spots of diversity on the landscape.

The attributes of a plot can't be assessed by plot size alone. The contribution of each plot design can be compared by quantity and the quality or usefulness of information returned for the total effort of sampling with that design. The Extensive plots do not contain subplots for collecting estimates of species cover essential for tracking information such as the spread of exotic species or the progress of restoration efforts. This omission makes the Extensive plots quicker to sample, but less useful to managers. The modified Whittaker and Intensive plots must be used if managers require cover estimates of the species detected. Sampling the modified Whittaker and Intensive plots required similar effort, approximating a ratio that allows fewer than two Intensive plots to be placed on the landscape for every one modified Whittaker plot (Table 17.1). With similar effort, the eight modified Whitaker plots provided greater species richness and cover for more species than the 15 Intensive plots (Table 17.1 and 17.2). The quality and quantity of information provided by the modified Whitaker plots dispels the popular notion that many small plots are more effective than a few large plots, at least for plant diversity studies such as this one.

The species accumulation curves that describe the contribution of each plot type to the total species detected elucidates the bias of the smaller plot sizes in terms of taxonomic completeness (Figure 17.3). The flattening of the sections of the curve contributed by both the Intensive and Extensive plots suggests further sampling would detect few new species. However, regardless of the order included, the steep curve defined by the modified Whittaker plots indicates these large plots detect species missed by the smaller plots (Figure 17.3). Small plots underestimate species richness. In fact, if a manager were to iteratively assess the degree of completeness of an inventory either with statistical techniques (see Bunge and Fitzpatrick 1993; Schreuder et al. 1999) or looking at the slope of the species accumulation curve, the inventory might be halted prematurely if sampling were only conducted with plots even of the 100 m^2 size (Barnett and Stohlgren 2003).

Other Applications for Nested Intensity Designs

There are many other applications of nested intensity sampling strategies to efficiently address time, funding, and spatial and temporal accuracy concerns. For example, pilot studies (Krebs 1989; Ludwig and Reynolds 1988; Reed et al. 1993) have long been recommended for the determination of sample size. The number of study plots required to account for variability may often be underestimated, as the small number of pilot plots may not sample enough locations to accurately describe the environmental variability across the study area. The use of this nested intensity design may help an investigator better understand the extent of the variation on the landscape without the cost and time associated with the inclusion of many detailed plots on the landscape. The number of plots and the proportion of the techniques used depend on the landscape, the number of vegetation types and rare habitats, specific questions of interest, and the time and money available for the work. The inventory must be an iterative process, with the number and location of plots informed and directed as the data are collected.

The nested intensity design could contribute to this iterative approach. The number of species in an Extensive plot, or the predicted species-area curve for that plot, might indicate that the plot location is important, perhaps due to high species richness or because of the presence of a particular exotic species. Such an area may require monitoring efforts more frequently than areas of less interest. The information obtained from Extensive plots may spawn the establishment of a modified Whittaker or Intensive plot at that location in future monitoring efforts to gain greater detail about species cover at finer scales and species composition at larger scales.

Plant ecologists with access to computer and statistical expertise can use spatial models to predict species distributions across the landscape to continue to improve their inventory programs (see chapter 14) (Chong et al. 2001; Reich and Bravo 1998). The Extensive plots used in this design provide a means for improving and testing the accuracy of such maps without the expense of placing many new large plots on the landscape.

Those ecologists who may not have adequate funding or scientific expertise to pursue such a thorough inventory system may also benefit from the nested intensity design. Simplified monitoring can be practiced between periodic intensive investigations, creating a nesting of knowledge and time. The ability to predict the number of species in a 100 m^2 plot with data from 1 m^2 is just one example. Those who do not have extensive taxonomy skills can simply count the number of different morphological species (based on different appearance) in Extensive plots to identify hot spots for later detailed research by trained taxonomists. One can learn to identify a small number of species and search Extensive plots or even the 100 m² plot of a modified Whittaker plot. For instance, we found the easily identified and highly competitive nonnative grass *Bromus inermis* at several plots. If a manager learned this species and could quantify its spread in study plots, the resulting data could direct control efforts. Early detection of this species would greatly facilitate control efforts. Meanwhile, the plant ecologists could monitor just the 1 m² subplots of a modified Whittaker or Intensive plot for the cover of such species in an effort to track subtle changes over time. An initial nested intensity inventory could direct monitoring to the highest priority areas (Stohlgren et al. 1997c). The data obtained from the abridged monitoring with 1 m² or Extensive plots could not only direct management, but also register thresholds that suggest the need for another intensive inquiry with modified Whittaker and Intensive plots.

The principles of the nested intensity inventory method described here could be adapted for use in any landscape, and to address a variety of management questions. Exploration of the methods used in this study demonstrated several benefits to inventory programs:

- Nested intensity designs are well suited to document local and landscape patterns of plant diversity, while increasing time efficiency and reducing costs.
- A subset of fairly large plots (e.g., the modified Whittaker plot or Keeley plot) are needed to capture more complete information on species richness and unique species in some areas—methods with smaller plot sizes may miss locally rare species.
- A moderate set of smaller Intensive plots are needed to increase the spatial extent of the study, while increasing the accuracy and representativeness of species-area curves.
- A greater set of Extensive plots is needed to further increase sampling intensity in heterogeneous landscapes, thus adding unique species and locating hot spots of diversity and patchy resources of special concern.

Nested intensity systems represent creative alternatives to single-type inventory techniques commonly used by managers of both public and private lands. Newer, innovative approaches could increase the effectiveness and spatial extent of established monitoring programs.

18

Quantifying Trends in Space and Time

The Issue

Much of this book has focused on accurately quantifying the spatial patterns of plant diversity. This is often attempted with a combination of remote sensing data, plot sampling techniques, and interpolations from spatial models, which provide a "snapshot" of plant diversity patterns (with some quantified levels of uncertainty) (see chapter 14). However, the ultimate challenge as plant ecologists is to measure and predict (model) changes in plant diversity in space and time—more like a "movie" (Stohlgren 1999a). There are a growing number of textbooks to help us understand spatial variation (Cressie 1991; Cullinan and Thomas 1992), and there are accepted techniques for time series analysis (Pole et al. 1994; West and Harrison 1997), but there are no well-accepted textbooks on spatiotemporal analyses. The "Holy Grail" in ecological research is to develop field and modeling techniques to detect and quantify patterns in space and time and to explicate underlying mechanisms (Adler and Lauenroth 2003; Lubchenco et al. 1991).

Paleoecologists and plant geographers have provided us with a basic understanding that vegetation does in fact change in space in time. Through pollen records, pack rat middens, macrofossils, and current vegetation data, they present coarse-scale descriptions of how dominant tree species and subdominant vegetation have "moved" across regions since the last ice age (Davis 1991; Pielou 1991). As discussed in chapter 15, the paleorecord is often spotty at best (discontinuous in time and space) and contains information on only a few species in any local flora (i.e., locally dominant species). It is possible to prove that a species was certainly "present" at some general location in the past, but it is difficult or impossible to prove that other species were truly absent. Some species leave poorer records, while others may have been hiding in small refugia or microsites nearby. Note that this situation is not unlike current vegetation sampling issues, where some plant species occur just outside the plot and are not recorded, some are in the plot as seeds and go undetected, and some may have lingered on the site (in a favorable microsite) after establishing under different climatic conditions. This creates lag effects in temporal series of plant species distributions in response to climate change and other factors.

Models of the effects of potential climate change on forest distributions have also shown that, given current species-environment relationships, some vegetation types in the central Rocky Mountains will quickly migrate to Canada (Romme and Turner 1991). Such models often make simplifying assumptions that current "average" species-environment relationships of a species describe the extreme behavior of individuals and various genotypes in the population and that there are no lag effects, such as individuals lingering in favorable microsites within inhospitable climate regions. Obviously such analyses are greatly limited by the grain and resolution of habitat maps, and such maps are not intended to show high resolution of all individuals within a species.

Paleoecological studies and climate-vegetation models hint at what we would like to be able to do as plant ecologists: measure and monitor changes in plant diversity in space and time, usually to address general and specific questions about species-environment relationships, species persistence, and species migrations and invasions at multiple scales. I use selected questions and topical issues to intertwine with the discussion that follows:

1. A small meadow on a cattle ranch contains a population of a rare (but not federally or state-protected) lily. What would happen to the population in the long term if grazing were curtailed?
2. An invasive, nonnative plant species was recently observed just outside a large national park. Which habitat types are most vulnerable to invasion and how quickly will it likely spread?
3. Relict populations of an endemic and now rare species are showing population declines associated with tree canopy closure during natural succession. If a natural fire regime replaced the current management practice of complete fire suppression, would the rare populations flourish and spread?
4. Bountiful, colorful flowers on nonnative thistles are attracting pollinators away from less-colorful, less statuesque native plant species. How will this alter the abundance, isolation, and fecundity of the native plant species over time?
5. Hot spots of native plant diversity appear associated with rare landscape features such as wetlands, tropical forests, riparian zones, and rare forest or grassland communities. How are these

hot spots threatened by land use changes, habitat loss, and fragmentation and isolation?

6. Land use changes, agriculture, and domestic livestock grazing have left only a few, small "natural" areas in many landscapes and regions—usually on unconquerable mesas or mountaintops, steep canyons, or designated protected areas. How well are most native species persisting under these pressures, and what are the long-term effects of land use on native plant diversity?

7. With greatly increased modern transportation, new plant species and genotypes, and plant pathogens from other countries are routinely entering the United States in packing materials, via horticultural trade and landscaping supplies, and by other means. How can we monitor the cumulative effects of these uninvited guests on native plant diversity?

Background and Sampling Considerations

The reader should now have a strong understanding of many sampling design issues concerning the measuring and monitoring of plant diversity. Certain questions should leap to mind when presented with a question. Question 1, on the fate of the lily population in the meadow, for example, raised many additional questions. Does the timing, seasonality, or intensity of grazing directly affect the lily population in the meadow? This requires an experimental approach and perhaps greenhouse studies to substantiate alternative explanations. Does grazing interact with other factors to negatively affect the fitness (reproductive success and persistence) of the lily? How is persistence linked to genetic, species, and habitat diversity? These questions require factorial experiments, genetics research, and long-term monitoring and modeling in the meadow. Is the species successfully reproducing, or expanding or contacting it range in the meadow? This requires monitoring of a larger area, including potential, but not realized, lily habitat. Are there viable seeds in the soil? This requires soil cores, seed bank investigations, and greenhouse experiments. How rare is the species on the ranch, in the surrounding landscape, and on neighboring ranches—that is, are there other seed sources to replenish the population if it were extirpated? This requires a high-resolution survey of habitats and careful survey of lily subpopulations. How rare are the habitats in which it lives, and how plastic is the species for surviving under less-than-optimal habitats? That is, can reproduction be stimulated over broader spatial scales (larger populations generally have greater persistence than smaller populations)? This requires remote sensing work, outplanting experiments, and long-term monitoring. Are there unforeseen external threats to the lily (a new pathogen arriving, such as sudden oak death in California)? This requires vigilance and links to broader-scale research on invasive organisms and their fate.

It is obvious that sampling design challenges increase with the scope of the problem. Monitoring the cover and abundance of a lily in a small meadow is far easier than landscape and regional surveys of lily habitat, detailed demographic studies of lily populations (and metapopulations), controlled yet realistic experiments on the effects of grazing and other factors on lily persistence, and population modeling over time in complex environments. Our first major challenge is in evaluating cost constraints versus the scope of the issues.

Cost Constraints versus the Scope of the Issues

In the lily example above, as with all other plant diversity questions, the issue of cost cannot be ignored. The general attributes of effective vegetation monitoring efforts have been discussed by several ecologists (Krebs 1989; MacDonald et al. 1991; Risser 1993; Stohlgren 1994; Strayer et al. 1986). The attributes always include clearly articulating the goals and objectives of the studies and identifying the scope and constraints of the problem. As discussed in chapter 3, these issues must be considered within the context of appropriate sampling designs (the number, size, and pattern of sampling plots) and an evaluation of the strengths and weaknesses of alternative study designs and field techniques. The lily example also points out the need for detailed experiments and developing a strong understanding of species biology and demography, and the potential internal and external threats to population persistence. The costs of any one part of this study could be substantial, and receiving funding to conduct all aspects of the study reasonably well would be unlikely.

In question 2, similar issues of scope come into play. Information is needed on the biology of the invading species; rates of spread in similar habitats; potential competitiveness of co-occurring species; the effects of current and future land use changes; the effects of manual, chemical, and biological control measures; and other factors that influence the spread and success of the invading species. As in most cases, information is needed from past research, carefully designed experiments, surveys, monitoring, and various predictive modeling approaches. Each aspect of the study is costly.

As the temporal scale of the study increases, the costs of collecting detailed data for modeling may rise significantly. In question 3, fire cycles can be decades to centuries long in many forests, and careful long-term monitoring is usually beyond the scope of a single investigator. Great lengths must be taken to ensure data precision and accuracy over time, with clearly defined sampling protocols, well-marked relocatable plots, and a strong commitment to data management (Figure 18.1) (Stohlgren 1994).

In most cases, long-term plant diversity studies take many measurements at relatively few sites. Most natural experiments using fire include very few true replicate burns. Thus time series analyses focus on trends in plant di-

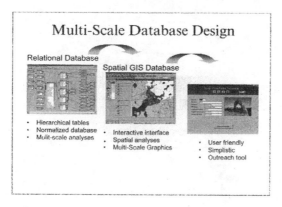

Figure 18.1. Combining a fully relational database linked to a geographic information system and web-based interface provides a useful way to integrate field data and modeling tools. Lesson 25. All biodiversity studies at landscape-scales and larger require extensive data management systems. Do not underestimate data management tasks and the links to surveys, monitoring, and modeling efforts. Befriend a data manager.

versity, perhaps precisely assessed but restricted spatially. Unique events and peculiar site characteristics and land use histories are fused into the analyses so temporal trends may not be extrapolated confidently to other areas (Stohlgren 1994).

As the biological complexity of the issue increases, again, costs rise. In question 4, about the colorful flowers on nonnative thistles that are attracting pollinators away from native plant species, several detailed investigations are needed. Information is needed on the host specificity of the pollinators, whether pollination is inadequate for persistence at the population or species level for the native plant species, and whether the observations and study conclusions are limited to a peculiar site and time (e.g., this situation may only occur under current, temporary conditions of severe grazing or atypically small populations of native species). A costly mix of long-term monitoring, extensive surveys, controlled experiments, and the population modeling of pollinators and host plants is needed to assess changes in the abundance, isolation, fecundity, and persistence of the native plant species over time.

Often basic field information is needed on the spatial distributions of multiple biological groups and environmental variables prior to designing a long-term monitoring program of plant diversity (Figure 18.2). Once basic patterns are quantified and information is available on sources of spatial and temporal variation, then integrated spatial and temporal studies can be appropriately designed.

The most obvious and least appreciated challenge in addressing changes in spatial and temporal patterns of diversity is spatial scale. As in question 5,

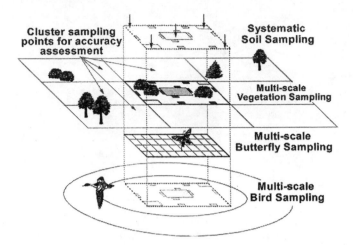

Figure 18.2. Plant ecologists must develop integrated skills in assessing the effects of multiple biological groups in monitoring spatial and temporal change in plant diversity. For example, birds and butterflies may be important in the seed dispersal, pollination, or persistence of particular plant species.

the daunting task of quantifying the hot spots of native biodiversity around the globe has been attempted by a few, brave scientists, such as Meyers et al. (2000). This study often substituted expert opinion where data were lacking, and for obvious reasons. Few of the hot spots and non-hot spots of diversity have complete (or even moderately complete) species lists, and less is known about the abundance, distribution, and uniqueness of most of the species of plants and animals on our planet since fewer than a few million species of the 10–50 million species have been cataloged to date (May 1999). Thus assessing the importance of rare landscape features such as wetlands, tropical forests, riparian zones, and rare forest or grassland communities in preserving hot spots of native plant diversity is a formidable challenge (Lavoie et al. 2003). The second part of question 5 brings in various real threats to these hot spots, including land use changes, habitat loss, and fragmentation and isolation. It is clear that such an evaluation requires detailed maps and monitoring of major human activities at very local scales. What is subtle is that information is required on the intensity, frequency, and spatial patterning of the threats alone or in combination, with a detailed understanding of the effects of multiple stresses on the patterns and processes of the threats in each ecosystem (or a fair number of replicate ecosystems).

Like question 5, question 6 raises issues of the sources and sinks of populations over time in a matrix of landscapes that generally includes only a small proportion of "more protected" environments in a sea of "less protected" environments (Pulliam 1988). Complex species movement patterns are difficult and costly to study, but there are some examples in the literature to guide the development of these studies (Harrison et al. 1988). The cost constraints increase as the number of factors influencing population movement increases, and the number of factors increases with the spatial and temporal scale of the study.

The global nature of the cost constraints is illustrated in question 7. The magnitude and direction of invasions of plants, animals, and diseases among countries will greatly alter the way we think about measuring and monitoring native plant diversity (Dark 2004; Sax and Gaines 2003; Stohlgren et al. 2005a). Invasive diseases such as Dutch elm disease, chestnut blight, white pine blister rust, and sudden oak death have altered our forests for the foreseeable future. Invasive plant species genotypes can quickly spread and overtake native genotypes, as shown by the spread of the common reed (*Phragmites australis*) throughout North America (Saltonstall 2002). Wildlife and insects that depend on those trees are likely directly affected, while co-occurring species may be affected indirectly. Local environments such as undercanopy light, nutrient cycling, and fire regimes are also directly and indirectly affected by invasive species (Keeley et al. 2003; Mack et al. 2000). Question 7 speaks to the potential increased rate of future changes in plant diversity, as new pathogens are shared more frequently with greatly increased modern transportation and increasing global trade. How we monitor the cumulative effects of these uninvited guests is perhaps the greatest environmental challenge of the 21st century—and it will not be cheap or easy!

For all the questions above, sampling designs must consider costs and information gain simultaneously. You often get what you pay for—so very inexpensive designs and very rapid assessments may not provide the information needed to make wise management decisions. Many other long-term monitoring designs spend too much money assessing very tiny areas of the landscape. Costs can be reduced by stratifying the landscape effectively and by using multiphase, multiscale, and nested intensity designs, as mentioned in earlier chapters. Still, cost greatly affects the size of the area surveyed or monitored, the number and size of plots, the amount of ancillary data collected, and the utility of data for modeling. Remembering that only a very small percentage of the landscape (or region, or globe) can be affordably sampled, there is an undeniable trade-off between the allocation of funds for field data collection versus modeling.

Modeling versus Data Collection

Funders of plant diversity studies often begin the planning process with a statement such as, "We have x dollars allocated for this study, what can you do for us?" If a research project is awarded a limited amount of funding (as is always the case), then decisions must be made about the appropriate allocation of resources to planning, equipment, travel, field work, data management, and modeling (and publication costs if all goes well). Predictive modeling remains only a minor component of most plant ecology studies, but this is changing as the scope of the ecological questions expands (e.g., questions 1–7 above). Predictive modeling capabilities for landscape-scale and larger studies assume some expertise in remote sensing, mathematical modeling, and computer use that goes beyond the training of many plant ecologists. These skills are as useful as plant taxonomy and orienteering skills are for field ecologists, and as statistics classes and software training are for data managers. The issue here is that there are significant trade-offs when funds are allocated among the field, data management, and modeling components of most integrated studies of plant diversity.

We learned from chapter 14 that the accuracy, precision, and certainty of spatial models are dependent on the sample size and placement, and on the natural variation, heterogeneity, and complexity of the landscape or region of concern. However, running different types and suites of models with the same dataset can greatly improve the amount of variation explained by the model (see Table 14.6). Thus funding either buys you more field sampling or greater modeling efforts. Other factors also come into play. Often new, improved remote sensing information becomes available after the fieldwork is completed. Alternatively, model outputs often show the need for additional sampling in areas of high heterogeneity and variability (i.e., where uncertainty is greater).

There is no easy way to decide in advance how much funding should be allocated to field sampling versus modeling. This illustrates the need for an iterative approach to plant diversity studies. As the field study progresses, careful attention is paid to the dispersion of plots to ensure that large areas of the landscape are not missed, that the full ranges of environmental gradients are covered, and that rare habitats are not overlooked.

Theoretical and Analytical Challenges

The primary theoretical and analytical challenges involve learning how to couple spatial and temporal analyses (the chicken) and develop realistic (or real) datasets that allow plant ecologists to adequately test new spatiotemporal models (the egg). Which is needed first?

Coupling Spatial and Temporal Analyses

I began the chapter by pointing out that there were several textbooks on spatial analysis (e.g., Cressie 1991) and several textbooks on time series analysis (e.g., Pole et al. 1994; Powell and Steele 1994; West and Harrison 1997), but no practical textbooks on spatiotemporal analyses. In addition, Powell and Steele (1994) pointed out that further progress on coupling spatial and temporal data in physical and biological systems is difficult to conceive without a substantial infusion of theoretical guidance (Adler and Lauenroth 2003).

Many analytical tools are in the development and testing phase. For example, upscaling information from plots (or points) to landscape levels involves spatial interpolation. Yet there is disagreement whether kriging (e.g., Legendre and Fortin 1989) or a Bayesian alternative to kriging (Le and Zidek 1992) is the best geostatistical tool for the interpolation of environmental data. Where repeated measures are taken on a few trees or plots, treatment of the data requires a full consideration of the statistical issues involved (see Moser et al. 1990, Zar 1996).

Haslett and Raftery (1989) provide an excellent example of space-time modeling of environmental data. They estimated the long-term average wind power output at a site in Ireland for which few data were available. The synthesis included deseasonalization, kriging, autoregressive moving average modeling, and fractional differencing in a way that can be applied directly in terrestrial ecology (e.g., to estimate spatiotemporal patterns in species richness). But these statistical models are not for novices.

Modeling changes in species establishment, growth, reproduction, and spread may present a challenge to mathematical modelers. Each phase may be determined by a different suite of factors. Migration of seed to a site and establishment may be influenced by highly stochastic elements involving disturbance, wind speed and direction, or animal movements, and past and near-term precipitation and temperature patterns. Growth is influenced by the availability of resources, competition, herbivory, climate, pathogens, and many other factors. Reproduction is influenced by as many, if not more factors when you consider pollinators and their populations. The science of mathematical modeling is in its infancy relative to the age-old complexity of the patterns and processes of natural systems at multiple spatial and temporal scales (Figure 18.3).

In the short-term, we might have to set low goals in modeling plant diversity. Recently colleagues have made important inroads in spatial modeling (Chong et al. 2001; Kalkhan and Stohlgren 1999). These models may be the precursor to accurate spatiotemporal models, because the first step in the process is accurately describing spatial resources. One potential approach to spatiotemporal modeling is to create "difference maps" from repeated spatial analyses. This approach might be ideal for tracking the spread and potential spread of invasive plant species. Current distributions can be mapped and modeled relative to environmental factors in the first phase. Then potential distributions or "probability of occurrence" models can be created to develop future scenarios. Finally, new field data on invading individuals and populations can be used to validate the scenarios. Implicit in this example is the need for frequent, reliable monitoring data.

Figure 18.3.   Lesson 26. Species-environment relationships may be different for seedlings, saplings, and mature stages of the same species, and for different genotypes within species. Slight changes in topography may drastically alter processes such as competition, herbivory, pollination, and resulting plant survivorship. Modeling species distributions, abundances, immigrations, and extirpations will require a blending of high-resolution spatial-temporal statistical models with individual-based and process-based models. (Photograph by author)

The Need for Realistic Datasets to Test Spatiotemporal Models

In many cases, new datasets are needed to design and test spatiotemporal models of changes in plant diversity at landscape and larger spatial scales. Extensive field surveys and long-term monitoring must be in position to provide accurate, precise (repeatable), and fairly complete information on an annual basis. A systematic inventory of biotic resources at landscape scales will require detailed, high-resolution, remotely sensed data (Bian and Walsh 1993). Far more attention must be paid to fine-scale patterning of natural resources (Fortin et al. 1989).

Test datasets are needed to develop optimum sampling strategies for monitoring changes in plant diversity in space and time (Stohlgren and Quinn 1992). Landscape scale surveys and monitoring will require combinations of systematically placed plots, stratified random sampling, targeted plots to cover extreme environmental gradients, and searching techniques to produce data that are geographically, ecologically, and taxonomically near-complete.

New generations of spatiotemporal models must be developed and tested on comparable, long-term datasets to evaluate the behavior of various models on the same standardized datasets. The models must be able to "estimate" spatially and temporally variable resources and complex ecosystem processes that work at large spatial scales (e.g., Davey and Stockwell 1991; Haslett and Raftery 1989; Twery et al. 1991). Again, spatially dispersed permanent plots and long-term monitoring will be needed to validate the new models and to quantify the feedbacks between biotic and abiotic components of ecosystems and landscapes (Lubchenco et al. 1991). Finally, encompassing alternative land management strategies (Matson and Carpenter 1990) and human values and economics (National Research Council 1990) will be the key to attracting funding to obtain the necessary datasets for more model development and testing.

Institutional Challenges

The institutional challenges facing future plant ecologists are substantial. Accurately quantifying long-term changes in plant diversity at landscape and larger spatial scales will take greater effort than in the past. Our past successes have been minor. The vast majority of ecological studies last only a few years, and the sizes and numbers of most vegetation plots are very small. Attracting greater funding for such endeavors is the second greatest challenge we face: a fair amount of funding has been spent on all previous plant diversity studies with only limited and usually local success. Attracting greater funding will only be successful if we take a "team approach" to plant diversity studies, incorporating new technologies and taking an experimental approach to the science of measuring plant diversity.

Taking a Team Approach
to Plant Diversity Studies

Early plant diversity studies remained the domain of taxonomists and naturalists typically working alone, usually by searching for and collecting plants in subjectively selected areas. This approach is ideal for developing fairly complete species lists and capturing rare species, but it is woefully lacking for quantifying changes in plant diversity over large areas through time. Later, plant ecologists added plot-based techniques, usually placed subjectively or quasi-randomly in homogeneous stands of vegetation. This approach was needed to accurately quantify changes in relatively common species, but it is limited for assessing changes in very rare species. A blending of these two approaches is needed to quantify changes in plant diversity, and taxonomists and plant ecologists can't do it alone. It is now necessary to add remote sensing specialists, spatial modelers, data managers, computer programmers, and landscape ecologists to the research team.

This hypothetical example from Rocky Mountain National Park, Colorado (Figure 18.4) is based on sampling common and rare habitats with nested intensity plots (see chapter 17), long-term vegetation plots and tran-

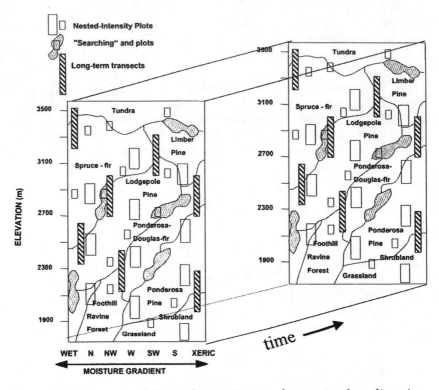

Figure 18.4. Conceptual design for monitoring changes in plant diversity at landscape scales in Rocky Mountain National Park, Colorado.

sects along elevation and moisture gradients (see chapter 16), and combining them with searching and targeted plot monitoring of very rare species populations and very rare habitats. The ultimate design is based on a blend of the theoretical approaches of Gleason, Clements, Whittaker, and others (see chapter 2). The plot sampling locations are based on stratified random sampling of dominant vegetation types in vegetation ecotones and heterogeneous areas, with additional plots in rare habitats and in sites with extreme environmental gradients. The stratification process is the responsibility of the remote sensing specialist in consultation with the landscape ecologist and statistician. The design is iterative. The samples sizes and optimal mix of plot and transect types and locations must be empirically determined and routinely tested. Sample data on spatial and temporal variation will ultimately determine sample size. Field crews contain well-trained taxonomists for immediate and accurate plant identifications (see Incorporating New Technologies below). The plots are not distributed in proportion to the area of the vegetation types, for that would oversample the common types and undersample very rare vegetation types. Sample sizes will be readjusted based on disturbance information gained and species and habitats added to the overall monitoring design. The programmer and data manager work with the statistician to provide a rule-based system for evaluating species richness and overlap, uniqueness, and patterns of species rarity as data are collected to ensure that some strata are not oversampled at the expense of missing additional rare/important strata. For example, rules would be established a priori to curtail sampling in large, homogeneous areas with few rare or unique species and sample additional areas of high diversity, high uniqueness, and high vulnerability to rapid or long-term change. The remote sensing specialists and spatial modelers have key responsibilities for initial stratification, modeling the data with levels of uncertainty, and identification of undersampled areas (and strata) for additional sampling. Everyone on the team helps to interpret and publish the findings.

Incorporating New Technologies

The next generation of plant ecologists will routinely incorporate new technologies in their work if recent progress is any indication. Palmtop computers are ideal for entering data in the field and checking species lists, and for preliminary analyses to guide iterative sample efforts (Figure 18.5).

A shortage of properly trained plant taxonomists may necessitate the increased development and use of species identification tools such as "polyclaves." Polyclaves are superior to unidirectional dichotomous keys in that they allow multiple identifying characteristics to be entered simultaneously to reach a rapid solution. Combined with palmtop and satellite communications capabilities, detailed species lists, line drawings, photographs, and assistance and advice from regional taxonomic experts could be downloaded

Figure 18.5.   Lesson 27. Recording data directly into palmtop computers saves time, improves the accuracy of data, and allows for immediate data analysis. Species lists, identification clues, and codes can be loaded on the palmtop in advance.

to aid in species identification. GPSs have been in use for many years to help field crews find sampling locations, rare plant populations, and long-term vegetation plots. Now GPSs can be combined with video mapping technologies (e.g., Figure 18.6; www.redhensystems.com) to allow for quick and accurate mapping of plant populations (Stohlgren et al. 2001).

Remote sensing and computer capabilities may become major expenses in future plant diversity studies. This points to the greater infrastructural support that may be required for plant diversity studies in the future, including more office and laboratory space and equipment, geographic information system and data management assistance, and modern herbarium facilities. Plant diversity studies have evolved beyond the wandering naturalist with a notebook and pocketknife-sharpened pencil.

There are also institutional challenges associated with rewarding team research. Most reward systems for promotion and tenure are based on the number of senior-authored publications. There are great incentives for conducting simple, single-investigator, short-term, small-scale experiments rather than complex, multi-investigator, long-term, large-scale studies. Reward systems must adapt and recognize the growing gap between "biocomplexity" and institutional simplicity.

Quantifying trends in plant diversity in space and time will be difficult and costly. The scales of studies will extend far beyond those of typical graduate degree programs in small study areas. Every effort should be made to extend the spatial and temporal scale of past and ongoing plant diversity studies, with a renewed focus on understanding the effects of increasing spatial and temporal scales on study results. This may be the only way to fully understand the link between ecological patterns and processes. The last, and greatest, institutional challenge in plant diversity studies remains—obtaining long-term commitments to research, surveys, monitoring, and modeling of plant diversity from local to global scales (Figure 18.7).

Since plant species are more freely moving around the globe with modern transportation and trade (Mack et al. 2000), we can expect rapid and continuing invasions of plants, animals, and diseases. It will become increasingly difficult to protect local and distinct assemblages of native species and genotypes.

Maintaining an Experimental Approach

Throughout this text I have espoused taking an experimental approach to plant diversity studies. This is especially true if the objectives of the study are to add data and information needed to understand the patterns and processes of plant diversity from local to global scales. Not all plant diversity studies will or should try to do this, but there would be significant advances in the science if data from multiple studies were comparable and complementary. Different strategies may be needed for different areas (see chapter 15), but in nearly every case, an experimental approach is likely warranted

Figure 18.6. Photograph (by author) of the camcorder and GPS system
tested in this study and a schematic of an application of video mapping
and data recording. A global position system provides latitude, longitude,
and elevation along side video from the camcorder. The data and video
can be downloaded directly to a geographic information system in a
computer to assess all sites with video taken. From Stohlgren et al. 2001.

337

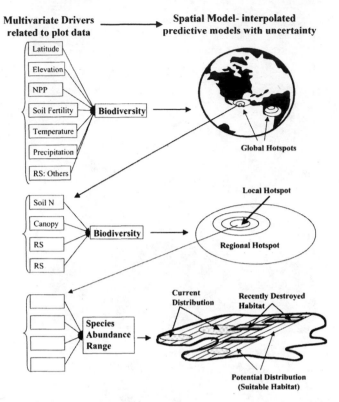

Figure 18.7. The future of plant diversity research is in linking patterns and processes from local to global scales.

for plant diversity studies. While many different approaches can yield valuable information, I propose that many landscape-scale plant diversity studies might clearly benefit if they contain at least some of the following attributes. These are suggestions rather than "lessons."

- Unbiased (or reduced biased) sample site selection for most plot-based studies should be complemented with purposive sampling of extreme or complex environmental gradients. Unbiased site selection allows for the generalization of results to similar habitats in the surrounding landscape. However, plot-based studies constrained by costs may preclude sampling along the full range of environmental gradients. It is important to establish some plots in extreme environments to better understand the limits of plant species distributions and patterns of diversity, uniqueness, and rarity.
- Plot-based surveys, monitoring, and modeling techniques, should be complemented with searching techniques. Plot-based surveys are needed to quantify patterns of plant diversity at large spatial scales with known precision and accuracy. Only a small portion of any landscape can be affordably sampled, usually less than 1%, so we

better be pretty good at modeling what is in the unsampled 99% of the landscape. Likewise, cost constraints limit our ability to establish plots in extremely rare habitats, so some searching techniques will always be needed to complement plot-based studies to evaluate species richness more completely.

- Studies should be designed to meet local objectives, but also for broad applications and synthesis with other data. There are great benefits to collecting compatible, comparable, and complementary databases on plant diversity to improve our regional and global understanding of spatial and temporal changes in plant diversity. Multiscale and nested intensity designs offer the greatest opportunities for comparable data.

- "Adequate replication" is a moving target and is difficult to achieve because of funding constraints and higher spatial and temporal variation. Except for very focused objectives, it is very difficult to determine appropriate sample sizes early in a study. Evaluating spatial and temporal variation is an endless task, so sampling must be an iterative process.

- Analysis of multiple parameters of plant diversity is the key. Not all aspects of plant diversity can be expected to change equally in space or time. Evaluating changes in species richness, cover and abundance, frequency and distribution, genetic diversity, habitat diversity, and all aspects of plant demography and evolution are important to fully understand patterns of plant diversity.

- Experiments aimed at isolating the underlying processes of plant diversity must be conducted in multiple vegetation types and biomes, with ranges of natural disturbances, and with many species and vegetation structures, and under a wide array of edaphic and biotic conditions. The experiments must also be complemented with surveys, monitoring, and modeling to confirm the links between patterns and processes of experimental units and natural landscapes.

- Monitoring genetic diversity will become increasingly important. This book has focused on plant *species* diversity, but the *genetic* diversity of many native plant species may be threatened by non-native genotypes, hybridization, and backcrosses with native types.

- Nonvascular plant diversity is an important component of plant diversity, but one poorly covered by this and other texts. Standardized field sampling protocols, taxonomic experts, and automated identification keys are generally lacking for landscape and regional studies of nonnative plant diversity in many areas of the globe. More work is needed in this area.

- Habitat loss, land use changes, contaminants, altered disturbance regimes, invasive species, and climate change will accelerate changes in plant diversity in large regions and countries. Monitoring the rapid changes in the quality, quantity, and connectedness of habitats and species distributions remains an important component of plant diversity studies.

- Integrated "team science" approaches will greatly advance plant diversity studies. Quantifying patterns of plant diversity, associated ecosystem processes, and associated environmental threats will require the coordinated efforts of taxonomists, remote sensing specialists, field ecologists, soil scientists, data managers, spatial modelers, and others.

In the final analysis, it will be the next generation of plant ecologists that provides us with the greatest advances in the study of plant diversity. These plant ecologists will experiment with new field techniques, remote sensing data, and computer technologies guided by more definitive theoretical frameworks. The more frequent use of multiscale sampling and nested intensity techniques, larger plots, and better replicated studies, and the collecting of detailed ancillary data integrated with improved geographic information system-based models will increase our ability to synthesize multiscale patterns of plant diversity from local to global scales. Most of our work lies ahead.

Glossary

Abundance (or frequency): The relative numbers of individuals in an area (Tansley and Chipp 1926).

Association: Stands of homogeneous vegetation in a community, similar to each other, yet different from other stands in the same community.

Biodiversity: The variety of all life, including genetic diversity, species diversity, and habitat or ecosystem diversity.

Biomass: The weight of a given individual or species in a given area (usually presented as the dry weight of living, aboveground foliage).

Climax community: Those communities for which there is no evidence of replacement (Daubenmire 1968, p. 216).

Community: A plant community can be understood as a combination of plants that are dependent on their environment and influence one another and modify their own environment (Mueller-Dombois and Ellenberg 1974, p. 27).

Competition: When one or more plants restrict the availability of necessary resources for the growth of another plant.

Connectivity: The degree to which patches of a given type are joined by corridors [adapted from Wiens (1993)].

Constancy: (1) The percentage of occurrences of a species in samples of a uniform size scattered over the geographic range of an association (Daubenmire 1968, p. 76)—it is equivalent to "frequency" among a series of stands rather than a single stand (i.e., describes a species that occurs in most (usually 50% or more) of the stands in a given vegetation type. (2) The percentage of times a species is recorded in specific plant associations growing in different areas (Tansley and Chipp 1926).

Density: The number of individuals of a species in a given area.

Dominant: The most conspicuous species in an area (usually determined by basal area, cover, or density).

Exclusiveness: The degree to which a given species is in one plant association and not in others (Tansley and Chipp 1926). This might also be called "uniqueness" (Stohlgren et al. 1997b).

Extent: The total area under study [adapted from O'Neill et al. (1996)].

Fidelity: The degree to which a species is restricted to one community (McIntosh 1985) or the extent to which a species occurs regularly in every stand of an association.

Flora: A list of species in an area (Tansley and Chipp 1926).

Frequency: The presence of a species in a given number of plots or the percentage of occurrences of a species in samples of a uniform size scattered throughout a single stand.

Grain: The spatial resolution of the data (O'Neill et al. 1996).

Ground cover (or foliar cover): The projected cover of foliage of a species or group of species.

Habitat: The sum of the effective environmental conditions under which the associations exist (Tansley and Chipp 1926).

Phenology: The study of biological periodicity in relation to the seasonal sequence of climatic factors (Daubenmire 1968, p. 71).

Quadrat: A small square (Pound and Clements 1898a) or rectangular (Daubenmire 1964) subplot for measuring small plant species (grasses, herbs).

Resolution: The finest level of features examined in a map, photograph, or study area.

Scale: The size or extent of the area being examined; the physical relationship of an area relative to a model (or map, or photograph) of the area being examined.

Sere community: A successional community (e.g., pioneer species) that is later replaced by a later successional community, and ultimately the climax community.

Species diversity: The number and kinds of species in an area and their genetic diversity.

Species richness: The number of species in a delineated area.

Vegetation: Small or large groupings of natural plants, types of forests, grasslands, etc. (Tansley and Chipp 1926).

References

Adler, P. B., and W. K. Lauenroth. 2003. The power of time: spatiotemporal scaling of species diversity. Ecology Letters 6:749–756.

Agee, J. K., and D. R. Johnson. 1988. *Ecosystem Management for Parks and Wilderness.* Seattle: University of Washington Press.

Ahmed, J., C. D. Bonham, and W. A. Laycock. 1983. Comparison of techniques used for adjusting biomass estimates by double sampling. Journal of Range Management 36:217–221.

Allen, R. B., R. K. Peet, and W. L. Baker. 1991. Gradient analysis of latitudinal variation in southern Rocky Mountain forests. Journal of Biogeography 18:123–139.

Arrhenius, O. 1921. Species and area. Journal of Ecology 9:95–99.

Ashby, E. 1948. Statistical ecology, II. A reassessment. Botanical Review 14:222–234.

Austin, M. P., and P. C. Heyligers. 1991. New approach to vegetation survey design: gradsect sampling. In: C. R. Margules and M. P. Austin, eds. *Nature Conservation: Cost Effective Biological Surveys and Data Analysis.* Clayton South, Victoria, Australia: CSIRO; pp. 31–36.

Ayres, J. M. C. 1993. *As matas de várzea do Mamirauá.* MCT-CNPq-Programa do trópico úmido, Sociedade civil de Mamirauá, Brazil.

Backéus, I. 1993. Ecotone versus ecocline: vegetation zonation and dynamics around a small reservoir in Tanzania. Journal of Biogeography 20:209–218.

Baker, W. L. 1989. Effect of scale and spatial heterogeneity on fire-interval distributions. Canadian Journal of Forest Research 19:700–706.

Baker, W. L. 1990. Species richness of Colorado riparian vegetation. Journal of Vegetation Science 1:119–124.

Baker, W. L., J. A. Munroe, and A. E. Hessl. 1997. The effects of elk on aspen in the winter range of Rocky Mountain National Park. Ecography 20:155–165.

Baker, W. L., and P. J. Weisberg. 1995. Landscape analysis of the forest-tundra ecotone in Rocky Mountain National Park, Colorado. Professional Geographer 47:361–375.

Barbour, M. G., J. H. Burk, W. D. Pitts, F. S. Gilliam, and M. W. Schwartz. 1999. *Terrestrial Plant Ecology,* 3rd ed. Menlo Park, CA: Benjamin/Cummings.

Barnett, D., and T. J. Stohlgren. 2001. Aspen persistence near the National Elk Refuge and Gros Ventre Valley elk feedgrounds of Wyoming, USA. Landscape Ecology 16:569–580.

Barnett, D. T., and T. J. Stohlgren. 2003. A nested-intensity design for surveying plant diversity. Biodiversity and Conservation 12:255–278.

Bartlein, P. J., C. Whitlock, and S. L. Shafer. 1997. Future change in the Yellowstone National Park region and its potential impact on vegetation. Conservation Biology 11:782–792.

Bashkin, M., T. J. Stohlgren, Y. Otsuki, M. Lee, P. Evangelista, and J. Belnap. 2003. Soil characteristics and exotic species invasions in the Grand Staircase-Escalante National Monument, Utah. Applied Soil Ecology 22:67–77.

Bazzaz, F. A. 1975. Plant species-diversity in old-field successional ecosystems in southern Illinois. Ecology 56:485–488.

Bazzaz, F. A. 1996. *Plants in Changing Environments.* New York: Cambridge Press.

Bell, G., and M. J. Lechowicz. 1991. The ecology and genetics of fitness in forest plants, I. Environmental heterogeneity measured by explant trials. Journal of Ecology 79: 663–685.

Bellehumeur, C., and P. Legendre. 1998. Multiscale sources of variation in ecological variables: modelling spatial dispersion, elaborating sampling designs. Landscape Ecology 13: 15–25.

Belnap, J. 1995. Soil surface disturbances: their role in accelerating desertification. Environmental Monitoring and Assessment 37:39–57.

Belnap, J. 1996. Soil surface disturbances in cold deserts: effects of nitrogenase activity in cyanobacterial-lichen soil crusts. Biology and Fertility of Soils 23:362–367.

Belnap, J. 1998. Impacts of trampling soils in southeast Utah ecosystems. In: L. M. Hill, ed. Proceedings: Learning from the Land: Science in the Grand Staircase-Escalante National Monument. November 4–5, 1997. U.S. Department of the Interior, Bureau of Land Management, Southern Utah University, Cedar City, Utah. BLM Technical Report BLM/UT/GI-98/006+1220, pp. 231–244.

Belnap, J., and D. A. Gillette. 1998. Vulnerability of desert soil surfaces to wind erosion: impacts of soil texture and disturbance. Journal of Arid Environments 39:133–142.

Belnap, J., and K. T. Harper. 1995. Influence of cryptobiotic soil crusts on elemental content of tissue in two desert seed plants. Arid Soil Research and Rehabilitation 9:107–115.

Belsky, A. J. 1983. Small-scale patterns in grassland communities in the Serengeti National Park, Tanzania. Vegetatio 55:141–151.

Belsky, A. J. 1986. Does herbivory benefit plants? A review of the evidence. American Naturalist 127:870–892.

Belsky, A. J., and D. M. Blumenthal. 1997. Effects of livestock grazing on stand dynamics and soils in upland forests of the interior west. Conservation Biology 11:315–327.

Berkowitz, A. R., K. Kolosa, R. H. Peters, and S. T. A. Pickett. 1989. How far in space and time can the results from a single long-term study be extrapolated? In: G. E. Likens, ed. *Long-Term Studies in Ecology: Approaches and Alternatives.* New York: Springer-Verlag; pp. 192–198.

Bian, L., and S. J. Walsh. 1993. Scale dependencies of vegetation and topography in a mountainous environment of Montana. Professional Geography 45:1–11.

Blackwood, L. G. 1991. Assurance levels of standard sample size formulas. Environmental Science and Technology 25:1366–1367.

Bock, C. E., J. H. Bock, W. R. Kenney, and V. M. Hawthorne. 1984. Responses of birds, rodents, and vegetation to livestock exclosure in a semidesert grassland site. Journal of Range Management 37:239–242.

Bock, C. E., J. H. Bock, and H. H. Smith. 1993. Proposal for a system of livestock exclosures on public rangelands in the western United States. Conservation Biology 7:731–733.

Bock, J. H., and C. E. Bock. 1995. The challenges of grassland conservation. In: A. Joern and K. H. Keeler, eds. *The Changing Prairie: North American Grasslands.* New York: Oxford University Press; pp. 199–222.

Bond, W. 1983. On alpha diversity and the richness of the Cape flora: a study in southern cape fynbos. In: F. J. Kruger, D. T. Mitchell, and J. U. M. Jarvis, eds. *Mediterranean-Type Ecosystems: The Role of Nutrients.* New York: Springer-Verlag; pp. 337–356.

Bonham, C. D. 1989. *Measurements for Terrestrial Vegetation.* New York: John Wiley & Sons.

Bonham, C. D., R. M. Reich, and K. K. Leader. 1995. A spatial cross-correlation of *Bouteloua gracilis* with site factors. Grasslands Science 41:196–201.

Boulinier, T., J. D. Nichols, J. R. Sauer, J. E. Hines, and K. H. Pollock. 1998. Estimating species richness: the importance of heterogeneity in species detectability. Ecology 79:1018–1028.

Bourdeau, P. F., and H. J. Oosting. 1959. The maritime live oak forest in North Carolina. Ecology 40:148–152.

Brady, W. W., J. W. Cook, and E. F. Aldon. 1991. A microplot method for updating loop frequency range trend data: theoretical considerations and a computer simulation. Research Paper RM-295. Fort Collins, CO: Rocky Mountain Forest and Range Experiment Station, U.S. Forest Service.

Bray, J. R., and J. T. Curtis. 1957. An ordination of the upland forest communities of southern Wisconsin. Ecological Monographs 27:326–349.

Brooks, M. L., and J. R. Matchett. 2003. Plant community patterns in unburned and burned blackbrush (*Coleogyne ramosissima* Torr.) shrublands in the Mojave Desert. Western North American Naturalist 63:283–298.

Brown, B. J., and T. F. H. Allen. 1989. The importance of scale in evaluating herbivory impacts. Oikos 54:189–194.

Brown, R. L., and R. K. Peet. 2003. Diversity and invasibility of southern Appalachian plant communities. Ecology 84:32–39.

Brubaker, L. B. 1986. Tree population responses to climate change. Vegetatio 67:119–130.

Brubaker, L. B. 1988. Vegetation history and anticipating future vegetation change. In: J. K. Agee and D. R. Johnson, eds. *Ecosystem Management for Parks and Wilderness*. Seattle: University of Washington Press; pp. 41–61.

Bruno, J. F., C. W. Kennedy, T. A. Rand, and M. B. Grant. 2004. Landscape-scale patterns of biological invasions in shoreline plant communities. Oikos 107:531–540.

Buchholtz, C. W. 1983. *Rocky Mountain National Park: A History*. Boulder: Colorado Associated University Press.

Bull, K. A., K. W. Stolte, and T. J. Stohlgren. 1998. *Forest Health Monitoring: Vegetation Pilot Field Method Guide*. Washington, DC: U.S. Forest Service.

Bunge, J., and M. Fitzpatrick. 1993. Estimating the number of species: a review. Journal of American Statistical Association 88:364–373.

Burkman, W. G., and G. D. Hertel. 1992. Forest Health Monitoring: a national program to detect, evaluate, and understand change. Journal of Forestry 90:26–27.

Burnham, K. P., and D. R. Anderson. 2002. *Model Selection and Inference: A Practical Information-Theoretic Approach*. New York: Springer-Verlag.

Busing, R. T., and P. S. White. 1993. Effects of area on old-growth forest attributes: implications for the equilibrium landscape concept. Landscape Ecology 8:119–126.

Byers, J. E., and E. G. Noonburg. 2003. Scale dependent effects of biotic resistance to biological invasion. Ecology 84:1428–1433

Cain, S. A. 1944. *Foundations of Plant Geography*. New York: Hafner Press.

Cambardella, C. A., and E. T. Elliott. 1993. Carbon and nitrogen distribution in aggregates from cultivated and native grassland soils. Soil Science Society of America Journal 57:1071–1076.

Campbell, D. G., J. L. Stone, and A. R. Junior. 1992. A comparison of the phytosociologic and dynamics of three (varzea) forests of known ages, Rio Jurua, West Brazilian Amazon. Journal of the Linnean Society 108:213–237.

Carrington, M.E. and J.E. Keeley. 1999. Comparison of postfire seedling establishment between scrub communities in mediterranean- and non–mediterranean-climate ecosystems. Journal of Ecology 87:1025–1036.

Case, T. J. 1990. Invasion resistance arises in strongly interacting species-rich model competition communities. Proceedings of the National Academy of Science USA 87:9610–9614.

Chaneton, E. J., and R. S. Lavado. 1996. Soil nutrients and salinity after long-term grazing exclusion in a flooding Pampa grassland. Journal of Range Management 49:182–187.

Cherrill, A. J., C. McClean, P. Watson, K. Tucker, S. P. Rushton, and R. Sanderson. 1995. Predicting the distributions of plant species at the regional scale: a hierarchical matrix model. Landscape Ecology 10:197–207.

Chew, R. M. 1982. Changes in herbaceous and suffrutescent perennials in grazed and ungrazed desertified grassland in southeastern Arizona, 1958–78. American Midland Naturalist 108:159–169.

Chiarucci, A., S. Maccherini, and V. De Dominicis. 2001. Evaluation and monitoring of the flora in a nature reserve by estimation methods. Biological Conservation 101 :305–314.

Chong, G. W. 2002. Multi-scale sampling of native and non-native plant diversity: examples of data analyses and applications. PhD dissertation. Fort Collins: Colorado State University.

Chong, G. W., R. M. Reich, M. A. Kalkhan, and T. J. Stohlgren. 2001. New approaches for sampling and modeling native and exotic plant species richness. Western North American Naturalist 61:328–335.

Christensen, N. L., and R. K. Peet. 1984. Convergence during secondary forest succession. Journal of Ecology 72:25–36.

Cid, M. S., J. K. Detling, A. D. Whicker, and M. A. Brizuela. 1991. Vegetational responses of a mixed-grass prairie site following exclusion of prairie dogs and bison. Journal of Range Management 44:100–105.

Clements, F. E. 1916. *Plant Succession: An Analysis of the Development of Vegetation.* Publication 242. Washington, DC: Carnegie Institution of Washington.

Clements, F. E. 1936. The nature and structure of the climax. Journal of Ecology 24:252–284.

Cline, S. P., S. A. Alexander, and J. E. Barnard, eds. 1995. *Environmental Monitoring and Assessment Program Forest Health Monitoring Program Quality Assurance Project Plan for Detection Monitoring Project.* Las Vegas, NV: U.S. Environmental Protection Agency.

Cochran, W. G. 1977. *Sampling Techniques,* 3rd ed. New York: John Wiley & Sons.

Coffin, D. P., and W. K. Lauenroth. 1989a. Disturbances and gap dynamics in a semiarid grassland: a landscape-level approach. Landscape Ecology 3:19–27.

Coffin, D. P., and W. K. Lauenroth. 1989b. Small scale disturbances and successional dynamics in a shortgrass community: interactions of disturbance characteristics. Phytologia 67:258–286.

Coffin, D. P., and W. K. Lauenroth. 1989c. The spatial and temporal variability in the seed bank of a semiarid grassland. American Journal of Botany 76:53–58.

Cole, R. G. 2001. Staggered nested designs to estimate hierarchical levels of variability. New Zealand Journal of Marine and Freshwater Research 35: 891–896.

Colebrook, J. M. 1982. Continuous plankton records: seasonal variations in the distribution and abundance of plankton in the North Atlantic Ocean and the North Sea (*Calanus finmarchicus*). Journal of Plankton Research 4:435–462.

Collins, S. L. 1987. Interaction of disturbances in tallgrass prairie: a field experiment. Ecology 68:1243–1250.

Collins, S. L., and D. E. Adams. 1983. Succession in grasslands: thirty-two years of change in a central Oklahoma tallgrass prairie. Vegetatio 51:181–190.

Collins, S. L., and S. C. Barber. 1985. Effects of disturbance on diversity in mixed-grass prairie. Vegetatio 64:87–94.

Collins, S. L., and S. M. Glenn. 1991. Importance of spatial and temporal dynamics in species regional abundance and distribution. Ecology 72: 654–664.

Collins, S. L., and S. M. Glenn. 1995. Grassland ecosystem and landscape dynamics. In: A. Joern and K. H. Keeler, eds. *The Changing Prairie: North American Grasslands.* New York: Oxford University Press; pp. 128–156.

Collins, S. L., and S. M. Glenn. 1997. Effects of organismal and distance scaling on analysis of species distribution and abundance. Ecological Applications 7:543–551.

Collins, S. L., S. M. Glenn, and D. J. Gibson. 1995. Experimental analysis of intermediate disturbance and initial floristic composition: decoupling cause and effect. Ecology 76:486–492.

Colwell, R. K. 1997. Estimate-S: statistical estimation of species richness and shared species from samples. Version 5.0.1. Available at: http://viceroy .eeb.uconn.edu/estimates. Last updated September 16, 2002.

Conkling, B. L., and G. E. Byers, eds. 1992. *Forest Health Monitoring Field Methods Guide.* Internal Report. Las Vegas, NV: U.S. Environmental Protection Agency.

Connell, J. H. 1978. Diversity in tropical rain forests and coral reefs. Science 199:1302–1310.

Connor, E. F., and E. D. McCoy. 1979. The statistics and biology of the species-area relationship. American Naturalist 113:791–833.

Conrad, H. S. 1951. *The Background of Plant Ecology.* Ames: Iowa State College Press.

Cook, J. W., W. W. Brady, and E. F. Aldon. 1992. Handbook for converting Parker loop frequency data to basal area. General Technical Report RM-212. Fort Collins, CO: Rocky Mountain Forest and Range Experiment Station, U.S. Forest Service.

Cooper, W. S. 1913. The climax forest of Isle Royale, Lake Superior and its development. Botanical Gazette 55:1–44, 115–140, 189–235.

Cornelius, J. M., and J. F. Reynolds. 1991. On determining the statistical significance of discontinuities within ordered ecological data. Ecology 72:2057–2070.

Cottam, G., and J. T. Curtis. 1949. A method for making rapid surveys of woodlands by means of pairs of randomly selected trees. Ecology 30:101–104.

Coughenour, M. B. 1991. Biomass and nitrogen responses to grazing of upland steppe on Yellowstone's northern winter range. Journal of Applied Ecology 28:71–82.

Coughenour, M. B. 1993. *Savanna—Landscape and Regional Ecosystem Model: Documentation.* Fort Collins, CO: Natural Resource Ecology Laboratory, Colorado State University.

Coughenour, M. B., and F. J. Singer. 1996. The concept of overgrazing and its application to Yellowstone's northern winter range. In: Technical Report NPS/NRYELL/NRTR/96–01. Denver, CO: U.S. Department of the Interior, National Park Service, Natural Resource Program Center, Natural Resource Information Division; pp. 1–11.

Coughenour, M. B., F. J. Singer, and J. J. Reardon. 1991. The Parker transects revisited: long-term herbaceous vegetation trends on Yellowstone's northern winter range. In: D. Despain, ed. *Plants and Their Environments: 1st Bicentennial Science Conference on the Greater Yellowstone Ecosystem.* Washington, DC: U.S. Department of the Interior, National Park Service; pp. 73–84.

Crawley, M. J. 1983. *Herbivory: The Dynamics of Plant-Animal Interactions.* Oxford: Blackwell.

Cressie, N. 1985. Fitting variogram models by weighted least squares. Mathematical Geology 17:563–586.

Cressie, N. 1991. *Statistics for Spatial Data.* New York: John Wiley & Sons.

Crumley, C. L. 1993. Analyzing historic ecotonal shifts. Ecological Applications 3:377–384.

Cullinan, V. I., and J. M. Thomas. 1992. A comparison of quantitative methods for examining landscape pattern and scale. Landscape Ecology 7:221–227.

Currie, D. J. 1991. Energy and large-scale patterns of animal- and plant-species richness. American Naturalist 137:27–49.

Curtis, J. T., and R. P. McIntosh. 1951. An upland forest continuum in the prairie-forest border region of Wisconsin. Ecology 32:476–498.

Czaplewski, R. L., and R. M. Reich. 1993. *Expected Value and Variance of Moran's Bivariate Spatial Autocorrelation Statistic Under Permutation.* Research Paper RM-309. Fort Collins, CO: U.S. Department of Agriculture, Rocky Mountain Experimental Range Station.

Dallmeier, F. 1992. *Long-Term Monitoring of Biological Diversity in Tropical Forest Areas: Methods for Establishment and Inventory of Permanent Plots.* MAP Digest 11. Paris, France: United Nations Educational, Scientific, and Cultural Organization.

Dallmeier, F., C. M. Taylor, J. C. Mayne, M. Kabel, and R. Rice. 1992. Effects of Hurricane Hugo on the Bisley Biodiversity Plot, Luquillo Biosphere Reserve, Puerto Rico. In: F. Dallmeier, ed. *Long-Term Monitoring of Biological Diversity in Tropical Forest Areas: Methods for Establishment and Inventory of Permanent Plots.* MAP Digest 11. Paris, France: United Nations Educational, Scientific, and Cultural Organization.

D'Antonio, C. M., and P. M. Vitousek. 1992. Biological invasions by exotic grasses, the grass/fire cycle, and global change. Annual Review of Ecology and Systematics 23:63–87.

Dark, S. J. 2004. The biogeography of invasive alien plants in California: an application of GIS and spatial regression analysis. Diversity and Distributions 10:1–9.

Darwin, C. 1859. *The Origin of Species.* London: Penguin Books [reprint].

Daubenmire, R. 1959. A canopy-coverage method of vegetational analysis. Northwest Science 33:43–64.

Daubenmire, R. F. 1940a. Exclosure technique in ecology. Ecology 21:514–515.

Daubenmire, R. F. 1940b. Plant succession due to overgrazing in the *Agropyron* bunchgrass prairie of southeastern Washington. Ecology 21:55–64.

Daubenmire, R. F. 1968. *Plant Communities: A Textbook of Plant Synecology.* New York: Harper and Row.

Daubenmire, R. F. 1974. *Plants and Environment,* 3rd ed. New York: John Wiley & Sons.

Davey, S. M., and D. R. B. Stockwell. 1991. Incorporating wildlife habitat into an AI environment: concepts, theory, and practicalities. AI Applications 5:59–104.

Davies, K. E., P. Chesson, S. Harrison, B. D. Inouye, B. A. Melbourne, and K. J. Rice. 2005. Spatial heterogeneity explains the scale dependence of the native-exotic diversity relationship. Ecology 86:1602–1610.

Davis, M. B. 1983. Quaternary history of deciduous forest of eastern North America and Europe. Annals of the Missouri Botanical Garden 70:550–563.

Davis, M. B. 1989. Retrospective studies. In: G. E. Likens, ed. *Long-Term Studies in Ecology: Approaches and Alternatives.* New York: Springer-Verlag; pp. 71–89.

Davis, M. B. 1991. Insights from paleoecology on global change. Ecological Society of America Bulletin 70:222–228.

Day, R. W., and G. P. Quinn. 1989. Comparisons of treatments after an analysis of variance in ecology. Ecological Monographs 59:433–463.

DeByle, N. V., and R. P. Winokur, eds. 1985. *Aspen: Ecology and Management in the Western United States.* Fort Collins, CO: Rocky Mountain Forest and Range Experiment Station, U.S. Forest Service.

DeFalco, L. A. 1995. *Influence of Cryptobiotic Crusts on Winter Annuals and Foraging Movements of the Desert Tortoise.* Fort Collins: Department of Biology, Colorado State University.

DeFerrari, C. M., and R. J. Naiman. 1994. A multi-scale assessment of the occurrence of exotic plants on the Olympic Peninsula, Washington. Journal of Vegetation Science 5:247–258.

Dewey, D. R., and K. H. Lu. 1959. A correlation and path-coefficient analysis of components of crested wheatgrass seed production. Agronomy Journal 51:515–518.

Diaz, S., J. G. Hodgson, K. Thompson, M. Cabido, J. H. C. Cornelissen, A. Jalili, G. Montserrat-Marti, J. P. Grime, F. Zarrinkamar, Y. Asri, S. R. Band, S. Basconcelo, P. Castro-Diez, G. Funes, B. Hamzehee, M. Khoshnevi, N. Perez-Harguindeguy, M. C. Perez-Rontome, F. A. Shirvany, F. Vendramini, S. Yazdani, R. Abbas-Azimi, A. Bogaard, S. Boustani, M. Charles, M. Dehghan, L. de Torres-Espuny, V. Falczuk, J. Guerrero-Campo, A. Hynd, G. Jones, E. Kowsary, F. Kazemi-Saeed, M. Maestro-Martinez, A. Romo-Diez, S. Shaw, B. Siavash, P. Villar-Salvador, and M. R. Zak. 2004. The plant traits that drive ecosystems: evidence from three continents. Journal of Vegetation Science 15:295–304.

Diersing, V. E., R. B. Shaw, and D. J. Tazik. 1992. US Army land condition trend analysis (LCTA) program. Environmental Management 16:405–414.

DiTommaso, A., and L. W. Aarssen. 1989. Resource manipulations in natural vegetation: a review. Vegetatio 84:9–29.

Dix, R. L. 1960. The effects of burning on the mulch structure and species composition of grasslands in western North Dakota. Ecology 41:49–56.

Dix, R. L., and J. E. Butler. 1954. The effects of fire on a dry, thin-soil prairie in Wisconsin. Journal of Range Management 7:265–268.

Dobzhansky, T. 1950. Evolution in the tropics. American Scientist 38:209–221.

Docters van Leeuwen, W. M. 1936. *Krakatau, 1883 to 1933. A. Botany.* Jardin Botanique de Buitenzorg (Buitenzorg, Java), Annales vols. 46–47.

Dormaar, J. F., B. W. Adams, and W. D. Willms. 1994. Effect of grazing and abandoned cultivation on a Stipa-Bouteloua community. Journal of Range Management 47:28–32.

Drude, O. 1902. Grundzüge der Pflanzenverbreitung. Vol. VI. In: A. Engler and O. Drude, eds. *Die Vegetation der Erde.* Leipzig: Wilhelm Engelmann; p. 671.

Duffy, D. C., and A. J. Meier. 1992. Do Appalachian herbaceous understories ever recover from clearcutting? Conservation Biology 6:196–201.

Dukes, J. S. 2002. Species composition and diversity affect grassland susceptibility and response to invasion. Ecological Applications 12:602–617.

Ellenberg, H. 1956. *Aufgaben und Methoden der Vegetationskunde.* Stuttgart: Eugen Ulmer.

Ellenberg, H. 1958. Bodenreaktion (einschlieblich Kaltfrage). In: W. Ruhland, ed. *Handbuch der Pflanzenphysiologie.* Berlin: Springer-Verlag; pp. 638–708.

Elliott, K. J., and D. L. Loftis. 1993. Vegetation diversity after logging in the southern Appalachians. Conservation Biology 7:220–221.

Ellison, L. 1960. Influence of grazing on plant succession of rangelands. Botanical Review 26:1–78.

Elton, C. S. 1958. *The Ecology of Invasions by Animals and Plants.* London: Methuen.

Elzinga, C. L., D. W. Salzer, and J. W. Willoughby. 1998. *Measuring and Monitoring Plant Populations.* Report BLM/RS/ST-98/005+1730. Denver, CO: U.S. Department of Land Management.

ESRI. 1997. ARC/INFO. Redlands, CA: Environmental Systems Research Institute.

Evanko, A. B., and R. A. Peterson. 1955. Comparisons of protected and grazed mountain rangelands in southwestern Montana. Ecology 36:71–82.

Facelli, J. M., R. J. C. Leon, and V. A. Deregibus. 1989. Community structure in grazed and ungrazed grassland sites in the flooding Pampa, Argentina. American Midland Naturalist 121:125–133.

Fairweather, P. G. 1991. Statistical power and design requirements for environmental monitoring. Australian Journal of Marine and Freshwater Research 42:555–567.

Fargione, J. E., and D. Tilman. 2005. Diversity decreases invasion via both sampling and complementarity effects. Ecology Letters 8:604–611.

Ferreira, L. V. 1997. Effects of the duration of flooding on species richness and floristic composition in three hectares in the Jau National Park in floodplain forests in central Amazonia. Biodiversity and Conservation 6:1353–1363.

Ferreira, L. V., and T. J. Stohlgren. 1999. Effects of river level fluctuation in species richness, diversity, and plant distribution in a floodplain in central Amazonia. Oecologia 120:582–587.

Fisser, H. G. 1970. *Exclosure Studies with Transects of Permanent Plots: 1969 Data.* Laramie, WY: Wyoming Agricultural Experiment Station, University of Wyoming.

Fleischner, T. L. 1994. Ecological costs of livestock grazing in western North America. Conservation Biology 8:629–644.

Floerl, O., T. K. Pool, and G. J. Inglis. 2004. Positive interactions between nonindigenous species facilitate transport by human vectors. Ecological Applications 14:1724–1736.

Foley, J. A. 1994. The sensitivity of the terrestrial biosphere to climate change: a simulation of the middle Holocene. Global Biogeochemical Cycles 8:505–525.

Fortin, M. 1994. Edge detection algorithms for two-dimensional ecological data. Ecology 75:956–965.

Fortin, M., P. Drapeau, and P. Legendre. 1989. Spatial autocorrelation and sampling design in plant ecology. Vegetatio 83:209–222.

Fox, J. F. 1979. Intermediate-disturbance hypothesis. Science 204:1344–1345.

Fox, M. D., and B. J. Fox. 1986. The susceptibility of natural communities to invasion. In: R. H. Groves and J. J. Burdon, eds. *Ecology of Biological Invasions*. Cambridge: Cambridge University Press; pp. 57–60.

Francis, R. E., R. S. Driscoll, and J. N. Reppert. 1972. *Loop-Frequency as Related to Plant Cover, Herbage Production, and Plant Density*. Research Paper RM-94. Fort Collins, CO: Rocky Mountain Forest and Range Experiment Station, U.S. Forest Service.

Frank, A. B., D. L. Tanaka, L. Hofmann, and R. F. Follett. 1995. Soil carbon and nitrogen of northern Great Plains grasslands as influenced by long-term grazing. Journal of Range Management 48:470–474.

Frank, D. A., and S. J. McNaughton. 1993. Evidence for the promotion of aboveground grassland production by native large herbivores in Yellowstone National Park. Oecologia 96:157–161.

Franklin, J. F. 1993. Preserving biodiversity: species, ecosystems, or landscapes? Ecological Applications 3:202–206.

Fridley, J. D., R. L. Brown, and J. F. Bruno. 2004. Null models of exotic invasion and scale-dependent patterns of native and exotic species richness. Ecology 85:3215–3222.

Fridriksson, S. 1975. *Surtsey: Evolution of Life on a Volcanic Island*. London: Butterworths.

Garcia-Pichel, F., and J. Belnap. 1996. Microenvironments and microscale productivity of cyanobacterial desert crusts. Journal of Phycology 32:774–782.

Gaston, K. J. 2000. Global patterns in biodiversity. Nature 405:220–227.

Gee, G. W., and J. W. Bauder. 1986. Particle size analysis. In: A. Klute, ed. *Methods of Soil Analysis. Part 1—Physical and Mineralogical Methods*. Madison, WI: American Society of Agronomy; pp. 383–411.

Gentry, A. H. 1988. Tree species richness of upper Amazonian forests. Proceedings of the National Academy of Sciences USA 85:156–159.

Gibson, C. W. D., and V. K. Brown. 1991. The effects of grazing on local colonization and extinction during early succession. Journal of Vegetation Science 2:291–300.

Giesler, R., M. Högberg, and P. Högberg. 1998. Soil chemistry and plant in Fennoscandian boreal forest as exemplified by a local gradient. Ecology 79:119–137.

Gilbert, B., and M. J. Lechowicz. 2005. Invasibility and abiotic gradients: the positive correlation between native and exotic plant diversity. Ecology 86:1848–1855.

Gillison, A. N. 1970. Structural and floristics of a montane grassland/forest transition, Doma Peaks Region, Papua. Blumea 18:71–86.

Gleason, H. A. 1917. The structure and development of the plant association. Bulletin of the Torrey Botanical Club 44:463–481.

Gleason, H. A. 1920. Some applications of the quadrat method. Bulletin of the Torrey Botanical Club 47:21–33.

Gleason, H. A. 1922. On the relation between species and area. Ecology 3:158–162.

Gleason, H. A. 1925. Species and area. Ecology 6:66–74.

Gleason, H. A. 1926. The individualistic concept of the plant association. Bulletin of the Torrey Botanical Club 53:1–20.

Glenn, S. M., and S. L. Collins. 1992. Effects of scale and disturbance on rates of immigration and extinction of species in prairies. Oikos 63:273–280.

Glenn-Lewin, D. C. 1977. Species diversity in North American temperate forests. Vegetatio 33:153–162.

Goodall, D. W. 1953a. Objective methods for the classification of vegetation, I. The use of positive interspecific correlation. Australian Journal of Botany 1:39–62.

Goodall, D. W. 1953b. Objective methods for the classification of vegetation, II. Fidelity and indicator value. Australian Journal of Botany 1:434–456.

Goodall, D. W. 1954. Objective methods for the classification of vegetation, III. An essay in the use of factor analysis. Australian Journal of Botany 2:302–324.

Goodall, D. W. 1957. Some considerations in the use of point quadrat methods for the analysis of vegetation. Australian Journal of Biological Science 5:1–41.

Gosz, J. R. 1991. Fundamental ecological characteristics of landscape boundaries. In: E. Holland, P. G. Risser, and R. J. Naiman, eds. *Ecotones: The Role of Landscape Boundaries in the Management and Restoration of Changing Environments.* New York: Chapman & Hall; pp. 8–30.

Gosz, J. R. 1993. Ecotone hierarchies. Ecological Applications 3:369–376.

Graber, D. M., S. A. Haultain, and J. E. Fessenden. 1993. Conducting a biological survey: a case study from Sequoia and Kings Canyon National Parks. In: *Proceedings of the Fourth Biennial Conference on Science in California's National Parks.* Davis, CA: Cooperative Park Studies Unit, University of California, Davis.

Grace, J. B. 1999. The factors controlling species density in herbaceous plant communities. Perspectives in Plant Ecology, Evolution, and Systematics 2:1–28.

Grace, J. B., and B. H. Pugesek. 1997. A structural equation model of plant species richness and its application to a coastal wetland. American Naturalist 149:436–460.

Green, D. G. 1982. Fire and stability in the postglacial forests of southwest Nova Scotia. Journal of Biogeography 9:29–40.

Green, D. G. 1983. The ecological interpretation of fine resolution pollen records. New Phytologist 94:459–477.

Green, D. G. 1989. Simulated effects of fire, dispersal, and spatial pattern on competition within forested mosaics. Vegetatio 82:139–153.

Greenberg, C. H., S. H. Crownover, and D. R. Gordon. 1997. Roadside soils: a corridor for invasion of xeric scrub by nonindigenous plants. Natural Areas Journal 17:99–109.

Greig-Smith, P. 1957. *Quantitative Plant Ecology.* London: Butterworths Scientific.

Greig-Smith, P. 1964. *Quantitative Plant Ecology.* London: Butterworths Scientific.

Grime, J. P. 1973. Competitive exclusion in herbaceous vegetation. Nature 242:344–347.

Grime, J. P. 1977. Evidence for the existence of three primary strategies in plants and its relevance to ecological and evolutionary theory. American Naturalist 11:1169–1194.

Grime, J. P. 1979. *Plant Strategies and Vegetation Processes.* New York: John Wiley & Sons.

Grubb, P. J. 1977. Maintenance of species-richness in plant communities—importance of regeneration niche. Biological Reviews of the Cambridge Philosophical Society 52:107–145.

Guenther, D., T. J. Stohlgren, and P. Evangelista. 2004. A comparison of a near-relict site and a grazed site in a pinyon-juniper community in the Grand Staircase-Escalante National Monument, Utah. In: C. van Riper III and K. L. Cole, eds. *The Colorado Plateau: Cultural, Biological, and Physical Research.* Tucson, AZ: University of Arizona Press; pp. 153–162.

Hadley, K. S., and T. T. Veblen. 1993. Stand response to western spruce budworm and Douglas-fir bark beetle outbreaks, Colorado Front Range. Canadian Journal of Forest Research 23:479–491.

Hansen, A. J., and F. di Castri, eds. 1992. *Landscape Boundaries: Consequences for Biotic Diversity and Landscape Flows.* New York: Springer-Verlag.

Hargrove, W. W., and J. Pickering. 1992. Pseudoreplication: a sine qua non for regional ecology. Landscape Ecology 6:251–258.

Harper, J. L. 1969. The role of predation in vegetational diversity. Brookhaven Symposium of Biology 22:48–62.

Harper, J. L. 1977. *Population Biology of Plants.* New York: Academic Press.

Harrison, S., D. D. Murphy, and P. R. Ehrlich. 1988. Distribution of the bay checkerspot butterfly, *Euphydryas editha bayensis:* evidence for a meta-population model. American Naturalist 132:360–382.

Hart, R. H., and J. A. Hart. 1997. Rangelands of the Great Plains before European settlement. Rangelands 19:4–11.

Haslett, J., and A. E. Raftery. 1989. Space-time modelling with long-memory dependence: assessing Ireland's wind power resource. Applied Statistics 38:1–21.

Haury, L. R., J. A. McGowen, and P. H. Wiebe. 1978. Patterns and processes in the time-space scales of plankton distributions. In: J. H. Steele, ed. *Spatial Pattern in Plankton Communities.* New York: Plenum; pp. 277–327.

Havesi, J. A., J. D. Istok, and A. L. Flint. 1992. Precipitation estimation in mountainous terrain using multivariate geostatistics. Part I: structural analysis. Journal of Applied Meteorology 3:661–676.

Hawkins, C. P. 1986. Pseudo-understanding of pseudoreplication: a cautionary note. Bulletin of the Ecological Society of America 67:185.

Heady, H. F. 1957. The measurement and value of plant height in the study of herbaceous vegetation. Ecology 38:313–320.

Heady, H. F. 1968. Exclosures. In: F. N. Golley and H. K. Buechner, eds. *A Practical Guide to the Study of Productivity of Large Herbivores*. Oxford: Blackwell Press: pp. 250–252.

Heltshe, J. F., and N. E. Forrester. 1983. Estimating species richness using the jackknife procedure. Biometrics 39:1–12.

Herben, T., B. Mandák, K. Bímova, and Z. Münzbergova. 2004. Invasibility and species richness of a community: a neutral model and a survey of published data. Ecology 85:3223–3233.

Herrick, J. E., W. W. Whitford, A. G. de Soyza, and J. Van Zee. 1996. Soil and vegetation indicators for assessment of rangeland ecological condition. In: *North American Workshop on Monitoring for Ecological Assessment of Terrestrial and Aquatic Ecosystems*. General Technical Report RM-GTR-284. Fort Collins, CO: Rocky Mountain Forest and Range Experiment Station, U.S. Forest Service.

Hickman, J., ed. 1993. *The Jepsen Manual: Higher Plants of California*. Berkeley: University of California Press.

Higgins, S. I., D. M. Richardson, and R. M. Cowling. 1996. Modeling invasive plant spread: the role of plant-environment interactions and model structure. Ecology 77:2043–2054.

Hinds, W. T. 1984. Towards monitoring of long-term trends in terrestrial ecosystems. Environmental Conservation 11:11–18.

Hobbs, R. J., and L. F. Huenneke. 1992. Disturbance, diversity, and invasion: implications for conservation. Conservation Biology 6:324–337.

Holland, M. M., P. G. Risser, and R. J. Naiman, eds. 1991. *Ecotones*. New York: Chapman & Hall.

Hongo, A., S. Matsumoto, H. Takahashi, H. Zou, J. Cheng, H. Jia, and Z. Zhao. 1995. Effect of exclosure and topography on rehabilitation of overgrazed shrub-steppe in the loess plateau of northwest China. Restoration Ecology 3:18–25.

Hormay, A. L. 1949. Getting better records of vegetation changes with the line interception method. Journal of Range Management 2:67–69.

Horn, H. S. 1975. Markovian properties of forest succession. In: M. L. Cody and J. M. Diamond, eds. *Ecology and Evolution of Communities*. Cambridge, MA: Belknap Press; pp. 196–211.

Howell, W. 1998. *Germination and Establishment of Bromus tectorum L. in Relation to Cation Exchange Capacity, Seedbed, Litter, Soil Cover and Water*. Prescott, AZ: Prescott College.

Hubbell, S. P. 2001. *The Unified Neutral Theory of Biodiversity and Biogeography*. Princeton, NJ: Princeton University Press.

Hughes, L. E. 1980. Six grazing exclosures with a message. Rangelands 2:17–18.

Hughes, L. E. 1983. Is no grazing really better than grazing? Rangelands 5:159–161.

Hughes, L. E. 1990. Twenty years of rest-rotation grazing on the Arizona strip—an observation. Rangelands 12:173–176.

Hughes, L. E. 1996. What's in an exclosure? Rangelands 18:201–203.

Hulbert, L. C. 1969. Fire and litter effects in undisturbed bluestem prairie in Kansas. Ecology 50:874–877.

Hulbert, L. C. 1988. Causes of fire effects in tallgrass prairie. Ecology 46:58.

Hulme, P. E. 1996. Herbivores and the performance of grassland plants: a comparison of arthropod, mollusc and rodent herbivory. Journal of Ecology 84:43–51.

Hunt, C. B. 1975. *Death Valley: Geology, Ecology, and Archeology.* Berkeley: University of California Press.

Hurlbert, S. H. 1984. Pseudoreplication and the design of ecological field experiments. Ecological Monographs 54:197–211.

Huston, M. A. 1979. A general hypothesis of species diversity. American Naturalist 113:81–101.

Huston, M. A. 1994. *Biological Diversity: The Coexistence of Species in Changing Landscapes.* New York: Cambridge University Press.

Huston, M. A. 1999. Local processes and regional patterns: appropriate scales for understanding variation in the diversity of plants and animals. Oikos 86:393–401.

Huston, M. A., and D. L. DeAngelis. 1994. Competition and coexistence: the effects of resource transport and supply rates. American Naturalist 144: 954–977.

Huston, M. A., and A. C. McBride. 2002. Evaluating the relative strengths of biotic versus abiotic controls on ecosystem processes. In: M. Loreau, S. Naeem, and P. Inchausti, eds. *Approaches to Understanding Biodiversity and Ecosystem Function.* New York: Oxford University Press; pp. 47–60.

Hutchings, S. S., and R. C. Holmgren. 1959. Interpretation of loop-frequency data as a measure of plant cover. Ecology 40:668–677.

Hutchinson, G. E. 1953. The concept of pattern in ecology. Proceedings of the Philadelphia Academy of Natural Science 105:1–12.

Isaaks, E. H., and R. M. Srivastiva. 1980. *An Introduction to Applied Geostatistics.* New York: Oxford University Press.

Jaccard, P. 1901. Etude comparative de la distribution florale dans une portion des Alpes et du Jura. Bulletin Societe Vaudoise des Sciences Naturelles 37:547–579.

Jacobson, G. L., T. Webb III, and E. C. Grimm. 1987. Patterns and rates of vegetation change during the glaciation of eastern North America. In: W. F. Ruddiman and E. H. Wright, eds. *North America and Adjacent Oceans During the Last Deglaciation.* Boulder, CO: Geological Society of America; pp. 277–288.

Jensen, J. R. 1996. *Introductory Digital Image Processing: A Remote Sensing Perspective,* 2nd ed. Upper Saddle River, NJ: Prentice-Hall.

Joern, A., and K. H. Keeler, eds. 1995. *The Changing Prairie: North American Grasslands.* New York: Oxford University Press.

Johnston, A. 1957. A comparison of the line interception, vertical point quadrat, and loop methods as used in measuring basal area of grassland vegetation. Canadian Journal of Plant Science 37:34–42.

Jones, J. R., and N. V. DeByle. 1985. Fire. In: *Aspen: Ecology and Management in the Western United States.* General Technical Report RM-119. Fort Collins, CO: U.S. Forest Service; pp. 77–81.

Jones, K. B. 1986. The inventory and monitoring process. In: A. Y. Cooperrider, R. J. Boyde, and H. R. Stuart, eds. *Inventory and Monitoring of Wildlife Habitat.* Denver, CO: Bureau of Land Management Service Center; pp. 1–10.

Junk, W. J. 1989. The use of Amazonian floodplains under an ecological perspective. Interciencia 14:317–322.

Kalkhan, M. A., R. M. Reich, and T. J. Stohlgren. 1998. Assessing the accuracy of Landsat Thematic Mapper classification using double sampling. International Journal of Remote Sensing 19:2049–2060.

Kalkhan, M. A., and T. J. Stohlgren. 2000. Using multi-scale sampling and spatial cross-correlation to investigate patterns of plant species richness. Environmental Monitoring and Assessment 64:591–605.

Kalkhan, M. A., T. J. Stohlgren, and M. B. Coughenour. 1995. An investigation of biodiversity and landscape-scale gap patterns using double sampling: a GIS approach. In: *Proceedings of the Ninth Conference on Geographic Information Systems for Natural Resources, Environment, and Land Information Management,* Vancouver, British Columbia; pp. 708–712.

Kallas, M. 1997. Armillaria root rot disease in the Black Hills National Forest. Master's thesis. Fort Collins: Colorado State University.

Kareiva, P. M., and M. Anderson. 1988. Spatial aspects of species interactions: the wedding of models and experiments. In: A. Hastings, ed. *Community Ecology.* Lecture Notes in Biomathematics 77. Berlin: Springer-Verlag; pp. 35–50.

Kaye, M. 2002. Population and age distributions of aspen (*Populus tremuloides*) in Rocky Mountain National Park, Colorado. PhD dissertation. Fort Collins: Colorado State University.

Kaye, M. W., K. Suzuki, D. Binkley, and T. J. Stohlgren. 2001. Landscape-scale dynamics of aspen in Rocky Mountain National Park, Colorado. In: *Sustaining Aspen in Western Landscapes: Symposium Proceedings. 13–15 June 2000; Grand Junction, CO.* Proceedings RMRS-P-18. Fort Collins, CO: Rocky Mountain Research Station, U.S. Forest Service.

Keel, S. H., and G. T. Prance. 1979. Studies of the vegetation of a black water igapó (Rio Negro-Brazil). Acta Amazonia 9:645–655.

Keeley, J.E., M. Carrington, and S. Trnka. 1995. Overview of management issues raised by the 1993 wildfires in southern California. In: J. E. Keeley and T. Scott, eds. *Brushfires in California: Ecology and Resource Management.* Fairfield, WA: International Association of Wildland Fire; pp. 83–89.

Keeley, J. E., D. Lubin, and C. J. Fotheringham. 2003. Fire and grazing impacts on plant diversity and alien plant invasions in the southern Sierra Nevada. Ecological Applications 13:1355–1374.

Keeley, J. E., and Fotheringham, C. J. 2005. Plot shape effects on plant species diversity measurements. Journal of Vegetation Science 16:249–256.

Kempton, R. A. 1979. Structure of species abundance and measurement of diversity. Biometrics 35:307–321.

Kerner von Marilaun, A. 1863/1951. *Plant Life of the Danube Basin*. Translated by H. S. Conrad. Ames: Iowa State College Press.

Kershaw, K. A. 1963. Patterns in vegetation and its causality. Ecology 44:377–388.

Kieft, T. L., C. S. White, S. R. Loftin, R. Aguilar, J. A. Craig, and D. A. Skaar. 1998. Temporal dynamics in soil carbon and nitrogen resources at a grassland-shrubland ecotone. Ecology 79:671–683.

Kikkawa, J., and E. E. Williams. 1971. Altitudinal distribution of land birds in New Guinea. Search 2:64–69.

Kindschy, R. R. 1987. Sagehen exclosure: a history of bitterbrush reproduction. Rangelands 9:113–114.

Knight, D. H. 1994. *Mountains and Plains: The Ecology of Wyoming Landscapes*. New Haven, CT: Yale University Press.

Kot, M., M. A. Lewis, and P. van den Driessche. 1996. Dispersal data and the spread of invading organisms. Ecology 77:2027–2042.

Kozlowski, T. T. 1982. Water supply and tree growth. Part II. Flooding. Forest Abstracts 43:145–161.

Kozlowski, T. T. 1984. Responses of woody plants to flooding. In: T. T. Kozlowski, ed. *Flooding and Plant Growth*. New York: Academic Press; pp. 129–164.

Krebs, C. J. 1989. *Ecological Methodology*. New York: Harper & Row.

Krebs, C. J. 1999 *Ecological Methodology*. Menlo Park, CA: Addison-Wesley Educational Publishers.

Kullman, L. 1996. Rise and demise of cold-climate *Picea abies* forest in Sweden. New Phytologist 134:243–256.

Kupper, L. L., and K. B. Hafner. 1989. How appropriate are popular sample size formulas? American Statistician 43:101–105.

Lacey, C. A. 1989. Knapweed management: a decade of change. In: P. K. Kay and J. R. Lacey, eds. *Proceedings of Knapweed Symposium, April 4–5, 1989*. Extension Bulletin 45, Plant and Soil Science Department and Cooperative Extension Service. Bozeman: Montana State University; pp. 1–6.

Lacey, C. A., P. Husby, and G. Handl. 1990. Observations on spotted and diffuse knapweed into ungrazed bunchgrass communities in western Montana. Rangelands 12:30–32.

LaRoe, E. T. 1993. Implementation of an ecosystem approach to endangered species conservation. Endangered Species Update 10:3–6.

Larsen, K. D. 1995. Effects of microbiotic crusts on the germination and establishment of three range grasses. Unpublished thesis. Boise, ID: Boise State University.

Lavado, R. S., J. O. Sierra, and P. N. Hashimoto. 1996. Impact of grazing on soil nutrients in a Pampean grassland. Journal of Range Management 49:452–457.

Lavoie, C., M. Jean, F. Delisle, and G. Létourneau. 2003. Exotic plant species of the St. Lawrence River wetlands: a spatial and historical analysis. Journal of Biogeography 30:537–549.

Law, R., and R. D. Morton. 1996. Permanence and the assembly of ecological communities. Ecology 77:762–775.

Le, N. D., and J. V. Zidek. 1992. Interpolation with uncertain spatial covariances: a Bayesian alternative to kriging. Journal of Multivariate Analysis 43:351–374.

Leach, M. K., and T. J. Givnish. 1996. Ecological determinants of species loss in remnant prairies. Science 273:1555–1558.

Lee, M. 2001. Non-native plant invasions in Rocky Mountain National Park: linking species traits and habitat characteristics. Master's thesis. Fort Collins: Colorado State University.

Legendre, P., and M. J. Fortin. 1989. Spatial pattern and ecological analysis. Vegetatio 80:107–138.

Levin, S. A. 1992. The problem of pattern and scale in ecology. Ecology 73:1943–1967.

Levine, J. M. 2000. Species diversity and biological invasions: relating local process to community pattern. Science 288:852–854.

Levine, J. M., and C. M. D'Antonio. 1999. Elton revisited: a review of evidence linking diversity and invasibility. Oikos 87:15–26.

Li, J., A. Herlihy, W. Gerth, P. Kaufmann, S. Gregory, S. Urquhart, and D. P. Larsen. 2001. Variability in stream macroinvertebrates at multiple spatial scales. Freshwater Biology 46:87–97.

Likens, G. E., ed. 1989. *Long-Term Studies in Ecology: Approaches and Alternatives.* New York: Springer-Verlag.

Lister, A. J., P. P. Mou, R. H. Jones, and R. J. Mitchell. 2000. Spatial patterns of soil and vegetation in a 40-year-old slash pine (Pinus elliottii) forest in the Coastal Plain of South Carolina, U.S.A. Canadian Journal of Forest Research 30: 145–155.

Lonsdale, W. M. 1999. Global patterns of plant invasions and the concept of invasibility. Ecology 80:1522–1536.

Loope, L. L., and D. Mueller-Dombois. 1989. Characteristics of invaded islands, with special reference to Hawaii. In: J. A. Drake, F. DiCasti, R. H. Groves, F. J. Kruger, H. A. Mooney, M. Rejmánek, and M. H. Williams, eds. *Biological Invasions: A Global Perspective.* New York: John Wiley & Sons; pp. 257–280.

Lubchenco, J., A. M. Olson, L. B. Brubaker, S. R. Carpenter, M. M. Holland, S. P. Hubbell, S. A. Levin, J. A. MacMahon, P. A. Matson, J. M. Melillo, H. A. Mooney, C. H. Peterson, H. R. Pulliam, L. A. Real, P. J. Regal, and P. G. Risser. 1991. The sustainable biosphere initiative: an ecological research agenda. Ecology 72:371–412.

Ludwig, J. A., and J. F. Reynolds. 1988. *Statistical Ecology: A Primer on Methods and Computing.* New York: John Wiley & Sons.

MacArthur, R. H., and E. O. Wilson. 1967. *The Theory of Island Biogeography.* Princeton, NJ: Princeton University Press.

MacDonald, L. H., A. W. Smart, and R. C. Wissmar. 1991. Monitoring guidelines to evaluate effects of forestry activities on streams in the Pacific Northwest and Alaska. EPA/910/9–91–001. Seattle: U.S. Environmental Protection Agency.

Mack, R. N. 1981. Invasion of *Bromus tectorum* L. into western North America: an ecological chronicle. Agro-Ecosystems 7:145–165.

Mack, R. N. 1996. Understanding the processes of weed invasions: the influence of environmental stochasticity. In: C. H. Stirton, ed. *Weeds in a*

Changing World. Symposium Proceedings no. 64. Brighton, England: British Crop Protection Council; pp. 65–74.

Mack, R. N., D. Simberloff, M. Lonsdale, H. Evans, M. Clout, and F. Bazzaz. 2000. Biotic invasions: causes, epidemiology, global consequences and control. Issues in Ecology 5:1–20.

Mack, R. N., and J. N. Thompson. 1982. Evolution in steppe with few large, hooved mammals. American Naturalist 119:757–773.

Madole, R. F. 1976. Glacial geology of the Front Range, Colorado. In: W. C. Mahaney, ed. *Quaternary Stratigraphy of North America.* Stroudsburg, PA: Dowden, Hutchinson, and Ross; pp. 319–351.

Magnuson, J. J., T. K. Kratz, T. M. Frost, C. J. Bowser, B. J. Benson, and R. Nero. 1991. Expanding the temporal and spatial scales of ecological research and comparison of divergent ecosystems: roles for LTER in the United States. In: P. G. Risser, ed. *Long-Term Ecological Research: An International Perspective.* New York: John Wiley & Sons; pp. 45–70.

Magurran, A. E. 1988. *Ecological Diversity and Its Measurement.* Princeton, NJ: Princeton University Press.

Magurran, A. E. 2003. *Measuring Biological Diversity.* Malden, MA: Blackwell.

Malanson, G. P. 1993. *Riparian Landscapes.* Cambridge: Cambridge University Press.

Margules, C. R., and R. L. Pressey. 2000. Systematic conservation planning. Nature 405:243–253.

Markgraf, V., and L. Scott. 1981. Lower timberline in central Colorado during the past 15,000 yr. Geology 9:231–234.

Martinez, E., and E. Fuentes. 1993. Can we extrapolate the California model of grassland-shrubland ecotone? Ecological Applications 3:417–423.

MathSoft Inc. 1999. S-Plus 2000. Seattle: MathSoft, Inc.

Matlack, G. R. 1994. Vegetation dynamics of the forest edgetrends in space and successional time. Journal of Ecology 82:113–123.

Matson, P. A., and S. R. Carpenter. 1990. Special feature: Statistical analysis of ecological response to large-scale perturbations. Ecology 71:2037–2068.

May, R. M. 1973. *Stability and Complexity in Model Ecosystems.* Princeton, NJ: Princeton University Press.

May, R. M. 1999. Unanswered questions in ecology. Philosophical Transactions of the Royal Society of London Series B-Biological Sciences 354: 1951–1959.

McCune, B. 1997. Influence of noisy environmental data on canonical correspondence analysis. Ecology 78:2617–2623.

McGowan, J. A. 1990. Climate and change in oceanic systems: the value of time series data. Trends in Ecology and Evolution 5:293–299.

McIntosh, R. P. 1985. *The Background of Ecology: Concepts and Theory.* New York: Cambridge University Press.

McNaughton, S. J. 1979. Grassland-herbivore dynamics. In: A. R. E. Sinclair and M. Norton-Griffiths, eds. *Serengeti: Dynamics of an Ecosystem.* Chicago: University of Chicago Press; pp. 46–81.

McNaughton, S. J. 1983. Serengeti grassland ecology: the role of composite environmental factors and contingency in community organization. Ecological Monographs 53:291–320.

McNaughton, S. J. 1993. Biodiversity and function of grazing ecosystems. In: E. D. Schulze and H. A. Mooney, eds. *Biodiversity and Ecosystem Function.* Berlin: Springer-Verlag; pp. 361–408.

McNaughton, S. J., M. Oesterheld, D. A. Frank, and K. J. Williams. 1989. Ecosystem-level patterns of primary productivity and herbivory in terrestrial habitats. Nature 132:142–144.

Messer, J. J., R. A. Linthurst, and W. S. Overton. 1991. An EPA program for monitoring ecological status and trends. Environmental Monitoring and Assessment 17:67–78.

Metzger, K. 1997. Modeling forest stand structure to a ten meter resolution using Landsat TM data. Master's thesis. Fort Collins: Colorado State University.

Meyers, N., R. A. Mittermeier, C. G. Mittermeier, G. A. B. da Fonseca, and J. Kent. 2000. Biodiversity hotspots for conservation priorities. Nature 403:853–858.

Milchunas, D. G., and W. K. Lauenroth. 1993. Quantitative effects of grazing on vegetation and soils over a global range of environments. Ecological Monographs 63:327–366.

Milchunas, D. G., W. K. Lauenroth, P. L. Chapman, and M. K. Kazempour. 1990. Community attributes along a perturbation gradient in a shortgrass steppe. Journal of Vegetation Science 1:375–384.

Milchunas, D. G., O. E. Sala, and W. K. Lauenroth. 1988. A generalized model of the effects of grazing by large herbivores on grassland community structure. American Naturalist 132:87–106.

Miller, R. I., and R. G. Wiegert. 1989. Documenting completeness, species-area relations, and the species-abundance distribution of a regional flora. Ecology 70:16–22.

Milne, B. T., A. R. Johnson, T. H. Keith, C. A. Hatfield, J. David, and P. T. Hraber. 1996. Detection of critical densities associated with piñon-juniper woodland ecotones. Ecology 77:805–821.

Mistry, S., J. Comiskey, and T. J. Stohlgren. 1999. In: A. Alonso and F. Dallmeier, eds. *Biodiversity Assessment and Monitoring of the Urumbaba Region, Peru.* SI/MAB Series 3, Smithsonian Institution Monitoring and Assessment of Biodiversity Program. Washington, DC: Smithsonian Institution.

Monk, C. D. 1967. Tree species diversity in the eastern deciduous forest with particular reference to north central Florida. American Naturalist 101: 173–187.

Montaña, C., J. Lopez-Portillo, and A. Mauchamp. 1990. The response of two woody species to the conditions created by a shifting ecotone in an arid ecosystem. Journal of Ecology 78:789–798.

Moser, E. B., A. M. Saxton, and S. R. Pezeshki. 1990. Repeated measures analysis of variance: application to tree research. Canadian Journal of Forest Research 20:524–535.

Mueggler, W. F. 1985. Forage. In: *Aspen: Ecology and Management in the Western United States.* General Technical Report RM-119. Fort Collins, CO: U.S. Forest Service; pp. 129–134.

Mueller-Dombois, D., and H. Ellenberg. 1974. *Aims and Methods of Vegetation Ecology.* New York: John Wiley & Sons.

Mulder, B. S., B. R. Noon, T. A. Spies, M. G. Raphael, C. J. Palmer, A. R. Olsen, G. H. Reeves, and H. H. Welsh. 1999. The strategy and design for the effectiveness monitoring program for the Northwest Forest Plan. General Technical Report PNW-GTR-437. Portland, OR: U.S. Forest Service.

Munz, P. A., and D. D. Keck. 1959. *A California Flora and Supplement.* Berkeley: University of California Press.

Myers, N., R. A. Mittermeier, C. G. Mittermeier, G. A. B. da Fonseca, and J. Kent. 2000. Biodiversity hotspots for conservation priorities. Nature 403: 853–858.

Nasri, M., and P. S. Doescher. 1995. Effect of competition by cheatgrass on shoot growth of Idaho fescue. Journal of Range Management 48:402–405.

National Research Council. 1990. *Forest Research: A Mandate for Change.* Washington, DC: National Academies Press.

National Research Council. 1994. *Rangeland Health: New Methods to Classify, Inventory, and Monitor Rangelands.* Washington, DC: National Academies Press.

Neilson, R. P. 1991. Climatic constraints and issues of scale controlling regional biomes. In: M. M. Holland, R. J. Naiman, and P. G. Risser, eds. *Role of Landscape Boundaries in the Management and Restoration of Changing Environments.* New York: Chapman & Hall; pp. 31–51.

Neilson, R. P. 1995. A model for predicting continental-scale vegetation distribution and water balance. Ecological Applications 5:362–385.

Nero, R. W., and J. J. Magnuson. 1992. Effects of changing spatial scale on acoustic observations of patchiness in the Gulf Stream. Landscape Ecology 6:279–291.

Neter, J., W. Wasserman, and M. H. Kutner. 1990. *Applied Linear Statistical Models: Regression, Analysis of Variance, and Experimental Designs,* 3rd ed. Homewood, IL: Irwin.

Nevah, Z., and R. H. Whittaker. 1979. Structural and floristic diversity of shrublands and woodlands in northern Israel and other Mediterranean areas. Vegetatio 41:171–190.

Nichols, H. 1982. Review of late Quaternary history of vegetation and climate in the mountains of Colorado. In: J. C. Halfpenny, ed. *Ecological Studies in the Colorado Alpine: A Festschrift for John W. Marr.* Occasional Paper 37. Boulder: Institute of Arctic and Alpine Research, University of Colorado; pp. 27–33.

Nichols, J. D. 1992. Capture-recapture models: using marked animals to study population dynamics. BioScience 42:94–102.

Niklas, K. J., B. H. Tiffney, and A. H. Knoll. 1983. Patterns in vascular land plant diversification. Nature 303:293–299.

Noss, R. 1983. A regional landscape approach to maintain diversity. BioScience 33:700–706.

Noss, R. F., and A. Y. Cooperrider. 1994. *Saving Nature's Legacy: Protecting and Restoring Biodiversity.* Washington, DC: Island Press.

Nusser, S. M., and J. J. Goebel. 1997. The National Resources Inventory: a long-term multi-resource monitoring programme. Environmental and Ecological Systematics 4:181–204.

Nusser S. M., F. J. Breidt, and W. A. Fuller. 1998. Design and estimation for investigating the dynamics of natural resources. Ecological Applications 8:234–245.

Oosting, H. G. 1948. *The Study of Plant Communities.* San Francisco: Freeman.

Opdam, P., R. Van Apeldoorn, A. Schotman, and J. Kalkhoven. 1993. Population responses to landscape fragmentation. In: C. C. Vos and P. Opdam, eds. *Landscape Ecology of a Stressed Environment.* New York: Chapman & Hall; pp. 147–171.

Orr, D. M., and C. J. Evenson. 1991. Effects of sheep grazing *Astrebla* grasslands in central western Queensland III. Dynamics of *Astrebla* spp. under grazing and exclosure between 1975 and 1986. Rangeland Journal 13:36–46.

Pacala, S. W., and M. J. Crawley. 1992. Herbivores and plant diversity. American Naturalist 140:243–260.

Palmer, C. J., K. H. Ritters, J. Strickland, D. C. Cassell, G. E. Byers, M. L. Papp, and C. I. Liff. 1991. *Monitoring and Research Strategy for Forests—Environmental Monitoring and Assessment Program (EMAP).* EPA/600/4–91/ 012. Washington, DC: U.S. Environmental Protection Agency.

Palmer, M. W. 1990. The estimation of species richness by extrapolation. Ecology 71:1195–1198.

Palmer, M. W. 1993. Putting things in even better order: the advantages of canonical correspondence analysis. Ecology 74:2215–2230.

Palmer, M. W. 1994. Variation in species richness: toward a unification of hypotheses. Folia Geobotany Phytotax Praha 29:511–530.

Palmer, M. W., and P. M. Dixon. 1990. Small-scale environmental heterogeneity and the analysis of species distributions along gradients. Journal of Vegetation Science 1:57–65.

Palmer, M. W., and G. Rusch. 2001. How fast is the carousel? Direct indices of species mobility in an Oklahoma grassland. Journal of Vegetation Science 12:305–318.

Parker, K. W. 1951. *A Method for Measuring Trends in Range Condition in National Forest Ranges.* Washington, DC: U.S. Forest Service.

Parsons, A. J., A. Harvey, and I. R. Johnson. 1991. Plant-animal interactions in a continuously grazed mixture. II. The role of differences in the physiology of plant growth and of selective grazing on the performance and stability of species in a mixture. Journal of Applied Ecology 28:635–647.

Peet, R. K. 1981. Forest vegetation of the Colorado Front Range: patterns of species diversity. Vegetatio 52:129–140.

Peet, R. K. 1988. Forests of the Rocky Mountains. In: M. G. Barbour and W. D. Billings, eds. *North American Terrestrial Vegetation.* New York: Cambridge University Press; pp. 64–103.

Penfound, W. T. 1963. A modification of the point-centered quadrat method for grassland analysis. Ecology 44:175–176.

Petchey, O. L., A. Hector, and K. J. Gaston. 2004. How do different measures of functional diversity perform? Ecology 85:847–857.

Peters, A., D. E. Johnson, and M. R. George. 1996. Barb goatgrass: a threat to California rangelands. Journal of Range Management 18:8–10.

Peters, R. L., and J. D. Darling. 1985. The greenhouse effect and nature reserves. BioScience 35:707–717.

Peters, R. L., and T. E. Lovejoy, eds. 1992. *Global Warming and Biological Diversity*. New Haven, CT: Yale University Press.

Peterson, D. L., and V. T. Parker, eds. 1998. *Ecological Scale: Theory and Applications*. New York: Columbia University Press.

Petterson, E. S. 1999. Prescribed fire effects on plant communities in Rocky Mountain bighorn sheep habitat. Master's thesis. Fort Collins: Colorado State University.

Pianka, E. 1980. Guild structure in desert lizards. Oikos 35:194–201.

Pickett, S. T. A. 1989. Space-for-time substitution as an alternative to long-term studies. In: G. E. Likens, ed. *Long-Term Studies in Ecology: Approaches and Alternatives*. New York: Springer-Verlag; pp. 110–135.

Pickett, S. T. A., and P. S. White, eds. 1985. *The Ecology of Natural Disturbance and Patch Dynamics*. New York: Academic Press.

Pielou, E. C. 1969. *An Introduction to Mathematical Ecology*. New York: John Wiley & Sons.

Pielou, E. C. 1977. *Mathematical Ecology*. New York: John Wiley & Sons.

Pielou, E. C. 1991. *After the Ice Age: The Return of Life to Glaciated North America*. Chicago: University of Chicago Press.

Pitelka, L. F., R. H. Gardner, J. Ash, S. Berry, H. Gitay, I. R. Noble, A. Saunders, R. H. W. Bradshaw, L. Brubaker, J. S. Clark, M. B. Davis, S. Sugita, J. M. Dyer, R. Hengeveld, G. Hope, B. Huntley, G. A. King, S. Lavorel, R. N. Mack, G. P. Malanson, M. McGlone, I. C. Prentice, and M. Rejmanek. 1997. Plant migration and climate change. American Scientist 85:464–473.

Planty-Tabacchi, A.-M., E. Tabacchi, R. J. Naiman, C. DeFerrari, and H. Decamps. 1996. Invasibility of species-rich communities in riparian zones. Conservation Biology 10:598–607.

Podani, J., T. Czaran, and S. Bartha. 1993. Pattern, area and diversity: the importance of spatial scale in species assemblages. Abstracta Botanica 17: 35–51.

Pole, A., M. West, and P. J. Harrison. 1994. *Applied Bayesian Forecasting and Time Series Analysis*. New York: Chapman & Hall.

Polley, H. W., and S. L. Collins. 1984. Relationships of vegetation and environment in buffalo wallows. American Midland Naturalist 112:178–186.

Polley, H. W., and L. L. Wallace. 1986. The relationship of plant species heterogeneity to soil in buffalo wallows. Southwestern Naturalist 31:493–501.

Pollock, M. M., R. J. Naiman, and T. A. Hanley. 1998. Plant species richness in riparian wetlands: a test of biodiversity theory. Ecology 79:94–105.

Post, W. M., and S. L. Pimm. 1983. Community variability and food-web stability. Mathematical Biosciences 64:169–192.

Pound, R., and F. E. Clements. 1898a. A method of determining the abundance of secondary species. Minnesota Botanical Gardens 2:19–24.

Pound, R., and F. E. Clements. 1898b. *A Phytogeography of Nebraska*. New York: Arno Press.

Powell, T. M. 1989. Physical and biological scales of variability in lakes, estuaries, and the coastal ocean. In: J. Roughgarden, R. M. May, and S. A.

Levin, eds. *Perspectives in Ecological Theory.* Princeton, NJ: Princeton University Press; pp. 157–176.

Powell, T. M., and J. H. Steele, eds. 1994. *Ecological Time Series.* New York: Chapman & Hall.

Preston, F. W. 1960. Time and space and the variation of species. Ecology 41:611–627.

Preston, F. W. 1962a. Canonical distribution of commonness and rarity: part I. Ecology 43:185–215.

Preston, F. W. 1962b. Canonical distribution of commonness and rarity: part II. Ecology 43:410–432.

Preston, F. W. 1969. Diversity and stability in the biological world. Brookhaven Symposia in Biology 22:1–12.

Pugesek, B. H., A. Tomer, and A. Von Eye, eds. 2003. *Structural Equation Modeling: Applications in Ecological and Evolutionary Biology.* New York: Cambridge University Press.

Pulliam, H. R. 1988. Sources, sinks, and population regulation. American Naturalist 132:652–661.

Puyravaud, J., J. Pascal, and C. Dufour. 1994. Ecotone structure as an indicator of changing forest-savanna boundaries (Linganamakki Region, southern India). Journal of Biogeography 21:581–593.

Ramensky, L. G. 1924. Die Grundgesetzmassigkeiten im Aufbau der Vegetationsdecke. Botan. Centralbl. N. F. 7:453–455.

Randall, J. M. 1996. Weed control and the preservation of biological diversity. Weed Technology 10:370–383.

Raunkiaer, C. 1934. *The Life Forms of Plants and Statistical Plant Geography: Being the Collected Papers of C. Raunkiaer.* Oxford: Clarendon Press.

Reardon, J. J. 1996. Changes in grazed and protected plant communities in Yellowstone National Park. In: F. J. Singer, ed. *Effects of Grazing by Wild Ungulates in Yellowstone National Park.* Washington, DC: U.S. Department of the Interior, National Park Service, Natural Resource Program Center, Natural Resource Information Division,; pp. 115–125

Reed, R. A., R. K. Peet, M. W. Palmer, and P. S. White. 1993. Scale dependence of vegetation-environment correlations: a case study of a North Carolina piedmont woodland. Journal of Vegetation Science 4:329–340.

Reich, R. M., and V. A. Bravo. 1998. Integrating spatial statistics with GIS and remote sensing in designing multi-resource inventories. Presented at the North American Symposium on Toward a Unified Framework for Inventorying and Monitoring Forest Ecosystem Resources, Guadalajara, Mexico, November 1–6, 1998.

Reich, R. M., R. L. Czaplewski, and W. A. Bechtold. 1994. Spatial cross-correlation in growth of undisturbed natural shortleaf pine stands in northern Georgia. Journal of Environmental and Ecological Statistics 1:201–217.

Reichard, S. H., and C. W. Hamilton. 1997. Predicting invasions of woody plants introduced into North America. Conservation Biology 11:193–203.

Reichard, S. H., and P. White. 2001. Horticulture as a pathway of invasive plant introductions in the United States. BioScience 51:103–113.

Reid, W. V., and K. R. Miller. 1989. *Keeping Options Alive: The Scientific Basis for Conserving Biodiversity.* Washington, DC: World Resources Institute.

Rejmánek, M. 1996. Species richness and resistance to invasions. In: H. Orians, R. Dirzo, and J. H. Cushman, eds. *Biodiversity and Ecosystem Processes in Tropical Forests*. Berlin: Springer-Verlag; pp. 153–172.

Reynolds, R. T., and C. H. Trost. 1980. The response of native vertebrate populations to crested wheatgrass planting and grazing by sheep. Journal of Range Management 33:122–125.

Rice, B., and M. Westoby. 1983. Plant species richness at the 0.1 hectare scale in Australian vegetation compared to other continents. Vegetatio 52:129–140.

Rice, E. L. 1952. Phytosociologic analysis of a tall-grass prairie in Marshall County, Oklahoma. Ecology 33:112–115.

Riegel, G. M., S. E. Green, M. E. Harmon, and J. F. Franklin. 1988. Characteristics of mixed conifer forest reference stands at Sequoia National Park, California. Technical Report no. 32. Davis, CA: Cooperative National Park Resources Studies Unit, University of California, Davis.

Riitters, K. H., B. E. Law, R. C. Kucera, A. L. Gallant, R. L. DeVelice, and C. J. Palmer. 1992. A selection of forest condition indicators for monitoring. Environmental Monitoring and Assessment 20:21–23.

Risser, P. G. 1993. Ecotones: ecotones at local to regional scales from around the world. Ecological Applications 3:367–368.

Robertson, G. P. 1987. Geostatistics in ecology: interpolating with known variance. Ecology 68:744–748.

Robertson, J. H. 1971. Changes on a sagebrush-grass range in Nevada ungrazed for 30 years. Journal of Range Management 24:397–400.

Robinson, G. R., and J. F. Quinn. 1988. Extinction, turnover and species diversity in an experimentally fragmented California annual grassland. Oecologia 76:71–82.

Robinson, G. R., J. F. Quinn, and M. L. Stanton. 1995. Invasibility of experimental habitat in California winter annual grassland. Ecology 76:786–794.

Romme, W. H., and M. G. Turner. 1991. Implications of global climate change for biogeographic patterns in the Greater Yellowstone ecosystem. Conservation Biology 5:373–386.

Romme, W. H., M. G. Turner, L. L. Wallace, and J. S. Walker. 1995. Aspen, elk, and fire in northern Yellowstone National Park. Ecology 76:2097–2106.

Rosentreter, R. 1994. Displacement of rare plants by exotic grasses. In: S. B. Monsen and S. G. Kitchen, eds. *Proceedings—Ecology and Management of Annual Rangelands*. General Technical Report INT-GTR-313. Ogden, UT: Intermountain Research Station, U.S. Forest Service; pp. 170–175.

Rosenzweig, M. L. 1995. *Species Diversity in Space and Time*. Cambridge: Cambridge University Press.

Rosenzweig, M. L., and Y. Ziv. 1999. The echo pattern of species diversity: pattern and processes. Ecography 22:614–628.

Rummel, D. J., and J. Roughgarden. 1983. Some differences between invasion-structured and co-evolutionary-structured competitive communities. Oikos 41:477–486.

Rummell, R. S. 1951. Some effects of grazing on ponderosa pine forest and range in central Washington. Ecology 32:594–607.

Rusek, J. 1993. Air-pollution-mediated changes in alpine ecosystems and ecotones. Ecological Applications 3:406–416.

Salo, J., R. Kalliola, Y. Makinem, P. Niemela, M. Puhakka, and P. D. Coleu. 1986. River dynamics and the diversity of Amazon lowland forest. Nature 322:254–258.

Salt, G. W. 1957. An analysis of avifaunas in the Teton Mountains and Jackson Hole, Wyoming. Condor 59:373–393.

Saltonstall, K. 2002. Cryptic invasion by a non-native genotype of the common reed, *Phragmites australis,* into North America. Proceedings of the National Academy of Science USA 99:2445–2449.

SAS Institute. 1998. SAS for Windows. Cary, NC: SAS Institute.

Sax, D. F., and S. D. Gaines. 2003. Species diversity: from global decreases to local increases. Trends in Ecology and Evolution 18:561–566.

Schreuder, H. T., M. S. Williams, and R. M. Reich. 1999. Estimating the number of tree species in a forest community using survey data. Environmental Monitoring and Assessment 56:293–303.

Schulz, T. T., and W. C. Leininger. 1990. Differences in riparian vegetation structure between grazing areas and exclosures. Journal of Range Management 43:295–299.

Scott, C. T. 1998. Sampling methods for estimating change in forest resources. Ecological Applications 8:228–233.

Scott, J. M., F. Davis, R. Csuti, R. Noss, B. Butterfield, C. Groves, H. Anderson, S. Caicco, F. D'Erchia, T. C. Edwards, Jr., J. Ulliman, and R. G. Wright. 1993. GAP analysis: a geographic approach to protection of biological diversity. Wildlife Monographs 123:1–41.

Seastedt, T. R. 1995. Soil systems and nutrient cycles of the North American prairie. In: A. Joern and K. H. Keeler, eds. *The Changing Prairie: North American Grasslands.* New York: Oxford University Press; pp. 157–174.

Shafer, C. L. 1990. *Nature Reserves: Island Theory and Conservation Practice.* Washington, DC: Smithsonian Institution Press

Sheley, R. L., B. E. Olson, and L. L. Larson. 1997. Effect of weed seed rate and grass defoliation on diffuse knapweed. Journal of Range Management 50:39–43.

Shmida, A. 1984. Whittaker's plant diversity sampling method. Israel Journal of Botany 33:41–46.

Shmida, A., and M. V. Wilson. 1985. Biological determinants of species diversity. Journal of Biogeography 12:1–20.

Short, H. L., and J. B. Hestbeck. 1995. National biotic resource inventories and GAP analysis: problems of scale and unproven assumptions limit a national program. BioScience 45:535–539.

Shultz, L. M. 1998. The flora and fauna of the Colorado Plateau: what do we know? In: L. M. Hill, ed. *Learning from the Land, Grand Staircase-Escalante National Monument Science Symposium.* Salt Lake City, UT: Paragon Press; pp. 203–210.

Simonson, S. 1998. Rapid assessment of butterfly diversity: a method for landscape assessment. Master's thesis. Fort Collins, CO: Colorado State University.

Simonson, S., G. Chong, P. Opler, and T. J. Stohlgren. 2001. Rapid assessment of butterfly diversity: a method for landscape evaluation. Biodiversity and Conservation 10:1369–1386.

Singer, F. J. 1995. Effects of grazing by ungulates on upland bunchgrass communities of the northern winter range of Yellowstone National Park. Northwest Science 69:191–203.

Singer, F. J. 1996. Differences between willow communities browsed by elk and communities protected for 32 years in Yellowstone National Park. In: F. J. Singer, ed. *Effects of Grazing by Wild Ungulates in Yellowstone National Park.* Washington, DC: U.S. Department of the Interior, National Park Service, Natural Resource Program Center, Natural Resource Information Division; pp. 279–290.

Smeins, F. E., T. W. Taylor, and L. B. Merrill. 1976. Vegetation of a 25-year exclosure on the Edwards Plateau, Texas. Journal of Range Management 29:24–29.

Smith, D. A., and E. M. Schmutz. 1975. Vegetative changes on protected versus grazed desert grassland ranges in Arizona. Journal of Range Management 28:453–458.

Smith, D. R. 1960. Description and response to elk use of two mesic grassland and shrub communities in the Jackson Hole region of Wyoming. Northwest Science 34:25–36.

Smith, M. D., J. C. Wilcox, T. Kelly, and A. K. Knapp. 2004. Dominance not richness determines invasibility of tallgrass prairie. Oikos 106:253–262.

Sneva, F. A., L. R. Rittenhouse, P. T. Tueller, and P. Reece. 1984. Changes in protected and grazed sagebrush-grass in eastern Oregon, 1937 to 1974. Station Bulletin 663. Corvallis: Agricultural Experiment Station, Oregon State University; pp. 3–11.

Sokal, R. R., and F. J. Rohlf. 1981. *Biometry: The Principles and Practice of Statistics on Biological Research*, 2nd ed. San Francisco: W. H. Freeman.

Sokal, R. R., and J. D. Thomson. 1987. Applications of spatial autocorrelation in ecology. In: P. Legendre and L. Legendre, eds. *Developments in Numerical Taxonomy.* NATO ASI Series, vol. 14. Berlin: Springer-Verlag; pp. 431–466.

Soulé, M. E., and K. A. Kohm. 1989. *Research Priorities for Conservation Biology.* Washington, DC: Island Press.

SPSS Inc. 1997. Systat, version 7.0. Chicago: SPSS Inc.

SPSS Inc. 2001. Systat, version 10.0. Chicago: SPSS Inc.

Stapanian, M. A., S. P. Cline, and D. C. Cassell. 1993. Vegetation structure. In: B. L. Conkling and G. E. Byers, eds. *Forest Health Monitoring Field Methods Guide.* Internal Report. Las Vegas: U.S. Environmental Protection Agency.

Stebler, F. G., and C. Schröter. 1892. Versuch einer übersicht über die Wiesentypen der Schweiz. Landwirtschaftliches Jahrbuch der Schweiz 6:95.

Steele, J. H. 1989. Discussion: scale and coupling in ecological systems. In: J. Roughgarden, R. M. May, and S. A. Levin, eds. *Perspectives in Ecological Theory.* Princeton, NJ: Princeton University Press; pp. 177–180.

Stohlgren, T. J. 1992. Resilience of a heavily logged grove of giant sequoia (*Sequoiadendron giganteum*) in Sequoia and Kings Canyon National Parks. Forest Ecology and Management 54:115–140.

Stohlgren, T. J. 1993. Bald eagle winter roost characteristics in Lava Beds National Monument, California. Northwest Science 67:44–54.

Stohlgren, T. J. 1994. Planning long-term vegetation studies at landscape scales. In: T. M. Powell and J. H. Steele, eds. *Ecological Time Series*. New York: Chapman & Hall; pp. 208–241.

Stohlgren, T. J. 1999a. Measuring and monitoring biodiversity in forests and grasslands in the United States. In: C. Aguirre and C. R. Franco, eds. *Proceedings of the North American Symposium Toward a Unified Framework for Inventory and Monitoring Forest Ecosystem Resources*. General Technical Report RMRS-P-12. Fort Collins, CO: Rocky Mountain Forest and Range Experiment Station, U.S. Forest Service; pp. 248–255.

Stohlgren, T. J. 1999b. The Rocky Mountains. In: M. J. Mac, P. A. Opler, C. E. Puckett Haecher, and P. D. Doran, eds. *Status and Trends of the Nation's Biological Resources*. Reston, VA: Biological Resources Division, U.S. Geological Survey; pp. 473–504.

Stohlgren, T. J. 2001. Data acquisition for ecological assessments. In: M. Jensen and P. Bougeron, eds. *Ecological Assessments*. New York: Springer-Verlag; pp. 71–78.

Stohlgren, T. J. 2002. Beyond theories of plant invasions: lessons from natural landscapes. Comments on Theoretical Biology 7:355–379.

Stohlgren, T. J., and R. R. Bachand. 1997. Lodgepole pine (*Pinus contorta*) ecotones in Rocky Mountain National Park, Colorado, USA. Ecology 78:632–641.

Stohlgren, T. J., R. R. Bachand, Y. Onami, and D. Binkley. 1998a. Species-environment relationships and vegetation patterns: effects of spatial scale and tree life-stage. Plant Ecology 135:215–228.

Stohlgren, T. J., D. T. Barnett, C. Flather, J. Kartesz, and B. Peterjohn. 2005a. Plant species invasions along the latitudinal gradient in the United States. Ecology 86:2298–2309.

Stohlgren, T. J., J. Belnap, G. W. Chong, and R. Reich. 1998b. A plan to assess native and exotic plant diversity and cryptobiotic crusts in the Grand Staircase-Escalante National Monument. In: L. M. Hill, ed. *Learning from the Land, Grand Staircase-Escalante National Monument Science Symposium Proceedings*. Salt Lake City, UT: Paragon Press; pp. 269–276.

Stohlgren, T. J., D. Binkley, G. W. Chong, M. A. Kalkhan, L. D. Schell, K. A. Bull, Y. Otsuki, G. Newman, M. Bashkin, and Y. Son. 1999a. Exotic plant species invade hot spots of native plant diversity. Ecological Monographs 69:25–46.

Stohlgren, T. J., D. Binkley, T. T. Veblen, and W. L. Baker. 1995a. Attributes of landscape-scale, long-term studies: malpractice insurance for landscape ecologists. Environmental Monitoring and Assessment 36:1–25.

Stohlgren, T. J., K. A. Bull, and Y. Otsuki. 1998c. Comparison of rangeland vegetation sampling techniques in the Central Grasslands. Journal of Range Management 51:164–172.

Stohlgren, T. J., K. A. Bull, Y. Otsuki, C. A. Villa, and M. Lee. 1998d. Riparian zones as havens for exotic plant species. Plant Ecology 138:113–125.

Stohlgren, T. J., T. N. Chase, R. A. Pielke, Sr., T. G. F. Kittel, and J. S. Baron. 1998e. Evidence that local land use practices influence regional climate, vegetation, and stream flow patterns in adjacent natural areas. Global Change Biology 4:495–504.

Stohlgren, T. J., G. W. Chong, M. A. Kalkhan, and L. D. Schell. 1997a. Multi-scale sampling of plant diversity: effects of minimum mapping unit size. Ecological Applications 7:1064–1074.

Stohlgren, T. J., G. W. Chong, M. A. Kalkhan, and L. D. Schell. 1997b. Rapid assessment of plant diversity patterns: a methodology for landscapes. Environmental Monitoring and Assessment 48:25–43.

Stohlgren, T. J., G. W. Chong, L. D. Schell, K. A. Rimar, Y. Otsuki, M. Lee, M. A. Kalkhan, and C. A. Villa. 2002. Assessing vulnerability to invasion by nonnative plant species at multiple spatial scales. Environmental Management 29:566–577.

Stohlgren, T. J., M. B. Coughenour, G. W. Chong, D. Binkley, M. A. Kalkhan, L. D. Schell, D. J. Buckley, and J. K. Berry. 1997c. Landscape analysis of plant diversity. Landscape Ecology 12:155–170.

Stohlgren, T. J., C. Crosier, G. Chong, D. Guenther, and P. Evangelista. 2005c. Life-history habitat matching in invading non-native plant species. Plant and Soil 277:7–18.

Stohlgren, T. J., M. B. Falkner, and L. D. Schell. 1995b. A modified-Whittaker nested vegetation sampling method. Vegetatio 117:113–121.

Stohlgren, T. J., D. Guenther, P. Evangelista, and N. Alley. 2005b. Patterns of plant rarity, endemism, and uniqueness in an arid landscape. Ecological Applications 15:715–725.

Stohlgren, T. J., M. W. Kaye, A. D. McCrumb, Y. Otsuki, B. Pfister, and C. A. Villa. 2000a. Using new video mapping technology in landscape ecology. BioScience 50:529–536.

Stohlgren, T. J., Y. Otsuki, C. A. Villa, M. Lee, and J. Belnap. 2001. Patterns of plant invasions: a case example in native species hotspots and rare habitats. Biological Invasions 3:37–50.

Stohlgren, T. J., A. J. Owen, and M. Lee. 2000b. Monitoring shifts in plant diversity in response to climate change: a method for landscapes. Biodiversity and Conservation 9:65–86.

Stohlgren, T. J., and J. F. Quinn. 1992. An assessment of biotic inventories in western U.S. national parks. Natural Areas Journal 12:145–154.

Stohlgren, T. J., J. F. Quinn, M. Ruggiero, and G. S. Waggoner. 1995c. Status of biotic inventories in US National Parks. Biological Conservation 71:97–106.

Stohlgren, T. J., L. D. Schell, and B. Vanden Heuvel. 1999b. How grazing and soil quality affect native and exotic plant diversity in Rocky Mountain grasslands. Ecological Applications 9:45–64.

Stolte, K. W. 1997. *1996 National Technical Report on Forest Health.* Report FS-605. Asheville, NC: Southern Research Station, U.S. Forest Service.

Stoms, D. M. 1992. Effects of habitat map generalization in biodiversity assessment. Photogrammetric Engineering and Remote Sensing 58:1587–1591.

Strayer, D., J. S. Glitzenstein, C. G. Jones, J. Kolasa, G. E. Lichens, M. J. McDonnell, G. G. Parker, and S. T. A. Pickett. 1986. *Long-Term Ecological Studies: An Illustrated Account of Their Design, Operation, and Importance to Ecology.* Occasional Publication no. 2. Millbrook, NY: Institute of Ecosystem Studies.

Stubbendieck, J., and G. D. Willson. 1987. Prairie resources of national park units in the Great Plains. Natural Areas Journal 7:100–106.

Sukachev, V. 1945. Biogeocoenology and phytocoenology. Comptes Rendus Academie des Sciences de U.S.S.R. 47:429–431.

Sukachev, V., and N. Dylis. 1964. *Fundamentals of Forest Biogeocoenology.* Edinburgh: Oliver & Boyd.

Suzuki, K. 1997. Aspen regeneration in elk winter range of Rocky Mountain National Park and Roosevelt National Forest, Colorado. Master's thesis. Fort Collins: Colorado State University.

Suzuki, K., H. Suzuki, D. Binkley, and T. J. Stohlgren. 1999. Aspen regeneration in the Colorado Front Range: differences at local and landscape scales. Landscape Ecology 14:231–237.

Swengel, A. B., and S. R. Swengel. 1995. The tallgrass prairie butterfly community. In: *Our Living Resources: A Report to the Nation on the Distribution, Abundance, and Health of U.S. Plants, Animals, and Ecosystems.* Washington, DC: National Biological Service, U.S. Department of the Interior; pp. 174–176.

Sykes, M. T., and I. C. Prentice. 1996. Climate change, tree species distributions and forest dynamics: a case study in the mixed conifer northern hardwoods zone of northern Europe. Climatic Change 34:161–177.

Takeuchi, M. 1962. The structure of the Amazonian vegetation. VI. Igapó. Journal Of the Faculty of Science University of Tokyo III 8:297–301.

Tansley, A. G., and T. F. Chipp, eds. 1926. *Aims and Methods in the Study of Vegetation.* London: British Empire Vegetation Committee and Crown Agents for the Colonies.

ter Braak, C. J. F. 1986. Canonical correspondence analysis: a new eigen vector technique for multivariate direct gradient analysis. Ecology 67:1167–1179.

ter Braak, C. J. F. 1987a. The analysis of vegetation-environment relationships by canonical correspondence analysis. Vegetatio 69:69–77.

ter Braak, C. J. F. 1987b. CANOCO—a FORTRAN program for canonical community ordination by [partial] [detrended] [canonical] correspondence analysis, principal components analysis and redundancy analysis. Wageningen, The Netherlands: TNO Institute of Applied Computer Science.

ter Braak, C. J. F. 1991. CANOCO. Wageningen, The Netherlands: Agricultural Mathematics Group.

Thomas, L., and C. J. Krebs. 1997. A review of statistical power analysis software. Bulletin of the Ecological Society of America 78:128–139.

Tiedemann, A. R., and H. W. Berndt. 1972. Vegetation and soils of a 30-year deer and elk exclosure in central Washington. Northwest Science 46:59–66.

Tilman, D. 1982. *Resource Competition and Community Structure. Monographs in Population Biology.* Princeton, NJ: Princeton University Press.

Tilman, D. 1988. *Plant Strategies and the Structure and Dynamics of Plant Communities.* Princeton, NJ: Princeton University Press.

Tilman, D. 1997. Community invasibility, recruitment limitation, and grassland biodiversity. Ecology 78:81–92.

Tilman, D. 1999. The ecological consequences of changes in biodiversity: a search for general principles. Ecology 80:1455–1475.

Tilman, D., W. Wedin, and J. Knops. 1996. Productivity and sustainability influenced by biodiversity in grassland ecosystems. Nature 379:718–720.

Timmins, S. M., and P. A. Williams. 1991. Weed numbers in New Zealand's forest and scrub reserves. New Zealand Journal of Ecology 15:153–162.

Tobey, R. 1981. *Saving the Prairies: The Life Cycle of the Founding School of American Plant Ecology, 1985–1955.* Berkeley: University of California Press.

Turelli, M. 1981. Niche overlap and invasion of competitors in random environments. Models without demographic stochasticity. Theoretical Population Biology 20:1–56.

Turner, M. B. 1989. Landscape ecology: the effect of pattern on process. Annual Review of Ecological Systematics 20:171–197.

Twery, M. J., G. A. Elmes, and C. B. Yuill. 1991. Scientific exploration with an intelligent GIS: predicting species composition from topography. AI Applications 5:45–53.

U.S. Forest Service. 1985. *Range Analysis and Management Handbook.* FSH 2209.21. Amendment 15. Washington, DC: U.S. Forest Service.

U.S. Forest Service. 1996. *Range Analysis and Management Training Guide.* Lakewood, CO: Rocky Mountain Region, U.S. Forest Service.

U.S. General Accounting Office. 1991. Rangeland management: comparison of rangeland condition reports. GAO/RCED-91–191. Washington, DC: U.S. General Accounting Office.

U.S. Soil Conservation Service. 1976. *National Range Handbook,* Washington, DC: U.S. Soil Conservation Service.

van der Maarel, E., and M. T. Sykes. 1993. Small-scale plant species turnover in a limestone grassland: the carousel model and some comments on the niche concept. Journal of Vegetation Science 4:179–188.

Vestal, A. G. 1943. Unequal scales for rating species in communities. American Journal of Botany 30:305–310.

Vitousek, P. M. 1990. Biological invasions and ecosystem processes: towards an integration of population biology and ecosystem studies. Oikos 57:7–13.

Walker, J., and R. K. Peet. 1984. Composition and species diversity of pine-wiregrass savannas of the Green Swamp, North Carolina. Vegetatio 55:163–179.

Walter, H. 1964. *Die Vegetation der Erde in öko-physiologischer Betrachtung. Band I: Die tropischen und subtropischen Zonen,* 2nd ed. Jena, Germany: VEB Fischer.

Walter, H. 1971. *Ecology of Tropical and Subtropical Vegetation.* Translated by D. Mueller-Dombois, edited by J. H. Burnett. Edinburgh: Oliver & Boyd.

Weaver, J. E., and F. E. Clements. 1938. *Plant Ecology,* 2nd ed. New York: McGraw-Hill.

Weaver, J. E., and W. W. Hansen. 1941. *Native Midwestern Pastures—Their Origin, Composition, and Degeneration.* Bulletin 22. Lincoln: University of Nebraska Conservation and Survey Division.

Webster, D. B. 1992. Viewpoint: replication, randomization, and statistics in range science. Journal of Range Management 45:285–290.

Weisberg, P. J., and W. L. Baker. 1995. Spatial variation in tree seedlings and krummholz growth in the forest-tundra ecotone of Rocky Mountain National Park, Colorado, U.S.A. Arctic and Alpine Research 27:116–129.

Weitz, A., D. Bunte, and H. Hersemann. 1993. Application of nested sampling technique to determine the scale of variation in soil physical and chemical properties. Catena 20: 207–714.

Welsh, S. L., and N. D. Atwood. 1998. Flora of Bureau of Land Management Grand Staircase-Escalante National Monument. Kanab, UT: Bureau of Land Management, Kanab.

Wesser, S. D., and W. S. Armbruster. 1991. Species distribution controls across a forest-steppe transition: a causal model and experimental test. Ecological Monographs 61:323–342.

West, M., and P. J. Harrison. 1997. *Bayesian Forecasting and Dynamic Models.* New York: Springer-Verlag.

West, N. E., K. H. Rea, and R. O. Harniss. 1979. Plant demographic studies in sagebrush-grass communities of southeastern Idaho. Ecology 60:376–388.

Westbrooks, R. 1998. *Invasive Plants, Changing the Landscape of America: Fact Book.* Washington, DC: Federal Interagency Committee for the Management of Noxious and Exotic Weeds.

Whicker, A. D., and J. K. Detling. 1988. Ecological consequences of prairie dog disturbances. BioScience 38:778–785.

Whisenant, S. G., and D. W. Uresk. 1990. Spring burning Japanese brome in a western wheatgrass community. Journal of Range Management 43:205–208.

White, R. S., and P. O. Currie. 1983. Prescribed burning in the northern Great Plains: yield and cover responses of 3 forage species in the mixed grass prairie. Journal of Range Management 36:179–183.

Whittaker, R. H. 1962. Classification of natural communities. Botanical Review 28:1–239.

Whittaker, R. H. 1965. Dominance and diversity in land plant communities. Science 147:250–260.

Whittaker, R. H. 1967. Gradient analysis of vegetation. Biological Review 42:207–264.

Whittaker, R. H. 1970. *Communities and Ecosystems.* London: Macmillan.

Whittaker, R. H. 1977. Evolution of species diversity on land communities. Evolutionary Biology 10:1–67.

Whittaker, R. H., and W. A. Niering. 1975. Vegetation of Santa Catalina Mountains, Arizona. V. Biomass, production, and diversity along the elevation gradient. Ecology 56:771–790.

Whittaker, R. H., W. A. Niering, and M. O. Crisp. 1979. Structure, pattern, and diversity of a mallee community in New South Wales. Vegetatio 39:65–76.

Wiens, J. A., C. S. Crawford, and J. R. Gosz. 1985. Boundary dynamics: a conceptual framework for studying landscape ecosystems. Oikos 45:421–427.

Willems, J. H. 1980. Observations on northwest European limestone grassland communities. 5.A. An experimental approach to the study of species-diversity and above-ground biomass in chalk grassland. Proceedings of the Koninklijke Nederlandse Akademie Van Wetenschappen Series C-Biological and Medical Sciences 83:279–295.

Wilson, E. O. 1988. *Biodiversity.* Washington, DC: National Academies Press.

Winegar, H. H. 1977. Camp Creek channel fencing—plant, wildlife, soil, and water responses. Rangeman's Journal 4:10–12.

Wiser, S. K., R. K. Peet, and P. S. White. 1998. Prediction of rare-plant occurrence: a southern Appalachian example. Ecological Applications 8: 909–920.

Woods, K. D., and M. B. Davis. 1989. Paleoecology of range limits—beech in the Upper Peninsula of Michigan. Ecology 70:681–696.

Woodward, A., E. G. Schreiner, D. B. Houston, and B. B. Moorhead. 1994. Ungulate-forest relationships in Olympic National Park: retrospective exclosure studies. Northwest Science 68:97–110.

Woodward, F. I. 1993. The lowland-to-upland transition—modeling plant responses to environmental change. Ecological Applications 3:404–408.

Worster, D. 1977. *Nature's Economy: A History of Ecological Ideas.* New York: Cambridge University Press.

Young, V. A. 1943. Changes in vegetation and soil of palouse prairie caused by overgrazing. Journal of Forestry 41:834–838.

Zar, J. H. 1974. *Biostatistical Analysis.* Englewood Cliffs, NJ: Prentice-Hall.

Zar, J. H. 1996. *Biostatistical Analysis,* 3rd edition. Upper Saddle River, NJ: Prentice-Hall.

Zedaker, S. M., and N. S. Nicholas. 1990. *Quality Assurance Methods Manual for Forest Site Classification and Field Measurements.* EPA/600/3–90/082. Corvallis, OR: U.S. Environmental Protection Agency.

Index